Man And Nature in Delaware:

An Environmental History of the First State, 1631 – 2000

by

William Henry Williams

A Delaware Heritage Press Book

Published by the
Delaware Heritage Press
121 Duke of York Street
Dover, DE 19901

Telephone:
(302) 744-5077

Website:
http://heritage.delaware.gov/

ISBN (hardcover edition): 978-0-924117-36-7
ISBN (softcover edition): 978-0-924117-37-4

Library of Congress Control Number: 2008939314

Previous page:

Eroding Beaches— This photograph of two World War II-era U.S. Army towers in the surf north of Rehoboth Beach provides a perfect metaphor for man's changing relationship with nature in Delaware. When the towers were built about 1941, they were well back from the ocean. As late as the mid-1970s, they were still some forty to fifty feet behind the dune line. Today, waves routinely break along the towers' bases, *photo by Richard B. Carter, 1999.*

Dedicated to Dr. Robert L. Phyliky,
to the late William E. Lowe Jr.,
and to the late John A. Murray

When man interferes with the Tao,
the sky becomes filthy,
the earth becomes depleted,
the equilibrium crumbles,
creatures become extinct.

Tao Te Ching [1]
Circa 600 B.C.

Contents

Acknowledgments

I AM INDEBTED TO OVER ONE HUNDRED people for their contributions to the completion of this book. Most cannot be included in this formal thanks but their efforts are greatly appreciated by the author.

Those that I specifically recognize on this page include six people who read and commented on the entire manuscript, and helped in a number of other ways. They are, Larry Curtis, Sam Glasscock, Carol Hoffecker, Kathleen Jamison, and Ed Kee. I am particularly grateful to Richard B. Carter, who not only read the entire manuscript but then went on to direct this project to completion and provided a number of his photographs for the book. Those who read and commented on at least one or two chapters of the manuscript include Marshall Becker, Barbara Benson, Molly Murray, and David Small.

Others who supplied critical information or other significant help include Mike Apgar, John Bryson, Pat Canzanno, Wendy Carey, Lorraine Fleming, Kevin Connelly, Mary Sue Drugash, Shirin Jamasb, Aleta Esham, Georgia Ryan, Randy Goss, David Grettler, William Henderson, Mary Tylecki Dickson, Tim Kavil, Charlie Kopay, Daniel Leathers, Judith Showell Loeber, June MacArtor, Russell McCabe, Peter Martin, Richard Kiger, Kate Marvel, J.L. Miller, Governor Russell W. Peterson, Ellen Rendle, Connie Cooper, Jonathan Sharp, Harry Themal, and Jerry Vaughn.

Institutions and their respective staffs who were particularly helpful include the Betz Library, Delaware Technical and Community College, Georgetown; the Delaware Public Archives; the Historical Society of Delaware; and the Morris Library of the University of Delaware. Delaware Technical and Community College, Georgetown, supplied an office for the author's use during the research and writing phase of this book. Karen Druliner edited the original manuscript. Lastly, the Delaware Heritage Commission took on the major responsibility of publishing this study of Delaware's environmental history.

William H. Williams

Additional Acknowledgements

The Delaware Heritage Commission acknowledges with gratitude the efforts of our good friend and former Executive Director, Dr. Deborah Haskell, in working with Dr. Williams on proof-reading and later in helping to format the reference notes. Jason L. Burleson-Gibson, formerly of the Delaware Public Archives, created and enhanced maps and used his considerable graphics skills to assist with illustrations.

The commission would also like to thank the very talented nature photographer, Dolores S. Michels of Felton, for the use of several of her fine nature photographs. We also thank Ms. Lorraine Fleming, the Hon. Russell W. Peterson and the Hon. William B. Chandler, III, among others, for reading the completed manuscript in advance.

We are especially grateful to the Williams family for their financial contribution to the publication. In particular, the commission wishes to thank Bill Williams's son, Mark T. Williams, a faculty member at Boston University, both for his editorial assistance and for arranging for this work to benefit from the services of an outstanding editor and indexer, Louisa R. Mandarino of BU's Editorial Institute. Ms. Mandarino, a PhD candidate in the field of literary editing, went above and beyond the call of duty both in preparation of the index and in editing and proofreading the final camera-ready manuscript.

Richard B. Carter

Preface

Memories of William H. Williams

Bill Williams left us much too soon. He passed away on Saturday, April 7, 2007, at the age of seventy years. He had completed his work on this book just days before his passing. Bill had served on the Delaware Heritage Commission for some seven years and had served as chairman of our publications committee. He believed that because of the increasing cost of university press books and because the commission's publications are in large part underwritten by annual appropriations from the Delaware General Assembly, the Heritage Commission could play a unique role in publishing new scholarly works by Delaware historians at a cost well within the reach of people of average means. As he was nearing completion of this manuscript, he came to me and said that he would like the Heritage Commission to publish the book. He gave us this wonderful work with the understanding that he would receive nothing for it. You will understand, therefore, why I see Bill's environmental history of Delaware as his parting gift to the people of his adopted state.

Bill grew up in Southold, on Long Island's North Fork, where his father was the local high school principal. He earned a bachelor's degree in history from Drew University in 1958 and an M.S. in education from Yeshiva University the following year. After teaching social studies for a time in Pawling, New York, he decided to pursue a Ph.D. in history at the University of Delaware. Bill and his wife, Helen, moved here with their young family in the 1960s. To our everlasting good fortune, the Williamses adopted Delaware and made it their home. In 1967, the Williamses moved to the southern Delaware town of Georgetown so that Bill could teach history in the University of Delaware parallel program on the campus of the newly established Delaware Technical and Community College. He remained there for the rest of his career. During the course of those years Bill showed himself to be a wonderfully gifted teacher who inspired and challenged hundreds and hundreds of southern Delaware students.

Bill also emerged during those years as one of the three or four finest historians ever to live and work in Delaware. The first of his seven books, an expansion of his doctoral dissertation, *Anglo-America's First Hospital: The Pennsylvania Hospital, 1751–1841*, was published in 1976. In the years that followed, Bill published two works of popular history, *Delmarva's Chicken Industry: Seventy-five Years of Progress* (1998), and *Delaware, The First State — An Illustrated History* (1999). Along the way he published a history of Georgetown's Wesley United Methodist Church (1979) and two books which have become classics of Delaware historical scholarship, *The Garden of American Methodism: The Delmarva Peninsula, 1769–1820* (1984), and *Slavery and Freedom in Delaware, 1639–1865* (1996).

Man and Nature in Delaware is in many ways the summing up of Bill's life and work as an historian. In its pages, he gives us a magnificent and sweeping view of this state that he came to know and love so well. He takes us from man's earliest habitation here, some fifteen thousand years before the time of Christ, up to the present day, and shows us how man's use of Delaware's lands, waters and air has changed. His book is a testament to the fact that history must treat the totality of the human experience on this earth. It is not just a record of wars and exploration and political intrigues, but, in a broader sense, the story of how man interacts with his environment. He has left us with a clear warning of what the future may hold for us and for our children if we do not come to grips with the forces we have so thoughtlessly unleashed.

Bill always approached the study of history with enormous powers of perseverance, true intellectual integrity, and a courageous willingness to follow wherever the data led him. He did not shy away from tough subjects, and he always allowed his conclusions to be shaped by the facts as they emerged through his careful and wide-ranging research.

People like Bill Williams don't come along very often in life. My colleagues and I on the Heritage Commission and our staff are honored to have known and worked with this good and wise man. We are proud to have been able to assist him in bringing this, his last, and to my mind his finest, work to the people of Delaware whom he came to know and love so well. It is our hope that this book and its author will assume their rightful place among the brightest stars in the firmament of historical scholarship.

Richard B. Carter, Chairman
Delaware Heritage Commission,
September, 2008

Introduction

In late August of each year, many Delawareans return from their summer vacations to the more rapid tempo of the workplace and the schoolhouse. Despite the comfortable familiarity of this transition, the end of summer 2000 was marked by a certain uneasiness as many pondered the implications of recent articles in Delaware's largest daily newspaper, the *News Journal*.

On August 20, 2000, a *News Journal* headline spoke of a "Year-old Toxic Leak Made Public" in Delaware City. On August 27, the headline read "Del. Among Worst in Air Quality: Some Toxins Linked to Cancer, E.P.A Says." The next day's headline focused on the toxic menace of excess chicken manure in southern Delaware; and on August 29 a headline proclaimed that "Thousands of Menhaden Die, So Far This Summer 4 Million Fish Have Perished in [Delaware's] Inland Bays." What was going on? "It sounds," said *News Journal* reporter Molly Murray, "like Delaware is going to hell in a hand basket."[1]

In examining the past, the questions that we ask are shaped by the priorities of the present. The increasing public concern about the health of our natural environment and its implications for the well-being of Delaware's human population have been particularly evident in the last three decades of the twentieth century. Clearly, saving our natural world from

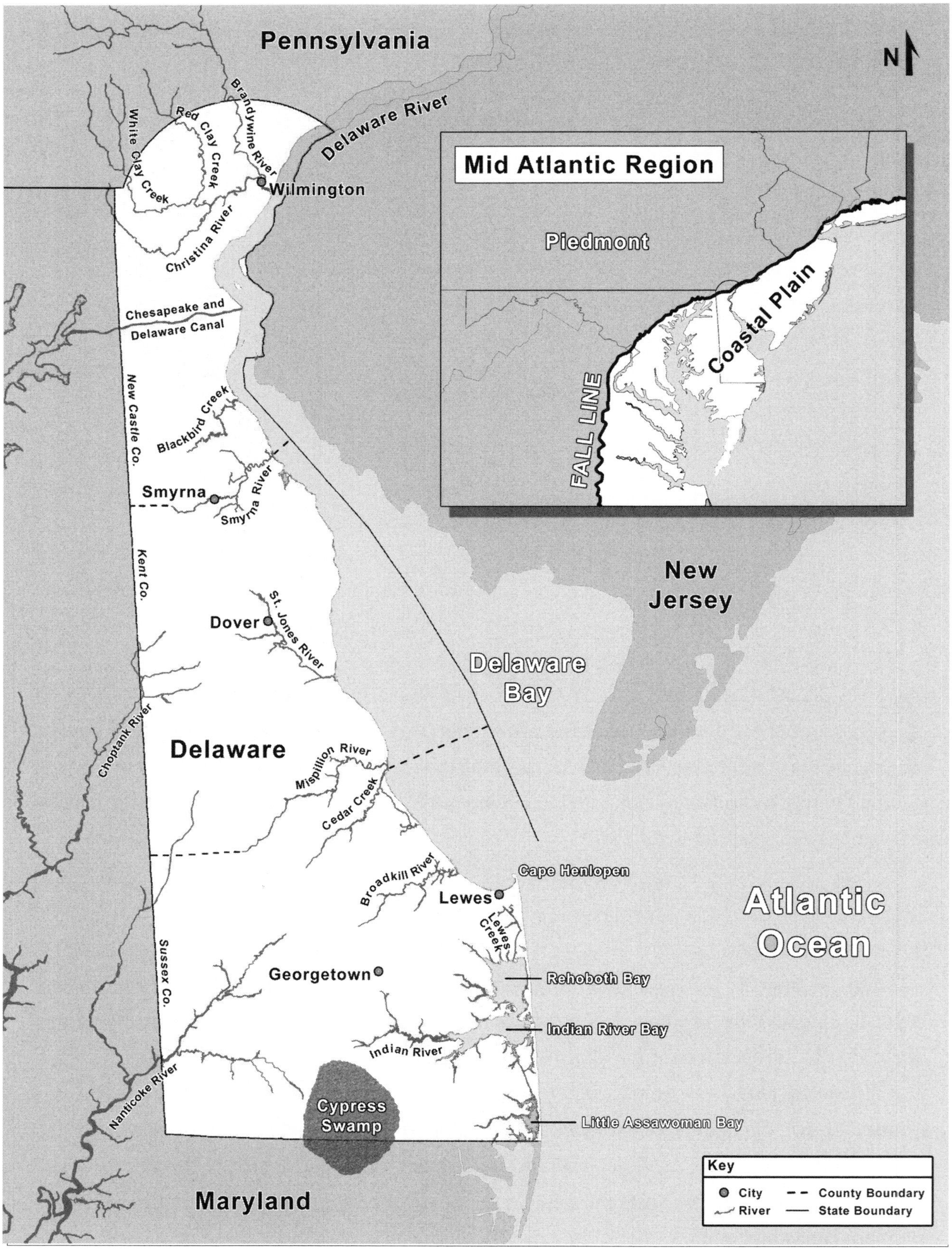

Figure 1: A topographical map of Delaware *(map created by Jason L. Burleson-Gibson).*

further decline and even restoring to health a land long victimized by industrial, agricultural, suburban, and urban-produced pollution, have become a state as well as a national priority. But the essential precondition to informed comment on environmental decline and restoration is some understanding of the historic roots of our present situation.

In examining Delaware's environmental past, this study attempts to provide a basic description of the extraordinary changes in the state's natural world over the last four centuries. It also attempts to focus some attention, in the Delaware context, on the old debate about the relative significance of the physical environment versus human choice in the evolution of the modern world. In short, does the natural world force serious constraints on human activity, or can humans rise above those constraints and, in the process, become the dominant force in the equation? Moreover, what are the implications of human agency breaking free of those constraints?

This study argues that throughout much, if not most, of Delaware's history, human actions and culture were limited by constraints imposed by the state's physical environment. However, when Delawareans reached a technological level that allowed them to throw off most of the shackles of nature-imposed constraints, they did so at their own peril. The rapid growth of industry and technology produced a much higher standard of living for the average Delawarean but at certain costs, including dirty air, toxic dumps, polluted water, diminution in flora and fauna, and, in some areas, aesthetically unpleasing landscapes. And yet, since 1970, a growing sensitivity to these environmental problems has led both the state and the federal government to address environmental issues in a manner that, by 2000, produced dramatic results in some areas. In a number of ways, Delaware's natural environment was healthier by the end of the twentieth century than it was fifty years earlier.

Although both environmental degradation and improvement are a significant part of this story, the primary focus is on the interaction of humans and their physical environment through four centuries. Indeed, the crucial nexus of nature and human agency is at the very core of this study because, as Donald Worster points out, wherever these two spheres "confront or interact with one another, environmental history finds its essential themes."[2]

It can be argued that in writing environmental history, one should take a global perspective because oceans, bays, rivers, insects, birds, mammals, reptiles, marine life, and wind currents are not respecters of state or national political boundaries. And yet a strong case can be made for environmental histories that have a far more restricted political or geographic focus. Quite simply, global environmental studies, like any other broad-based studies on an international scale, are largely dependent on information produced by in-depth research on the local, state, regional or national level. Moreover, many of the interactions between the natural world and human culture can best be understood when examined in a more narrowly circumscribed geographic or political landscape.

By focusing on a largely ignored but very vital aspect of Delaware's history, my hope is that this study will stimulate conversation and discussion of both our environmental past and our environmental future. If that conversation takes place, it will help to frame the public debate on how human forces and nature can best co-exist. The stewardship of Delaware's unique and finite natural resources will present clear challenges for ours and future generations, but with these can come tangible rewards.

William Henry Williams
Georgetown, Delaware — 2007

PROLOGUE:

Before The Arrival of Europeans

"This comes from Indians setting fire this time of year to the woods and thickets, in order to hunt...."

IN 15,000 B.C., DELAWARE HAD AN ARCTIC AND SUBARCTIC LANDSCAPE, A fascinating variety of animal life, and no people. The last Ice Age (Pleistocene Age), which stretched from approximately 1,000,000 B.C. to about 10,000 B.C. and featured huge, slow-moving glaciers from the north that scoured out, scraped away, and blanketed much of North America, was finally coming to an end. Although the closest that an ice sheet had come to Delaware was seventy-five miles to the northeast in what is now Belvidere, New Jersey, in 15,000 B.C., the massive Laurentide Glacier still covered a small part of northeastern Pennsylvania and northern New Jersey. Despite the fact that global temperatures were starting to rise, causing the southern edge of the Laurentide Glacier to begin a very gradual northward retreat into southern New York, even at such a considerable distance this huge ice sheet continued to exercise a chilling and dampening effect on Delaware's climate.[1]

Because Delaware, despite its rising temperatures, remained a relatively cold place in 15,000 B.C., its grasses, plants and trees most closely resembled those of today's central and northern Alaska, northern Canada, or northern Russia. In northern New Castle County, for example, treeless tundra dominated. Further south in New Castle County, and in Kent and Sussex, a predominantly spruce taiga (subarctic coniferous forest) and grasslands covered the

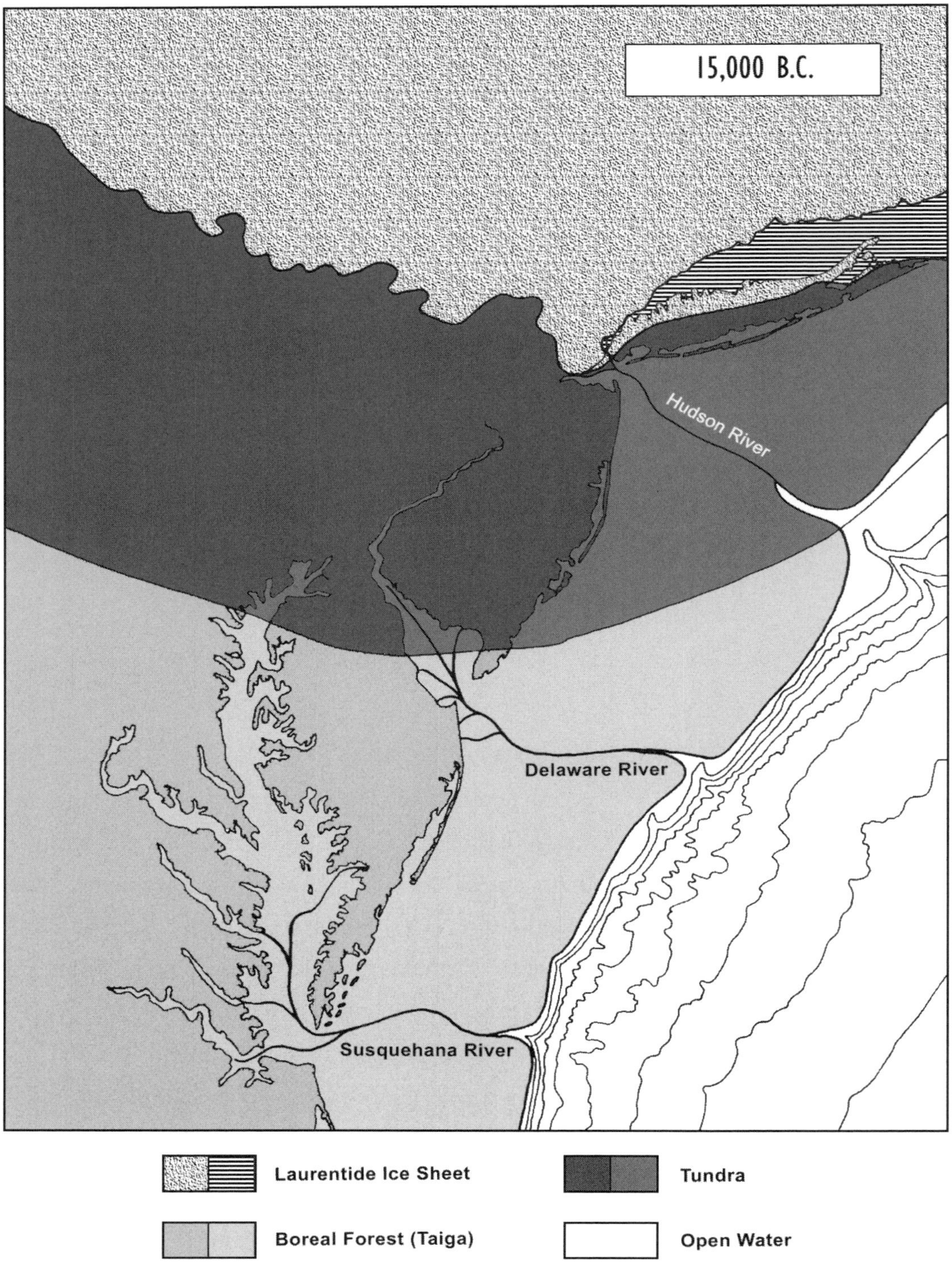

Figure 2: A map of what are now Delaware and the Delmarva Peninsula as they appeared in 15,000 B.C. *(courtesy of the University of Delaware Dept. of Geology with enhancements by Jason L. Burleson-Gibson).*

land. This varied natural habitat supported the greatest variety of wildlife species in the prehistory and history of Delaware. Probably roaming the landscape alongside elk, caribou, and white-tailed deer were far larger mammals such as grazing woolly mammoths and browsing mastodons.[2] Although no actual remains of these giant mammals have been uncovered within the state's present boundaries, the discovery of megafauna remains on the Atlantic Continental Shelf—once an eastern extension of the Delmarva Peninsula—and in adjoining states suggests the presence of these huge mammals in Delaware in 15,000 B.C. Other megafauna probably present included giant-sized musk oxen, moose, and bears.[3]

Forming the critical food base for Delaware's wildlife were grasses, shrubs and trees. Because of the region's cold climate, however, Delaware's flora grew too slowly to support large, fixed concentrations of grazing mammals. In short, a great variety of species did not translate into a great number of mammals. Over subsequent millennia, gradually warming temperatures produced more rapid annual growth in Delaware's flora which, in turn, supported a smaller variety but a significantly larger number of wildlife.[4]

By 9000 B.C., most of the megafauna in both Delaware and in the rest of North America were either extinct or fast on the way to extinction. The relatively sudden demise of woolly mammoths, mastodons, and other giant mammals is one of the great scientific puzzles of prehistoric times. Paleobiologists have advanced a number of unproven theories which blame, respectively, dramatic change in climate, migrating Native American hunters, contagious diseases, a host of other factors, and even combinations of some or all of the above. Particularly controversial is the view that Native Americans were "super predators" who, in a relatively short period of time, simply swept the North American landscape clean of megafauna.[5] In the future state of Delaware, however, the extinction of mammoths, mastodons and other very large mammals by Native American hunters is an unconvincing explanation. By the time the first humans arrived on the Delmarva Peninsula, the megafauna were probably gone.[6]

Delaware's first humans were, presumably, descendants of Asian immigrants who crossed to Alaska from Siberia. Although evidence for specific dates is hard to come by, perhaps from about 10,000 to about 9000 B.C., small numbers of these nomadic people pushed into the northeastern sector of the Delmarva Peninsula.[7] They found Delaware's extreme north dominated by rolling piedmont while, further south, stretched the Atlantic Coastal Plain. Although the geological topography of 9000 B.C. may have been similar to today's Delaware, the land area was significantly larger because southern Delaware extended a considerable distance into the Atlantic Ocean. Despite the fact that glaciers were slowly retreating in the face of warmer temperatures, they continued to cover much of North America. Because they still locked up so much of the planet's water supply, their continued presence caused sea levels to be dramatically lower than they are today. Consequently, much of

Figure 3: A map of what are now Delaware and the Delmarva Peninsula as they appeared in 9,000 B.C. *(courtesy of the University of Delaware Dept. of Geology with enhancements by Jason L. Burleson-Gibson).*

the sea bed of the future Delaware Bay was above sea level in 9000 B.C. The only water there was the Delaware River, which cut its narrow winding channel south and then east towards the Atlantic Ocean, carrying melted glacial runoff from the retreating Laurentide ice sheet. Indeed, the boulders at Big Stone Beach, in southeastern Kent County and those on the bottom of the Atlantic Ocean, approximately fourteen miles due east of the Indian River Inlet, were deposited there when ice flows carrying glacial debris finally melted in the Delaware River before it entered the Atlantic. Because sea level was then seventy-five to ninety feet below today's mark, the Atlantic coastline was probably fifteen to thirty miles due east of Rehoboth Beach.[8]

While familiarizing themselves with their new land's topography, the first Delawareans hunted and trapped such familiar sources of protein as white-tailed deer, elk, caribou, smaller mammals, birds and fish. Usually they killed their prey with stone-bladed spears which were hurled or driven by hand into the quarry's body. The use of the bow and arrow in Delaware was still almost 10,000 years away.[9] Although fish were also part of the Delaware diet in 9000 B.C., shellfish such as crabs, clams, mussels, and oysters were rarely eaten. The problem was that the initial meltdown of glaciers worldwide caused the sea level to rise at a comparatively rapid rate from 15,000 B.C. to 3000 B.C. As a result, estuaries, which provide the prime habitat for shellfish to flourish, did not have sufficient time to stabilize before being overwhelmed by constantly rising ocean waters. Only with the less rapid increase in sea level after 3000 B.C. were bodies of water such as the Delaware Bay, Rehoboth Bay, and Indian River Bay stable enough to become prime shellfish habitats. Although wild plants provided still another source of nutrition, they seem to have played a minor role in the diets of Delaware's earliest settlers.[10]

Delawareans of 9000 B.C. relied on stone tools for hunting. They had no clay pots or vessels, and there is little evidence that they were able to store food for long periods of time. The limited food storage capacity and the rapid exhaustion of local food sources by a hunter-gatherer life-style dictated that social-political organizations be restricted to small, nomadic or semi-nomadic bands with an extended family at the core. These bands spent most of the year in a variety of temporary camps while members scoured the nearby countryside for nutritious fauna and, to a lesser degree, flora. Because they provided almost every male member with a significant role as hunter and warrior, these bands were remarkably egalitarian organizations for males.[11] The social and political status of Delaware women in 9000 B.C., however, is less clear.

In addition to the ongoing search for food, a Native American band might locate temporarily near a site that provided specific types of rocks from which stone projectile points and other stone tools could be fashioned. At such lithic locations one band of very early Delawareans might encounter other bands equally intent on collecting and quarrying appropriate rocks. Because exogamic practices—breeding outside the band—were often employed by hunter-gatherers, these encounters provided opportunities for interband marriages as well as venues for bartering and exchanging goods. Such marriages helped stabilize kinship and political relationships between neighboring bands.

Like most people of their era, these first Delawareans found their economic and social options restricted by the sometimes capricious nature of their physical environment. Clearly their small-band, nomadic or semi-nomadic pattern of life was a logical response to the powerfully intrusive natural world that surrounded them. Indeed, although the cultures of prehistoric Delawareans were not entirely molded by their physical environment, their natural world did present serious constraints on how those cultures developed. As Helen Rountree and Thomas E. Davidson point out: "A critical factor in shaping Native American cultural traditions on Delmarva was the specific environmental conditions that existed" in the area.[12] Probably no prehistoric Delawareans were more constrained by nature than those first settlers of approximately 9000 B.C.

Choices

Almost eleven thousand years later, in 1600 A.D., Delaware's inhabitants, after some cultural and technological adaptation, were still dependent on stone tools and possessed a culture that continued to be impacted profoundly by the surrounding physical world. But in reacting to the natural climate and landscape, and to the surrounding flora and fauna, Delaware's Native Americans of the time affected their physical world in a manner unimaginable to those who had lived in the state about 9000 B.C.

Thanks to the gradual warming of the earth's climate and the subsequent northward retreat of the Laurentide Glacier, by 1600 A.D., Delaware's temperatures and humidity levels may have approximated those of today. In the rolling hills of northern Delaware's piedmont, long-term climatic change had led to the replacement of the flora of tundra by deciduous forests of oak, chestnut, beech, poplar, hickory, maple, ash, elm, and walnut. In that part of Delaware that was within the Atlantic Coastal Plain, comprising some ninety-four percent of the state's land mass, taiga forests had given way to hardwoods, with stands of pines scattered throughout the forests south of the St. Jones River. Pines, however, made up a smaller percentage of southern Delaware's trees than they do today. The state's hardwood forests of 1600 A.D. produced an understory of "mast" that included berries, tubers, edible greens, rootstocks, and seeds that dramatically increased their biotic carrying capacity for birds, other animals and humans.[13] This and the opening of occasional grasslands by natural or man-set fires provided food for larger numbers of birds and mammals than could be supported by the very different biotic base that existed in Delaware at the end of the Ice Age.

The melting glaciers caused a gradual but significant rise in sea level that produced, by 1000 B.C., a Delaware Bay and a Delaware Atlantic Coastline that approximated present coastal configurations. But even more important, beginning in 3000 B.C., the rise in sea level had begun to slow to the point where ecological stability would be established in Delaware's marine bodies and in the bordering wetlands, producing a far more congenial estuarine home for shellfish. Rising water temperatures also contributed to a better shellfish habitat.[14] By 1600 A.D., Delaware's saltwater bodies had long offered the state's Native Americans a smorgasbord of oysters, clams, scallops, crabs, mussels, and other mollusks. The more moderate rise in sea level after 3000 B.C. coincided with a gradual increase in herring, shad, and other anadromous fish that made annual runs up Delaware's freshwater streams.

By 1600 A.D., most of northern Delaware's human residents were probably Lenape—eventually labeled the "Delaware" by the English—a very loosely associated Indian people. According to anthropologist Marshall Becker, they occupied the west bank of the Delaware River from Pennsylvania's Lehigh Valley in the north to Delaware's Leipsic River Valley in the south, about midway between Smyrna and Dover. (A number of other scholars maintain that the Lenape homeland was far more extensive in size, stretching from the Delaware Valley eastward to the Atlantic Ocean and the lower Hudson Valley.) Some population estimates for the Lenape in Delaware during the early 1600s, which are only educated guesses, range from as few as 100 or 150 to more than 2,000. Other Native American peoples in Delaware during the same period included a handful of Nanticokes living along Broad Creek in southwestern Sussex and a limited number of a people later called Siconese (also spelled "Sekonese" or "Ciconicin") who may have been living northwest of Cape Henlopen. An estimated total of the combined population of the three Native American groups in Delaware in 1600 A.D., ranges from as low as 300 to as high or higher than 2,500. Of course the relatively low population numbers may have been impacted by the presence of deadly and contagious European diseases that often swept through Indian communities ahead of white colonization.[15]

The above population estimates of the total number of Native Americans in Delaware in 1600 A.D. roughly translate to between .2 and 1.3 people per square mile. Such estimates are less than or comparable to the 1.1 people per square mile in Alaska in 2000, but far below Delaware's 401 people per square mile in the same year.[16]

In addition to the advent of deadly Old World diseases, Delaware's very low population density in 1600 A.D. may be explained partially by the region's primary economic activities. Throughout prehistory and history, a predominantly agricultural economy has generally supported a much higher population density per square mile than a hunter-gatherer economy. One estimate places the figure from ten to one hundred times higher. The primary reason is that farming produces far more calories per acre than hunting-gathering. In addition, because agriculture demands a more sedentary life-style than hunting-gathering, it generally produces far more children per family. Unlike the case with agricultural societies, children among hunter-gatherers are consciously spaced through sexual abstinence, infanticide, and abortion, so that movement across the landscape would not be unreasonably hindered by large numbers of young children. By contrast, a more stationary agricultural people, "unconstrained by the problem of carrying young children on treks," could give birth and raise as many babies as it could feed. There are indications that most of Delaware's residents in 1600 A.D. were primarily hunter-gatherers, which meant that of necessity most families deliberately spaced and limited the number of their children.[17]

As early as 1000 A.D., some corn, beans and squash were grown across much of the Middle Atlantic region (Maryland to New York). Six hundred years later, most Native Americans to the south, west and northwest of Delaware continued to hunt and fish, and to gather shellfish, seeds, roots, berries and other fruits of the forest. But a significant number also depended heavily on the nutrition provided by the crucial triad of corn, beans and squash. Until recently, so little was known about the Indian peoples who inhabited Delaware in the pre-contact period before the arrival of Europeans that scholars filled in the blanks about their economic activities by assuming that all of the indigenous people of the Middle Atlantic region were "generally alike." Therefore, the agricultural model that was a good fit for some nearby Native Americans, was also applied to all of Delaware's Indians in 1600 A.D. Indeed, as late as 1991, a leading Delaware archaeologist wrote that raising corn "was the basic source of livelihood among Lenape groups" prior to the coming of Europeans.[18]

But the Lenape, according to more recent evidence, chose to plant and grow only limited quantities of corn, beans and squash in their temporary summer sites, and probably not enough to store significant amounts for winter. As one eighteenth century visitor to the lower Delaware Valley observed, their agriculture was "very trifling." Indeed, except from about 1640 until 1660, when the Lenape may have raised more corn than usual to sell to the first Swedish and Dutch settlers, Delaware's Lenape rejected a strong commitment to the agricultural model and continued, instead, to depend heavily on a hunter-gatherer economy that included only a modest agricultural component.[19]

Thanks to improved technology, Delaware's indigenous people of 1600 A.D. had become more efficient hunters and fishermen than their distant predecessors. The bow and arrow was introduced to the Delmarva Peninsula about 900 A.D. and proved so superior to the traditional spear in killing wild game that, 600 years later, the spear had been largely abandoned. Other technological advances included the development of more sophisticated snares for trapping small mammals, and dip nets, leisters (spear-like instruments with three or more barbed prongs), and fish weirs (traps) for harvesting herring, shad, and other anadromous fish as they made their annual spring runs up Delaware's creeks and rivers. Dugout canoes, carved from the trunks of swamp cypress and other trees, enabled Native Americans to place and then empty fish weirs close to shore.[20] But the hunting and fishing technology of 1600 A.D. was still primitive by modern standards. The type of bow used on much of the Delmarva Peninsula, for example, probably fired an arrow at a level trajectory for not much more than twenty yards.

This meant that it was important for hunters to get close to their quarry before drawing a bow string. Hunting techniques usually meant individual stalking of prey for most of the year followed by group drives and surround hunts in late fall and early winter. In all three activities, dogs may have been used for tracking and cornering deer and other game.[21]

Deer hunting was also made easier by the annual burning of the woods in November and December which removed underbrush and opened up a clear line of sight for shooting arrows at game. The practice of annually burn-ing parts of the countryside was noted in 1632 by Dutchman David de Vries. As he approached Cape Henlopen in early December of that year, but still out of sight of land, De Vries smelled the "sweet" aroma of burning bushes and herbs and wrote in his journal, "this comes from the Indians setting fire at this time of year to the woods and thickets, in order to hunt." Along New Castle County's White Clay Creek during the seventeenth century, Native Americans torched the woods to drive deer from the shelter of thickets into the shooting range of waiting bowmen. Burning away the forest understory also encouraged the growth of new grasses and tender shoots that drew deer to the same site the next year. Attracting deer to the same site was important because deer tended to be territorial and were generally content to remain in a specific area with a radius of approximately one mile.[22]

Some of Delaware's woods were better hunting venues than others. Deer and wild turkeys, the premier game of the pre-contact period, were drawn to the nut and seed-producing mixed forests of oak and hickory that were common in the rolling hills of the piedmont as well as along Delaware's slightly elevated spine that stretches from southwestern New Castle County to southern Sussex County. By contrast, the stands of mature pines that were scattered throughout southern Delaware attracted fewer deer, turkeys, and other wildlife because pine trees were less successful in producing the nutritious seeds and mast favored by most birds and mammals.[23]

Although Delaware's deer and most other mammals were more numerous in 1600 A.D. than at the end of the Ice Age, four-legged predators such as cougars (mountain lions) and wolves joined with Native Americans in controlling the numbers of grazing and browsing wildlife. Far less numerous than white-tailed deer, but an occasional source of protein nevertheless, were Delaware's elk and even, perhaps, a bison or two. With the introduction of European weapons came the extirpation of elk by the early eighteenth century and the state's handful of wild bison by the late eighteenth century.[24]

Shellfish were more significant in the diet of Delawareans in 1600 A.D. than in that of their predecessors at the end of the Ice Age. Indeed, because stable estuaries produced more bountiful harvests, shellfish in general and clams and oysters in particular probably were as significant as venison in the diet of Native Americans in central and southern Delaware. Because food was scarce in late winter and early spring, it was then that oysters were generally taken from their salt water settings despite the fact that, at this time of year, they had more limited nutritional value.[25]

Why the Lenape and, to a lesser extent, the other native peoples of Delaware rejected some or most aspects of an economic model primarily based on agriculture, and continued to adhere to a semi-nomadic, hunter-gatherer model with only a limited agricultural component is a mystery. While seventeenth century English political theorist Thomas Hobbes labeled the life of preagricultural man as "poor, nasty, brutish and short," the truth was that opting for farming over hunting-gathering did not guarantee a longer and healthier life. Rather, as archaeological studies from many locations throughout the world point out, first generation farmers were often "smaller and less well nour-ished, suffered from serious diseases, and died on the average at a younger age then the hunter-gatherers they replaced." [26]

White-tailed Deer at Kent County's Bombay Hook ational Wildlife Refuge, *photo by Dolores Michels.*

Contributing to this decline in general health was a shift in the diets of first generation farmers that replaced most of the nuts, greens, roots, seeds, berries, and fruit gathered from the wild with two or three cultivated crops. Quite simply, the more varied diet of hunter-gatherers may not have offered as many calories, but it was probably healthier for the human body than the limited varieties of grain and vegetables that fueled the bodies of most early agricultural people.

Perhaps native Delawareans of 1600 A.D. were reluctant to make the transition to a more sedentary, agricultural existence because they somehow sensed the nutritional shortcomings produced by abandoning a hunter-gatherer existence. Or perhaps for other reasons they found a primarily hunter-gatherer economy preferable to a primarily agricultural economy.

One important factor to keep in mind was that the pre-contact agricultural model available to Delaware's Native Americans was less complete and therefore less attractive than its counterpart across the Atlantic. The problem with the pre-contact Indian agricultural model was that it lacked large domestic animals to provide meat, milk, hides, and fertilizer, or pull plows, wagons and carts, and to turn millstones. Except for the wild ancestors of dogs, the varied and fascinating fauna of North America did not include large animals that were susceptible to domestication. Bison, elk, and deer, for example, seemed impossible to train for farm work. By contrast, certain large wild animals of the Eurasian land mass were quite amenable to domestication, and they subsequently became the ancestors of such important modern livestock as horses, cows, pigs, sheep, and goats.[27]

In addition to certain deficiencies in the available Native American agricultural model, Delawareans of 1600 A.D. may have been resistant to economic change because the natural resources crucial to supporting a hunter-gatherer existence were more plentiful throughout most of Delaware than further south on the Delmarva Peninsula, where a more significant com-mitment to agriculture was present among Indian peoples. Delaware's fruitful oak-hickory forests not only attracted a higher concentration of wild game than the piney woods further south, they also yielded a greater harvest of wild nuts, roots, seeds, greens, and berries to feed Native Americans. In short, Delaware's abundant oak-hickory forests made a primary dependence on wild flora as well as wild game a very viable way of life for Delaware's pre-contact human inhabitants.[28]

Just as the killing and sundering of game was made easier by the introduction of the bow and arrow and the development of sharper cutting implements made from stone, so too was the preparation, storage, and cooking of gathered plants enhanced by a more sophisticated technology such as the presence, as early as several thousand years B.C., of stone tools for grinding seeds, and, by 1200 B.C. or even earlier, the production and utilization of clay vessels. For the Indian peoples of 1600 A.D., ceramic containers "increased the efficiency of food preparation," because the boiling and simmering of stews and gruels in clay pots made possible the extraction of more calories from a specific quantity of meat, fish, shellfish, domestic and wild plants than earlier, less efficient methods of cooking. Clay pots also provided convenient receptacles for storing a limited amount of food.[29]

By 1600 A.D., the Lenape were organized into bands of perhaps twenty-five to thirty kin-related individuals whose periodic movements across the landscape were largely dictated by the changing seasons and the availability of food resources in specific places.[30] Because of their own fixed-site life-styles, Europeans would later label Lenape bands as nomadic. But the movements of these bands were predictable and usually limited to the same general locales visited the previous year at approximately the same date. The closing days of winter were usually marked by a shortage of food. By March, some Lenape band members would leave their winter camp for a temporary site along a creek or river, intent on harvesting herring, shad and other fish making their annual spawning runs upstream. In the next three months, foraging members of the band collected wild greens and berries that were ripening in forests and meadows. During

April or May the band would usually move to a summer site and plant limited quantities of maize (corn), beans, and squash.[31]

If it was a new summer site, males wielding stone axes cleared away trees and brush just enough to let the sun's rays reach the ground. Females loosened the soil with sharp-pointed sticks, and then hand-planted seeds on a staggered schedule so that harvests might stretch from August to October. Old fields required less effort in clearing but, after only three years or so of continuous planting, lost most of their nutrients so new fields had to be cleared and planted. Crop destruction by grazing and foraging wildlife such as deer, raccoons, squirrels, bears, beavers, rabbits, woodchucks, and birds must have been a serious problem. To keep mammals and birds away from garden plots, children probably were used as mobile scarecrows and the older boys may have cut their eye teeth as hunters by shooting or trapping unwanted four-footed visitors.[32]

In late summer and early fall, the limited crops of corn, beans and squash were harvested. Late fall was a period of intense foraging and hunting because it represented the last window of opportunity before winter for plant and nut gathering, and because wild game were now heavy with meat after eating well during summer and autumn. Summer sites were abandoned by late fall, because most of the band had to spend many days in distant woods and meadows where the numbers of deer and other wildlife had not yet been depleted. Once cold weather really set in, the band sought a more permanent encampment—perhaps a previous winter site—from which hunting parties would periodically venture forth to shoot and trap game during the days of moderate weather.[33]

In southern Delaware in 1600 A.D., where Native Americans were probably Nanticoke in southwestern Sussex and Siconese in eastern Sussex, there are indications suggesting that agriculture played a more important role than among the Lenape. Indeed, the greater importance of agriculture may have made Sussex's Indians more sedentary than the Lenape who lived in New Castle.[34]

Historically, the more sedentary the society, the less egalitarian it was. While a hunter-gatherer people, divided into small bands, generally functioned in a democratic manner, a more sedentary society with its higher population density, of necessity, depended on the directional will of ranked leaders. Of all of the native peoples of the Middle Atlantic area in 1600 A.D., the Lenape of the lower Delaware Valley and the Munsee of the upper Delaware Valley, were probably most committed to follow a primarily hunter-gatherer existence with small bands being the primary organizational unit. In Sussex, however, there are indications that the more sedentary Nanticokes and perhaps the Siconese were moving towards a tribal organization where rank was more evident. Although the semi-nomadic life of the Lenape may have produced a more egalitarian society than in some other Indian communities, it also continued to limit seriously the Lenape birthrate. When the Susquehannocks, a much more agricultural, sedentary and more populous tribe from the lower Susquehanna Valley dominated the Lenape in the early seventeenth century, the latter's inability to resist may have been due to the fact that they did not have enough warriors to defend their territorial rights.[35]

In the century prior to the coming of Europeans, Delaware's Indians continued to lead a life significantly shaped by the constraints, challenges and opportunities produced by the particulars of the state's natural environment. Delaware's climate, soil, water, flora and fauna combined to encourage certain economic and social choices and to discourage others. But to say this does not discount the important role played by human volition. Indeed, although nature continued to provide the raw materials necessary for a hunter-gatherer existence, the natural world also provided the soil and climate necessary to support a primarily agricultural existence. Presented with two economic options, Delaware's Lenape chose to remain primarily hunter-gatherers while evidence suggests that, by the end of the sixteenth century, the Nanticokes and Siconese in southern Delaware were moving toward a more agriculturally-based economy.

One might expect that by remaining primarily hunter-gatherers, most of Delaware's indigenous people continued to follow a way of life that had minimal impact on the natural landscape. The modest population produced by a semi-nomadic life-style of hunter-gatherers guaranteed that, unlike some other areas along the eastern seaboard, much of Delaware's landscape would remain relatively untouched by human economic and social activity prior to the arrival of Europeans. But even in the thinly populated Delaware of 1600 A.D., there was increasing evidence of the hand of man. Native Americans had developed a hunting technology that was far more efficient in killing and capturing birds, mammals, mollusks, and fish than that of their distant predecessors who first entered Delaware at the end of the Ice Age. By 1600 A.D., Delaware's Native Americans were also commonly firing the woods to improve hunting and to clear out a few fields for planting. But all of these activities had a very modest impact on Delaware's natural world compared to the environmental onslaught to come.

CHAPTER ONE:

European Ways, 1631 – 1683

"...a land flowing with milk and honey."

When Europeans first explored and colonized Delaware in the seventeenth century, they encountered a natural world that reminded them, in many ways, of their native lands across the Atlantic. It soon became clear, however, that the soil, climate, flora and fauna of Delaware were significantly different from that of their European homelands, and that these differences would have a considerable impact on the lives of first and second generation European pioneers.

Although there are at least fourteen different soil types in Delaware, in 1600 the land shared one common denominator: most of it was not recently cultivated by Native Americans. One observer noted that as late as 1654–55, Delaware from Fort Casimir (site of the modern town of New Castle) south to Cape Henlopen was not cleared and planted by either Native Americans or Europeans.[1] This meant that Delaware's soil enjoyed an initial virginal fertility that had been lost by European fields many centuries before. Beyond the commonality of virginal fertility, however, Delaware's soils varied dramatically depending on geographic location.

Stretching south from Pennsylvania to an imaginary line running northeast from present-day Newark to present-day Wilmington, the rolling piedmont covered approximately six percent of Delaware. Rising in Pennsylvania, freshwater streams cut ravines and gullies through this northern New Castle County hill country as they rushed southeast to connect with the fast-flowing Brandywine and the sluggish Christina which, after merging, emptied into the Delaware River. The soil in Delaware's piedmont, which was derived from crystalline rocks, was fairly fertile and well drained, and yet retained moisture reasonably well.[2] The sloping terrain and significant rainfall, however, would combine with the clearing of timber and the introduction of the European plow to produce significant soil erosion.

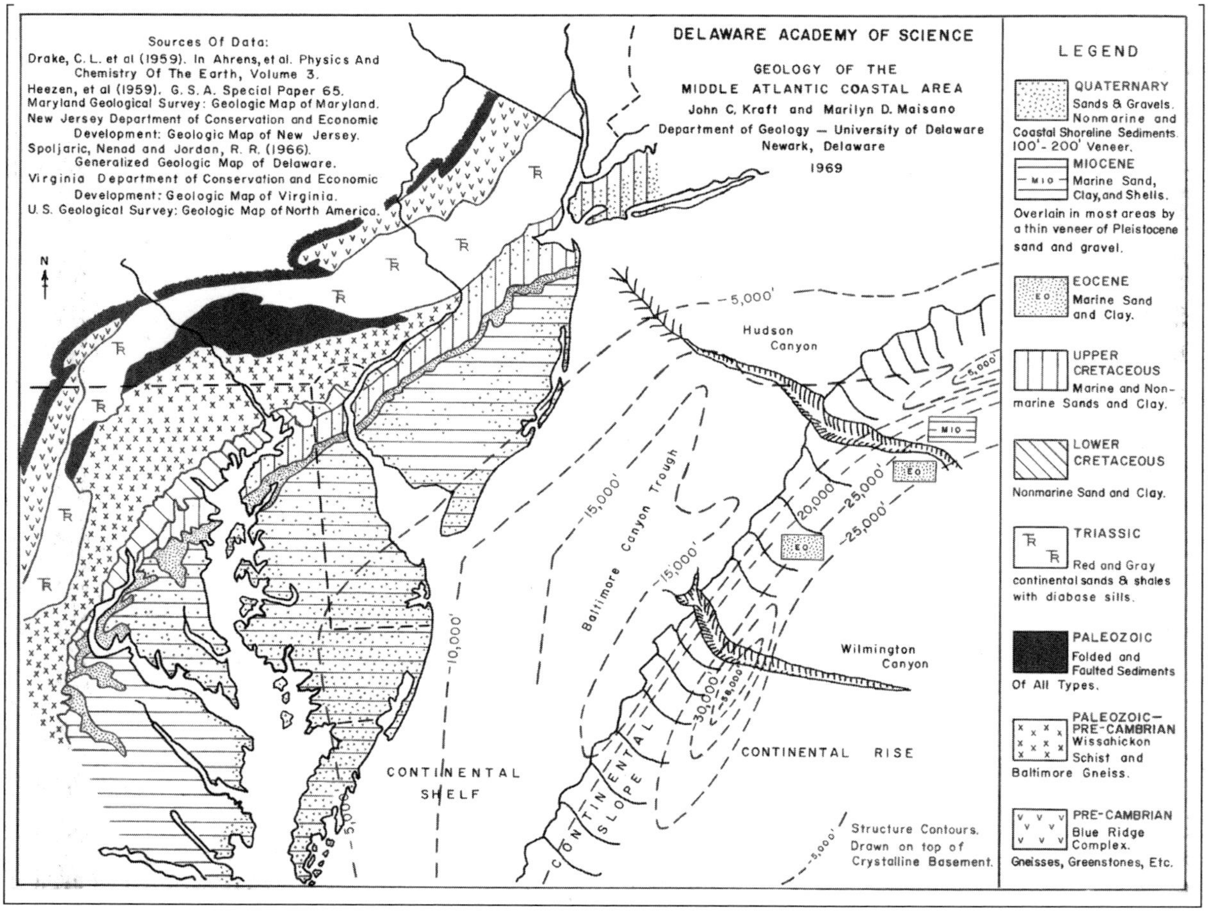

Middle Atlantic Geology— This map, produced in 1969 by John C. Kraft and Marilyn D. Maisano of the University of Delaware's Department of Geology shows the geology of Delaware and the Middle Atlantic coast, including the outer continental shelf, *courtesy of the University of Delaware.*

The remaining 94 percent of the state is part of the Atlantic Coastal Plain that stretches from New England to Florida. It is often called "flat" Delaware, and is divided into two slightly different landscapes. The northernmost sector of the Coastal Plain features slight surface undulations from the piedmont south to the Smyrna River (central and southern New Castle County). But from the Smyrna River south to the Maryland boundary (Kent and Sussex counties), the terrain really flattens out with only occasional swales forged by sluggish tidewater streams adding a little variety to the landscape's dominating horizontal lines.

Despite the monotony of the region's horizontal lines, the inhabitants of flat Delaware sensed that each individual place was, in its own way, unique thanks to the peculiar twists and turns of a local stream, the drainage patterns of a local marsh, the geographic proximity to bay or ocean, or the way that people had physically reshaped their part of the landscape. In short, most Delawareans understood that their economic and cultural lives were partly dependent on the subtle peculiarities of local geography and how those peculiarities would be altered by man. Clearly, no specific place in the First State was exactly like any other specific place.

The directional flow of Delaware's rivers

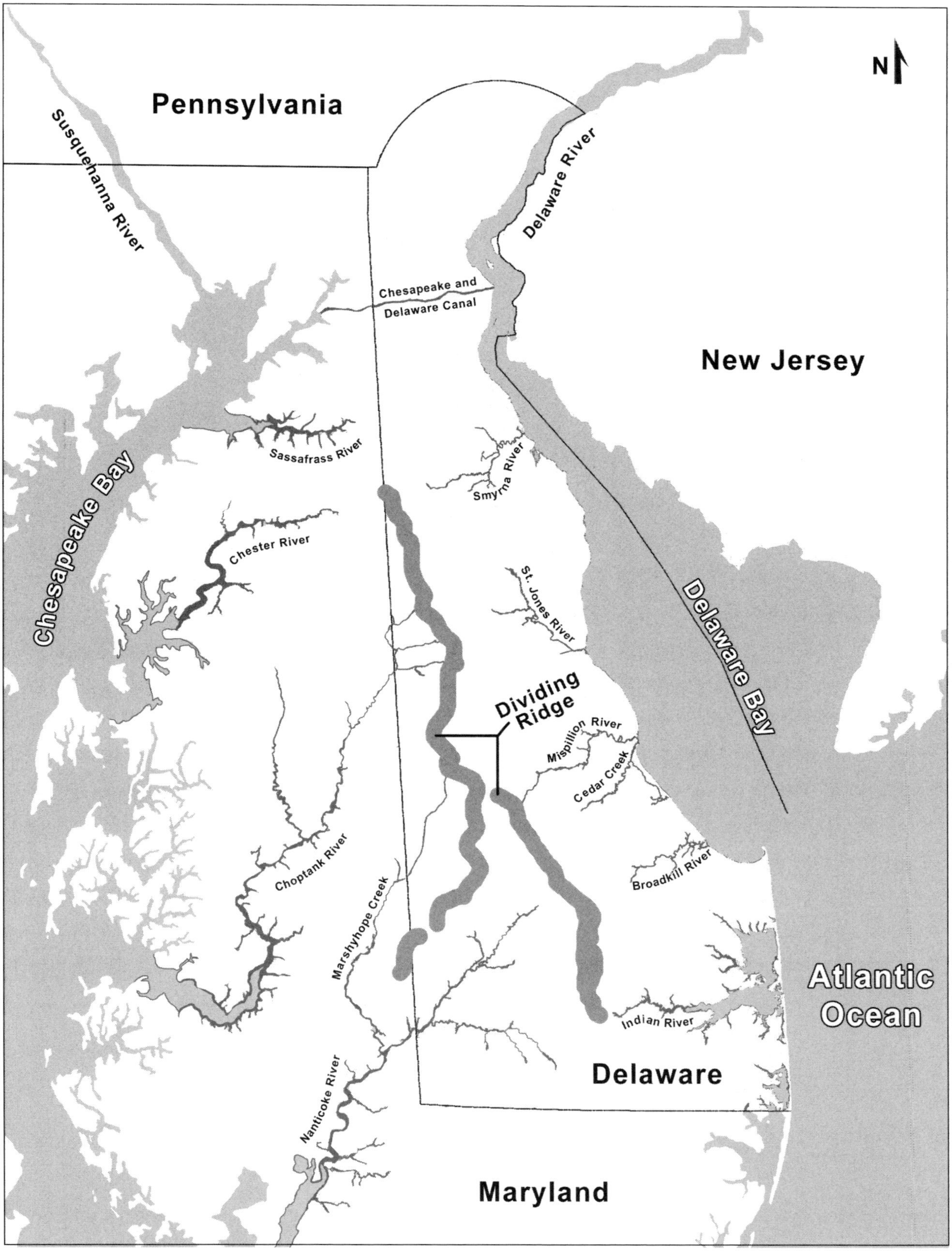

Figure 4— A map of Delaware showing the Dividing Ridge that runs generally from north to south through most of the state *(map created by Jason L. Burleson-Gibson)*.

south of the piedmont is dictated by Dividing Ridge,[3] a visually unimpressive land form that rises only a few feet above the surrounding countryside. It begins in Maryland's Upper Eastern Shore, runs southeast into southwestern New Castle County and then south into Kent County, Delaware. About seven miles southwest of Camden, Dividing Ridge turns southeast again and pushes into Sussex County where it finally disappears north of the Maryland line in what was, some three hundred years ago, the northern part of Cypress Swamp. Acting as the watershed of central and southern Delaware, Dividing Ridge bisects the state so that all of flat Delaware's significant river systems, except the Nanticoke, flow eastward into the Delaware River or Bay, or into the Atlantic Ocean. But unlike the fast-flowing streams of Delaware's piedmont, the rivers further south meander—often in serpentine fashion—across the level landscape, carving out shallow, marshy river valleys as they make their way to the sea. Missing in flat Delaware were the rushing freshets of the piedmont that could be easily harnessed for the future water power needs of European settlers.

The soil of flat Delaware, depending on location, was made up of differing combinations of sand, silt, and clay. Sand was the most abundant of the three components and proportionally increased as one moved south from the Christina River Valley to the southern border of Sussex County. The soil of eastern Kent and Sussex, from the seacoast to about ten miles inland, tended to have a higher clay and lower sand component than did the soil further west, particularly in those areas of Kent and Sussex flanking Dividing Ridge. Although a significant amount of sand improved soil drainage, it also made the land highly susceptible to the leaching of valuable nutrients over time. When annual rainfall exceeds 32 inches or 800 millimeters per year, leaching begins to have a significant impact on soil fertility. All of Delaware has an annual rainfall of more than 40 inches. The relationship between leaching and the increasing sand content found in Delaware's soil was demonstrated, in the early nineteenth century, by the fact that the presence of such natural nutrients as calcium, magnesium and potassium gradually decreased as one moved from northern New Castle County to southern Sussex County.[4] Moreover, the growing presence of sand in the soil, as one moved south, guaranteed that much of the land had difficulty holding and sustaining moisture. This, in turn, accentuated the impact of droughts on crops planted and cultivated by Europeans in much of central and southern Delaware. There were specific areas of Delaware south of the piedmont, however, where the soil had enough silt and clay to prevent the serious leaching of vital nutrients and to retain significant amounts of moisture. Nevertheless, as with most of the rest of the American South, the mediocre nature of most of the soil of central and southern Delaware would impose a low ceiling on the region's standard of living so long as most of the area's residents worked the land to support themselves. In addition, there was something about Delaware's soil that, except for the early colonial period, joined with climate in

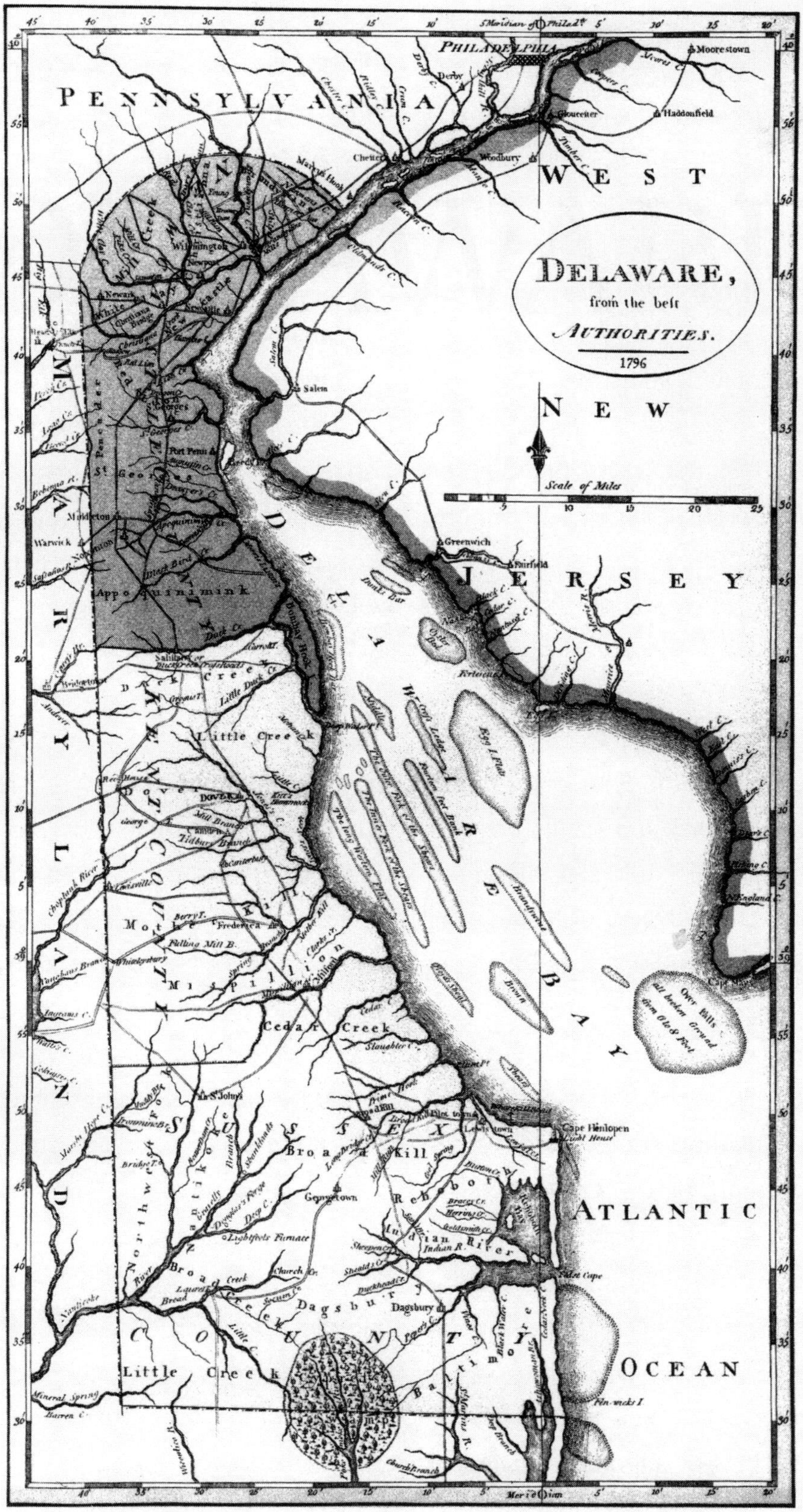

The Great Cypress Swamp— This circa 1796 map of the State of Delaware shows the size and importance of the Great Cypress Swamp, the oval-shaped area straddling Delaware's southern boundary with Maryland. At the time, the swamp extended over some 50,000 acres and gave rise to the headwaters of numerous tidal creeks and rivers, *courtesy of the Delaware Public Archives*.

precluding the commercial production of tobacco.[5]

The first Europeans found that Delaware's climate resembled that of their native lands in having four distinct seasons. Although exact comparative daily temperatures for the seventeenth and eighteenth centuries are not available, anecdotal information and today's statistics—London's mean temperature in 2001 is approximately six degrees warmer in January and sixteen degrees cooler in July than that of Wilmington, Delaware—indicate that summers on this side of the Atlantic were considerably hotter and winters at least as cold or colder than those in southern Sweden, The Netherlands, Great Britain and Ireland. But even in little Delaware, there were modest temperature differences between north and south. Present-day statistical comparisons and anecdotal information from the past suggest that, during the seventeenth century, the annual average temperature in Sussex was about three degrees warmer than in northern New Castle County. This difference translated into Sussex having approximately ten more frost-free days per year than the northern part of the state. The subtle, but real climatic differences between northern, particularly piedmont, Delaware and the rest of the state would be reflected in the choice of either horses or mules as the primary agricultural beasts of burden. In New Castle County, particularly in the piedmont, the horse was preferred. By the early nineteenth century, however, Kent and Sussex were mule country because mules performed better than horses in hot and humid climates.[6]

Despite Delaware's similar or slightly colder winters and considerably warmer summers, Europeans in general and Englishmen in particular continued to equate climate with latitude during the seventeenth and much of the eighteenth century. To them it was astonishing that Delaware, which was on the same general latitude as southern Spain and Portugal, would have winters colder than London, which was latitudinally aligned with Labrador. Johan Printz, Swedish Governor of northern Delaware from 1643–1653, declared the extreme cold of the lower Delaware Valley was unknown in southern Sweden. Peter Lindestrom, in visiting the Delaware Valley in the mid-seventeenth century, noted that the winters were as severe "as the most severe in Old Sweden." In 1719, a third Swede, Andreas Hesselius, wrote with surprise that a Delaware cold snap in late January was comparable to "the strongest which is ever experienced in Sweden."[7] Although recurrent experiences constantly contradicted them, Europeans were slow to alter their ideas concerning the relationship of latitude to climate. Even as late as the American Revolution, English Methodist preacher Francis Asbury was surprised that in Delaware, winters and summers were more extreme than in Great Britain.[8]

Delaware's temperature extremes caused Europeans to lay on heavier clothing in the winter—often furs and animal skins—and to be particular about textiles used to make summer clothing. Delaware's Swedes, for example, found it impossible to do physical labor in the middle of summer's hot and humid days or to "wear any other clothes but linen made quite wide (loose) on account of the oppressive heat."[9]

The hot and humid summers also encouraged the use of enslaved Africans in Delaware's fields and probably restricted sheep raising and the production of certain European root crops. Moreover, Delaware's hot summer temperatures and its often porous soil may have altered strategies for raising grain. But the oppressive heat also raised the stock of watermelons in the eyes of Delaware's first whites because, "during the hot summer...they refresh and cool off a person strongly." Despite having hotter summers than Europe, Delaware's growing season began too late in the spring for successful competition with areas further south in the commercial production of tobacco and cotton.[10]

Delaware's annual rainfall during the seventeenth century far surpassed that of northern and western Europe, and today Wilmington averages more than 45 inches per year while London averages only about 30 inches per year. Unlike northern and western Europe's gentle summer rains, however, most of Delaware's summer moisture tends to come in brief but heavy downpours that are largely wasted in runoff. Two hundred years ago, Delaware's heavy showers caused greater soil erosion in cultivated fields than did the gentler showers of northern and western Europe. The rainwater that did remain on Delaware's fields, quickly evaporated in the hot sun or rapidly filtered through the sandy topsoil into the substrata. Probably as disconcerting to Europeans as the uneven nature of Delaware's summer rainfall was the accompanying fireworks display that lit up Delaware's skies. Although electric storms occurred in Europe, the intensity and drama of thunder and lightning were modest by American standards. In the mid-seventeenth century, Peter Lindestrom observed that Europeans reacted to Delaware's summer storms "with great horror, for when it thunders the whole heaven appears to be on fire, that nothing can be seen but flames of fire and smoke."[11]

Trees, Marshes and Swamps

The earliest Dutch, Swedish and English explorers found most of Delaware covered by mature forests featuring oak and hickory, with some pine trees growing in southern Kent and Sussex. Of course natural and Indian-set fires gave fast-growing pines an initial advantage on the newly burned-over landscape, but soon they would be joined by sweet gum and red maple. Eventually, this natural succession led to other hard woods such as oak and hickory taking over and most of the original pine trees were crowded out. Among other sylvan varieties present in Delaware during the late pre-contact period were peach trees, which were originally introduced to the New World from Europe by the Spanish, probably at the end of the fifteenth century. By the mid-sixteenth century they had been brought north to the woods of the Middle Atlantic region by Native American cultivators or by natural propagation.[12]

Evidence suggests that, because of the large size of old growth trees, in many places underbrush couldn't get sufficient sunlight to grow. This gave a park-like appearance to much of Delaware's forested landscape. In 1739, English evangelist George Whitefield found part of his journey from Lewes to New Castle County as easy as "riding through Hyde Park." Ten years later on his trip from Philadelphia to Wilmington, Peter Kalm noted that

> the trees of the forest were tall but branchless below, so that it left a free view to the eye, and no underwood obstructed the passage between them. It would have been easy in some places to have gone under the branches with a carriage for a quarter of a mile, the trees standing at great distance from each other.

There were also sections of wooded Delaware where vines, particularly grapevines, grew in abundance. In 1616, for example, a Dutch explorer reported that the trees along Delaware Bay were often covered by vines. In fact the site for the town of New Castle was initially called Grapevine Point.[13] These park-like and vine-dominated sylvan settings combined with wetlands, meadows, streams, rivers, bays and the Atlantic Ocean to provide habitat for a variety of fauna, ranging from insects to large birds, fish and mammals.

Particularly crucial to survival of Delaware's fauna were the brackish and saltwater marshes that dominated the coast from the mouth of the Christina River in the north to Fenwick Island in the south. From northern Delaware to Cape Henlopen these marshes generally fronted on the Delaware Bay and River. South of Cape Henlopen, they nestled behind a line of sand dunes that protected them from the pounding surf of the Atlantic Ocean. These coastal marshes were vital to the ecological health of Delaware because they nourished the very base of the food chain that eventually fed so many of Delaware's fish, reptiles, birds and mammals.

Scattered across the landscape further inland were freshwater marshes and swamps—marshes were treeless wetlands while swamps were tree-covered wetlands—that also nourished food sources for concentrated populations of wildlife. In addition, these freshwater marshes and swamps served as headwaters for Delaware's tidal creeks and rivers. The largest freshwater swamp was the Great Cypress Swamp, which stretched from southern Sussex County south into Maryland. It was a primary source of the Nanticoke and Indian rivers in Delaware, and the St. Martins, Pocomoke, and Wicomico rivers on Maryland's Eastern Shore. Large numbers of bald cypress trees gave Cypress Swamp a landscape that was "truly characteristic of the deeper South."[14]

Fauna

Delaware's fresh and saltwater marshes and swamps produced swarms of mosquitoes that made life miserable for both Native Americans and Europeans during the warmer months. Although mosquitoes were present in Europe, the numbers and aggressive nature of the Delaware varieties were beyond the previous experiences of early white colonists. In the mid-eighteenth century, Sweden's Peter

Kalm reported that in the lower Delaware Valley, which included northern Delaware, the mosquitoes sucked "so much blood that they can hardly fly away." The resulting bite left a bigger red welt than mosquito bites in Sweden. Even if one closed all the doors and windows, the English style chimneys had no dampers, which afforded the insects "free entry into the houses."[15]

Mosquitoes of the lower Delaware Valley, according to the hyperbolic Earl of Carlisle in 1778, were "as large as sparrows." To protect himself he wore trousers in the summer, "which is the constant dress of this country." Ten years later, an anonymous observer found in the Cape Henlopen area of eastern Sussex,

> inconceivable swarms of mosquitoes and sand flies which infest every place, and equally interrupt the tranquility of the night and the happiness of the day. The attacks are intolerable upon man as well as beast.[16]

Particularly good breeding habitat for mosquitoes was provided by the many fresh and saltwater marshes and swamps along the sluggish tidal streams of flat Delaware. The result was that enormous numbers of potential disease vectors were in place, awaiting the introduction of serious contagion from the other side of the Atlantic such as malaria and yellow fever.

Although some other Delaware insects, and some of its smaller mammals and larger birds were less vexing than mosquitoes, they did possess traits that made them seem exotic to Europeans. Cicadas, tree frogs, opossums, American passenger pigeons, bald eagles, and fireflies, for example, were all encountered for the first time in the wild by Sweden's Andreas Hesselius while living in Delaware in the early eighteenth century.[17] The loud noises made by the first two, the opossum's peculiar pouch, the enormous flocks of passenger pigeons that filled the sky in the fall, the unusual size of the soaring bald eagles, and the light of the firefly, an insect found in some parts of Europe but evidently not a familiar sight to most Swedes, were genuine curiosities.

Lack of familiarity with some of Delaware's indigenous insects combined with mounting political rivalry between the Swedes and the Dutch to produce a brief military panic in northern Delaware. In the mid-seventeenth century, New Sweden's strongest military post was Fort Christina on the north bank of the Christina River, only two miles upstream from the Delaware. In 1651, the Dutch challenged Swedish hegemony in the region by building Fort Casimir on the west bank of the Delaware at the present site of the town of New Castle, a few miles south of Fort Christina.

One night in 1654, a newly arrived soldier from Sweden who, evidently, had never seen a firefly was on guard duty at Fort Christina. Suddenly, he shouted "Alarm, alarm, the enemy is about." The drummer beat a call to arms and the fort's inhabitants rushed to their defensive stations. When asked what he had seen, the soldier responded, "The enemy is here close to the fort, see how many burning torches there are." But when others, who were more

American Bison were once native to Delaware.

seasoned in the New World, looked out into the darkness, they were relieved to see that the enemy torches were nothing but fireflies.[18]

The first Europeans also were astonished at the numbers and varieties of fish in Delaware's bays and rivers. Nothing in their European experience had prepared them for the fecundity of nearby waters. In 1633, for example, a Dutch vessel in the mouth of the Delaware River caught in one throw of its nets as many "as thirty men could eat of perch, roach* and pike." William Penn wrote that the fish in the Delaware were numerous and included sturgeon, herring, rock, shad, catshead, sheepshead, eel, smelt, perch, roach; and in the streams flowing into the Delaware, freshwater trout were plentiful. In addition, whales were common in Delaware Bay as were great oyster banks.[19]

Of the largest indigenous mammals, the occasional bison was probably the least familiar to the Swedes, Dutch, English, Welsh, Irish and Scots-Irish who settled in southeastern Pennsylvania and Delaware. Of course a European bison or "wisent" had existed since time out of mind, but it had become a rare species by the seventeenth century, surviving in very limited numbers in the Bialowieski Forest of present-day eastern Poland near the Byelorussian border, and on the north slopes of the Caucasus Mountains. Except for natives of these areas, few Europeans had seen bison or even knew very much about them.

In North America, by contrast, perhaps as many as thirty million bison ranged the Great Plains by the seventeenth century. In the latter part of the sixteenth century some of the Great Plains' bison crossed to the east bank of the Mississippi and, in very small herds, moved inland towards the Atlantic Coast. Just how far these buffalo moved east before being killed by both Native Americans and westward advancing whites is unclear and a matter of dispute among historians. Nevertheless, there is indication that a small number reached eastern Pennsylvania and even Delaware before being hunted to extinction. In 1698, for example, Welsh Quaker Gabriel Thomas noted the presence of bison—he called them buffalo—in eastern Pennsylvania. In the late eighteenth century the last small herd of wild bison in Delaware was supposedly spotted on Murderkill Neck, southeast of Dover by Drummer Gray, an enslaved African-American. Although wild bison were gone from Delaware by the end of the eighteenth century, a few farmers raised

**Roach— a fish similar in appearance to a European variety of sunfish.*

buffalo—perhaps descendants of bison calves captured after their wild buffalo parents were slain—until at least 1843. Unlike cattle, however, bison as livestock were never numerous. The bison were difficult to domesticate and their size, strength and tendency to roam made them a constant threat to farm fences.[20]

Deer and elk, two ungulates found in many parts of Europe, were also indigenous to Delaware. Although elk were never numerous, white-tailed deer were present in considerable numbers and supplied European settlers with protein to eat and skins for commercial trade. In Sussex, deerskins were used for money during the late seventeenth century. The use of buckskin as currency in Delaware and across much of early America eventually led to "bucks" becoming a slang term for dollars. The slaughter of deer for both their skins and their meat was extraordinary. As early as 1657, a Dutch official reported that thousands of deer were annually killed in New Netherland on the Delaware. One witness reported that deer carcasses weighing from 60 to 100 pounds were such a glut on the market that one could be purchased at a reasonable price. Such fur-bearing animals as beaver, fox, otter, and bear were also found throughout seventeenth-century Delaware. Three large Delaware carnivores, the solitary cougar and the more gregarious gray wolf and red wolf, were eventually hunted to extinction because of their predatory nature rather than for their pelts.[21]

In 1634, an English explorer reported that in the lower Delaware Valley, "the quantity of fowl is so great as can hardly be believed." Perhaps no bird drew greater attention than the wild turkey, which could be found only in North America. Pre-Columbian Mexicans domesticated the wild turkey, and about 1517 the first tame birds were sent east across the Atlantic to Spain. By 1541, they could be found in Great Britain. In the next century, domesticated turkeys were carried back across the Atlantic to eastern North America, including Delaware, by Anglo-American colonists. But what an additional bonus for Europeans to find that considerable numbers of the domesticated turkey's wild cousin, eight- to more than twenty-pound packages of easy-to-hunt protein were obligingly perched on tree limbs in Delaware's forests, just waiting to be harvested.[22]

Of the variety of snakes found in seventeenth century Delaware, only the rattlesnake and the copperhead were poisonous. Although both snakes existed only in the New World, it was the rattlesnake that fascinated Europeans the most. In the lowlands of the Middle Atlantic region, the indigenous variety of rattlesnake was the timber rattler which, at three to four and one half feet, generally averaged a foot longer than the copperhead. In 1684, a German visitor to the west bank of the lower Delaware Valley reported that "there are more rattlesnakes (whose bite is fatal) in the land than is agreeable to us."[23] Some observers, however, took even more literary license in describing the rattler. In the mid-seventeenth century, Peter Lindestrom noted that rattlesnakes were "a kind of large, horrible and abominable snake," with "jaws like a dog, they can cut and bite off a person's leg, as if it had been cut by an ax."[24]

American Cougar

Value Systems

How much did the respective value systems of Delaware's Indians and whites influence their interaction with the state's flora and fauna in the seventeenth and eighteenth centuries? In recent years it has become fashionable to point to Native Americans as our first great ecologists. Their value system, it was said, prevented them from exploiting the natural world in a manner that threatened its health and stability.

But this may be too simplistic a view. After all, the preservation of the New World's flora and fauna probably depended at least as much on population density and on the level of available hunting technology as it did on the value system espoused by indigenous people. Indeed, there is some indication that when Native Americans possessed the necessary hunting technology and were present in comparable numbers, they were as great a threat to America's wildlife as were people of European ancestry. At the end of the last Ice Age, for example, Native American hunters to the west of Delaware may have played a key role in destroying a significant number of North American megafauna such as wooly mammoths because, at the time, they possessed the necessary weapons to do the job. Moreover, there are indications that, after the introduction of the bow and arrow, the deer population in the nearby Chesapeake region was significantly reduced by overhunting prior to the advent of Europeans. Shortly after the arrival of whites, the acquisition of European weapons by many Native Americans led to a dramatic increase in the slaughter of fauna. By the mid-seventeenth century, for example, Indians along the Delaware River had guns. In 1654, as he moved south along the Delaware's west bank, Peter Lindestrom noted that by the time he reached the Schuylkill River, the "hunting and shooting" by Native Americans had taken a devastating toll of local wildlife.[25]

Resurrecting and then ennobling the ideas and values of precapitalist peoples, such as Native Americans, is particularly attractive to those who believe that most of the world's contemporary ills, particularly its environmental problems, are rooted in the practices of capitalism. But as William Beinart and Peter Coates remind us, this is "a powerful ideological statement rather than good history."[26] What is better history is the observation that a higher level of technology, a more exploitative approach to the use of natural resources, and a rapidly expanding population base combined to cause Europeans to leave a far larger imprint on Delaware's natural landscape than the Lenape and the Nanticoke.

The Impact of Europeans

In any case, European diseases, and the economic, social, and political disruptions caused by the rapidly increasing white population killed off or drove most of Delaware's Native Americans away from the state by the mid-eighteenth century. But even as early as the late seventeenth century, the interaction between Delaware's humans and their natural world primarily featured the activities of men and women of European rather than of Native American ancestry.

Since time out of mind, Europeans had seen wilderness as both alien and threatening. But God, according to Genesis I:26-29, directed mankind to "rule the fish in the sea, the birds in the heaven, the cattle, all wild animals on earth and all reptiles that crawl upon the earth." To the European mind, God had given humans dominion over wilderness in order to transform and domesticate it. Driven for centuries by their own temporal desires and a clear mandate from Heaven, Europeans gradually turned their own continent's wilderness into cultivated fields and grassy meadows. As early as the mid-eleventh century, for example, woodlands now accounted for less than fifteen percent of the English landscape.

Timber Rattlesnake

By the seventeenth and eighteenth centuries, extensive forests had generally disappeared from all but certain peripheral areas of Europe such as northern and central Scandinavia.[27]

When Europeans arrived in the New World, however, they were confronted with a wilderness that was as alien and as threatening as Europe's dark forests had once been to their distant ancestors. Thus, the initial challenge in the New World was to beat back wilderness and, in the process, recreate in America a countryside reminiscent of the Europe left behind. In shaping landscape as well as in shaping certain aspects of Delaware's cultural life, early white settlers worked hard to replicate the European model.[28] Trees were cut down, fields cleared, streams dammed, marshes drained, and homes were built as wilderness was transformed into a semi-domesticated landscape. But in the process of beating back wilderness, white Delawareans were only partially successful in recreating a European landscape and culture on this side of the Atlantic. The problem was that certain physical conditions in Delaware and in the rest of the New World made exact replication impractical.

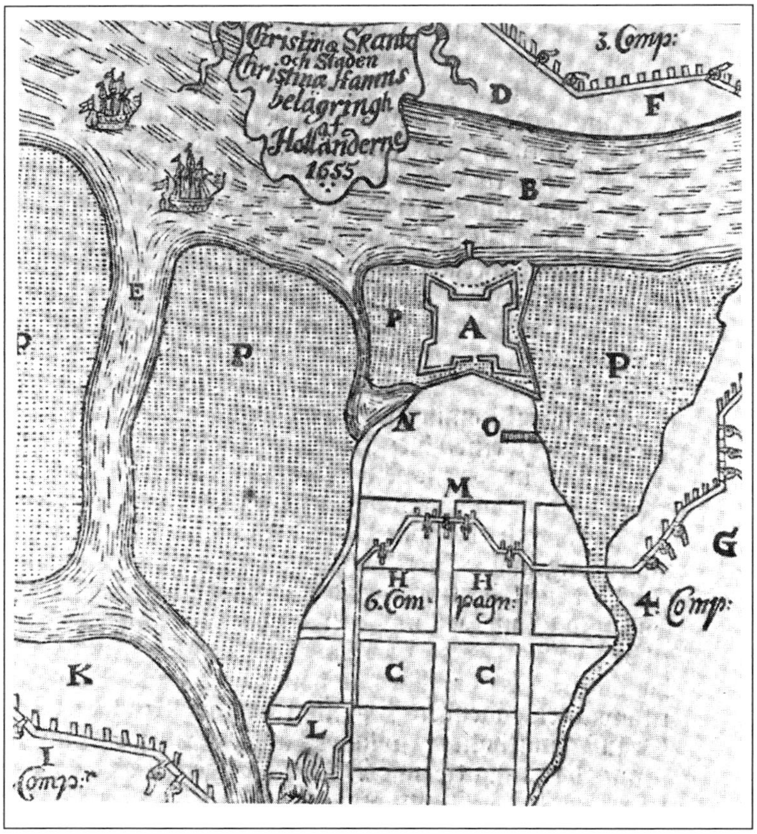

Fort Christina— Swedish soldiers and settlers built Fort Christina soon after their arrival and garrisoned it with troops. A wall surrounded the small square fortress which appears as "A" on the map. The village of Christinaham was laid out on a grid pattern nearby, *map courtesy of the Historical Society of Delaware.*

The earliest Europeans were first drawn to Delaware by the desire to exploit its fur-bearing animals and its oil-bearing whales. In 1631, the Dutch established the first European settlement in Delaware along the west bank of Lewes Creek and called it Swanendael because of the large number of swans in the area. Within a year, however, local Native Americans—perhaps "Sickonese" — successfully attacked the small outpost and killed all thirty-two of its white residents. In 1632, a second Dutch expedition arrived at Lewes Creek, found the skeletal remains of the slain colonists, made peace with the local Indians, and then remained in the area until 1633 to harvest the whales that seemed so plentiful in Delaware Bay. Disappointed that of seventeen whales harpooned, only seven were actually recovered, the Dutch abandoned Delaware for the next eighteen years because the Swanendael colony proved unprofitable.[29]

Sweden was attracted to Delaware because of Dutch commercial interests in the lower Delaware Valley. During the first half of the seventeenth century, Sweden was an important European military power but a second-rate commercial entity. In fact, most of the commerce on the Baltic Sea was carried on by Dutch merchants in Dutch-owned ships. For that reason, when the New Sweden Company was formed in 1637 to plant a colony in the Delaware Valley, the Swedes welcomed Dutch investment and invited Dutch entrepreneurs to join the board of directors.[30] Led by Dutch businessmen and Swedish government officials, the New Sweden Company established Delaware's first permanent European settlement in 1638 at Fort Christina on the north bank of the Christina River, about two miles upstream from the Delaware. Although the commander of the colonizing expedition, the captains of the expedition's two ships and half of their crews were Dutch, most of the colonists aboard were subjects of the Swedish crown.[31]

In subsequent years, Dutch interests in the

Company were bought out and "New Sweden"—as the colony was called—increasingly became a Swedish venture. With land holdings purchased from Native Americans, New Sweden stretched along the west bank of the Delaware Bay and River from Cape Henlopen in the south to the falls of the Delaware in the north, just opposite the site of the future city of Trenton. On the east bank of the Delaware Bay and River, New Sweden stretched from Cape May in the south to present-day Gloucester City, New Jersey in the north. Most of New Sweden's colonists lived on the west bank between the Christina River and the Schuylkill. Although the colony's capital was moved in 1645 from the north bank of the Christina to Tinicum Island, just south of the present site of the Philadelphia Airport, Fort Christina continued to be the colony's primary commercial center and chief port.[32]

Plans for a thriving trade in animal furs and skins, which had initially attracted the Dutch and the Swedes to the lower Delaware Valley, met with some initial success. The first two Swedish ships to return to Europe from Fort Christina, for example, carried a combined total of 2,200 beaver, bear, and otter pelts. By 1663, European fur traders in the Delaware Valley were annually purchasing up to 10,000 pelts from Native Americans, but now most of the fur-bearing animals were hunted or trapped some distance inland from the lower Delaware Valley. With the rapid depletion in the number of fur-bearing animals, it is not surprising that in the immediate area of New Sweden, the beaver trade had just about ended as early as 1648. By 1710, it had pretty much collapsed throughout the entire Delaware River Valley.[33]

Into the late seventeenth century, Delaware Valley colonists still believed the supply of whale oil was endless. William Penn perpetuated the idea in 1685 when he wrote, "the mighty whales roll upon the coast, near the mouth of the Bay of Delaware. Eleven caught and worked into oil [in] one season." It was easy to believe there would be a "considerable profit" from whaling because the whales seemed "so numerous." But Penn's statement also indicated that whales, which had once been plentiful inside the mouth of Delaware Bay were no longer a common sight north of Cape Henlopen. To find their quarry, whalers had to venture beyond the bay into the Atlantic.[34]

To the Swedes and to the Dutch, who had conquered New Sweden in 1655, it became clear that the lower Delaware Valley could not long maintain an economy primarily based on the annual harvest of animal pelts and whale oil. Clearly agriculture and other, more sustainable economic pursuits had to be encouraged to attract larger numbers of white colonists and to provide the economic foundation for a more prosperous future. But unlike the fur trade, which required the maintenance of wilderness, successful agriculture demanded the destruction of wilderness. In fact, no human activity in history has been as destructive of the planet's wilderness as agriculture.

The Finns

But what Old World peoples were skilled at cutting down trees and clearing new fields to prepare the way for agriculture? During the seventeenth and eighteenth centuries, Delaware was primarily settled by the Dutch, Swedes, Finns, English, Scots-Irish, and Africans. While Africans would provide much of the labor in transforming the Delaware landscape from forests to fields, their relatively late arrival in the colony—they were not brought to Delaware in large numbers until the early and mid-eighteenth century—their lack of familiarity with the tasks involved in clearing nontropical forests, and their enslaved status meant that the specific values, skills and strategies crucial to pushing back Delaware's wilderness were not, on the whole, African.[35] Because the Dutch and English had eliminated most of their own vast forests long before the seventeenth century, their descendants had lost the necessary skills and the distinctive value system that were prerequisites to the successful conquest of the New World's forests. Only on the periphery of northern and western Europe did there exist people who still had the skills to quickly turn much of the wooded wilderness of the Delaware Valley into semi-cleared fields.[36]

The Scots-Irish lived on this periphery but, by the mid-seventeenth century, most of the remaining forests in their homeland in northern Ireland were under royal or aristocratic protection and were, therefore, off limits to the ax. In short, the Scots-Irish no longer had the skills or values of a forest-clearing people. In time, after immigrating to the Delaware Valley in very large numbers in the early eighteenth century, the Scots-Irish would play a crucial role in the settling of the American frontier. But this would happen only after they learned the necessary skills and attitudes for a successful pioneer existence in the forest from another ethnic group already in the Delaware Valley.[37]

During the Swedish occupation of northern Delaware and the rest of the lower Delaware River Valley (1638–1655), the colony was under the leadership of the New Sweden Company. To make agriculture successful in its newly established foothold, the New Sweden Company sought out people who still possessed the requisite values and skills to quickly clear some of the land for crops. They found their pioneers among the Finns who had migrated from Finland to the untamed wilderness areas of central Sweden in the late sixteenth and early seventeenth centuries, a time when Finland was part of the Swedish Empire.

The first Finns arrived in the lower Delaware Valley in 1640, and by 1655 they probably represented at least one third of New Sweden's population. Immigration to the lower Delaware Valley by these "forest people" continued after the Dutch (1655) and English (1664) conquests of the area. Thanks to this continuing influx of Finns and their unusually high birth rate, by the 1670s they may have become a majority of the small number of Europeans living in the lower Delaware Valley. In the future state of Delaware, most Finns initially lived within a few miles of the Delaware River in a band that stretched from what is now the Pennsylvania line in the north to the town of New Castle in the south. Perhaps 30 percent of the European population of Delaware was Finnish by the 1670s.[38]

While living on the Finnish frontier and then in their new homes in the wilderness areas of central Sweden, these Finnish pioneers, according to Terry G. Jordan and Matti Kaups in their 1989 work, *The American Backwoods Frontier*, created a distinct culture that became the basis for "the most successful forest colonization process ever devised." But that culture also led to an extraordinarily wasteful stewardship of the land and its wild creatures. The immigration of Finns to the lower Delaware Valley just about guaranteed that the fertility of the region's soil and the fecundity of its flora and fauna would be seriously compromised.[39]

Why should we look so closely at the Finns who, during the seventeenth century, probably never numbered more than five or six hundred in the entire lower Delaware Valley? After all, they would soon be easily outnumbered by first English and then Scots-Irish immigrants? The answer is that the first cultural imprint is far more durable than subsequent cultural imprints, even if the latter are left by groups of far greater numerical size and economic power.[40]

Although the Swedes joined with the Finns in becoming the first Europeans to establish permanent homes in Delaware and in the rest of the lower Delaware Valley, it was the Finnish imprint rather than the Swedish one that was the most durable, not only in Delaware but across much of the nation. The Finnish forest culture won out because it was so well suited to meeting the challenges of life on the American colonial frontier. Understandably, many Swedes and many members of other ethnic groups, who subsequently moved to the lower Delaware Valley, quickly adopted much of the Finnish forest culture. By 1663, Dutch authorities in control of Delaware south of the Christina River were acknowledging this fact when they included Swedes along with Finns as "particularly fitted" for pioneer life in forested early Delaware.[41]

But not everyone admired the Finnish pioneers and their forest culture. Back in Sweden the Finns may have been successful in turning forests into fields, but their techniques created no end of controversy. Their slash and burn method of agriculture and their wasteful hunting practices were condemned for needlessly destroying trees, carelessly draining nutrients from the soil, and recklessly slaughtering wildlife. No wonder they were called "Forest destroying Finns," and were charged with shooting and trapping animals only for their hides and thus strewing Sweden's depleted forests with rotting carcasses.[42]

Understandably, the Finns were in perpetual conflict with Swedish villagers and royal officials for poaching, smuggling, tax evasion, and setting fire to the Swedish countryside. In view of their Old World behavior, it isn't surprising that many of the first Finns to arrive in Delaware had been forced by the Swedish government to leave Sweden and cross the Atlantic. Nor is it surprising that the Finnish population in Delaware and elsewhere in the lower Delaware Valley resented the Swedish Crown's strong-arm tactics, and harbored a strong antipathy to government controls of any sort.[43]

In addition to their visceral distrust of

government, which subsequently became a familiar trait of many American frontiersmen, the Finns brought with them specific forest-clearing techniques, experience with a subsistence-based economy, and certain unique values that would eventually characterize most American pioneers until they ran out of forested landscapes west of the Mississippi Valley. To begin with, they had generations of experience at chopping down trees, and they possessed a superbly designed steel ax that was superior to the axes and hatchets of the Dutch, English and Scots-Irish, and far superior to the stone implements of Native Americans. It was said that with his ax, one Finnish woodsman could clear-cut an acre of virgin forest in seven to ten days. Because most frontier farmers rarely cultivated more than five or six acres, and often only one acre, felling trees was not excessively time consuming for the Finns.[44]

They and their emulators either cut down trees, or killed them by chopping a ring pattern around the tree trunk, a process called girdling. From several months to a couple of years later, both the felled trees and the still-standing-but-now-dead ones were dry enough to burn. Torching the dead timber left a small section of cleared woodland, covered by a layer of ashes, that produced a burst of soil fertility lasting three to five years. By then, however, repeated planting without adding additional fertilizer drained the soil of nutrients. Long steeped in the practice of geographic mobility, the Finns simply cleared new land nearby or picked up what few belongings they had and moved deeper into the forest where they replicated their Old World homes by building log cabins.[45] In subsequent years the Finnish log cabin became a great American icon that conjured up the image of a restless frontier people repeatedly pushing back small portions of the forest, spending a few years depleting the land of its natural assets, and then pushing onward to conquer new wilderness areas.

In addition to corn and some tobacco, Finnish pioneers in Delaware planted only limited amounts of other grains and vegetables in their stump-laden fields. Because commercial grain production—which generally required plowed fields—was not important to them, the Finns did not involve themselves with the backbreaking work of pulling stumps out of the ground by their roots. In the fields of Finnish farmers, the hoe was the primary agricultural implement and the plow was used rarely because it was physically exhausting to plow fields filled with stumps. Not needing cattle and horses to pull plows, the Finns were more committed to raising pigs than to any other livestock. Swine matured quickly and seemed to thrive under very little supervision in nearby woods where they foraged for mast and roots on the forest floor.[46]

As in the Old World, the Finns depended heavily on hunting to put food on the table. Unlike other colonists both in Delaware and throughout most of America, they had considerable prior experience with firearms and used them to deplete the local deer and wild turkey populations.[47] Indeed, the decision by frontier Finns to pick up and move deeper into the forest may have been dictated as much by the diminished numbers of deer and wild turkey in their immediate neighborhoods as by the declining soil fertility of their fields.

In some ways the economic lifestyle of the Finnish pioneers was similar to that of the Lenape who predated them in the lower Delaware Valley. Hunting game and gathering nuts, fruits, seeds, and roots from both forest and meadow played a crucial role in the household economies of both peoples. Moreover, the Finns and their emulators also committed themselves to growing Indian corn, tobacco and a few other Native American crops. But in environmental impact and food yield, the economies of the two peoples did not produce similar results. As Jordan and Kaups point out, Finnish pioneers in the lower Delaware Valley produced a farming pattern that "was more complex, destructive and productive than the Indian type."[48] It was inevitable that the early presence of Finns and their white imitators, armed with a superior ax technology, a more proactive attitude towards nature, and a greater population density, would send shock waves through Delaware's natural landscape.

In 1655, a hyperbolic Peter Lindestrom waxed euphoric about nature's gifts to Delaware. The Christina River was "rich in fish," the entire Christina Valley "was a land flowing with milk and honey," and its forests were filled with "valuable trees and fruits." Indeed, nature had been so generous with the river valley that the pen was "too weak to describe, praise and extol it." An effusive Lindestrom noted that the rest of Delaware, from the future boundary of Pennsylvania in the north to Cape Henlopen in the south, was blessed with fertile soil. And yet, by 1655, one senses that these gifts provided by nature were already under siege by the Finns who now could be found as far south as Sandhock (future town of New Castle) and were mounting hunting expeditions into Delaware's unsettled regions further south and west.[49] Simply put, it was Delaware's fate to begin its European era under the influence of the most successful forest-conquering culture in history, but one that used fire carelessly, chopped down trees indiscriminately, practiced agriculture negligently, and destroyed wildlife enthusiastically.

Although the Finns may have left the most significant cultural imprint, other European practices and attitudes played a role in shaping Delaware's interaction with its natural landscape. The Dutch, for example, brought a hard nose for profit to Delaware which led at least some of them to favor commercial agriculture over the slash-and-burn subsistence farming practices of the Finns. To produce fields for their commercial crops, the Dutch preferred building dikes and digging drainage ditches to clearing forests. And why not? In seventeenth-century Europe, no people were as skilled as the Dutch in pushing back tidal waters and in draining marshlands and swamps in order to claim new lands for the plow. In fact, in 1675, slightly more than a decade after the English conquered the Dutch holdings along the west bank of the lower Delaware, Governor Edmond Andros demanded that the directors of a dike-building and trench-digging project, aimed at draining a marsh just north of the town of New Castle, be Dutch because there were few others "here who have the knowledge of such work."[50]

Soil exhaustion had long been a problem in the Old World. To deal with it, Dutch and other European farmers annually rested one third to one half of their fields so that the worn-out soil could regain some of its lost fertility. But fallow farming, as this practice was called, took a high percentage of cropland out of production each year and thus significantly reduced the potential harvest. Although the Dutch were constantly claiming new farmland from the North Sea, by the early seventeenth century they no longer had enough cropland to successfully feed their rapidly growing population without dramatically reducing the practice of fallow farming.

But how to abandon fallow farming and yet maintain a certain level of soil fertility? The Dutch solved this problem by heavily manuring their fields and constantly rotating their crops. So successful were these new methods of maintaining soil fertility that, by the late seventeenth century, fallow farming was no longer practiced in much of The Netherlands. The use of these innovative methods soon made Dutch farming the envy of Europe. The English took great notice of the agricultural revolution taking place across the North Sea and were soon applying Dutch practices to increase the fertility of their own fields and to drain their own marshes.

While both the Dutch and English were moving forward with many remarkable agricultural advances during the seventeenth and eighteenth centuries, these advances were not transferred en masse to colonial Delaware. Although, by the mid-seventeenth century, some of Delaware's Swedes and Dutch practiced a crude form of crop rotation, there is no evidence that it was practiced by a preponderance of the area's settlers. Despite the obvious advantages of the new farming model, most of Delaware's Dutch and English farmers stubbornly continued to follow the more traditional but wasteful agrarian practices of their more distant ancestors in Europe. All of this caused Peter Kalm to observe, in 1749, that in the lower Delaware Valley as well as in the Anglo-American colonies to the north that he visited, the advanced nature of English husbandry that was clearly visible in some parts of Great Britain, was "scarcely recognizable here."[51]

But even more significant than the survival in Delaware of ancient and wasteful Dutch and English farming practices was the availability of cheap land on this side of the Atlantic. Wherever farmland is inexpensive and easily accessible, pressure to make every acre very productive is considerably diminished. Cheap land in colonial Delaware not only made fallow farming feasible, it also encouraged a certain disregard for manuring and constant crop rotation because one could easily replace played-out fields by simply clearing nearby forests. In addition, the initial abundance of forests caused most farmers to turn their livestock loose to roam freely through the woods. This made it impossible to gather livestock manure to spread on fields and meadows. No wonder Delaware farmers preferred rotating fields to rotating crops.[52]

In summary, the example of the Finns proved to be very influential in shaping the forest-clearing and farming practices of subsequent waves of European colonists in Delaware. Despite a seventeenth and eighteenth century agricultural revolution in The Netherlands and in England, agricultural reform made little headway in early Delaware. Rather, the memory of more ancient and wasteful European agricultural practices and the presence of vast stretches of cheap virgin wilderness reinforced the tendency of some of Delaware's Dutch, and most of its Swedes, English and Scots-Irish to follow the extraordinarily successful, but environmentally destructive, Finnish model when interacting with the natural world.

CHAPTER TWO:

Chopping Down Trees and Decimating Wildlife, 1631–1867

"There were not nearly so many edible birds... as there used to be when they were children."

IN 1837, BRANDYWINE MILLER EDMUND CANBY SAT DOWN TO "AN ELEGANT supper" of "terrapins, venison, bear's meat and wild turkey..." The venison, bear's meat and wild turkey were most likely imported from outside Delaware because by this date deer, bear and wild turkey were either extinct in Delaware or very close to it. Only terrapins continued to be commonly caught in the First State and their numbers would drop dramatically during the nineteenth century. A century earlier, Delaware gentry could have sat down to the same elegant supper, but with the pride and satisfaction that the meal was a product of Delaware's woods, meadows, swamps, marshes and waterways.[1]

By the close of the eighteenth century, Delaware's once abundant natural resources were seriously depleted. An expanding white population had cut down or girdled most of the state's oldest and most valuable trees. Hunters and trappers had so decimated wildlife that, by 1800, Delaware's few remaining bears had taken refuge in Sussex County's Cypress Swamp, while elk, wild bison, cougars, and wolves were extinct in the state. Even the white-tailed deer, once such an important source of protein and pelts to both Indians and whites, was now an uncommon sight in Delaware's woods, fields and meadows. It was also apparent that the numbers of certain fish and wild fowl were considerably less than in earlier times. By 1800, wasteful farming practices had so drained much of Delaware's soil of nutrition that many First State farmers were either striking out for the virgin lands beyond the Appalachians, particularly the Ohio River Valley, or at least seriously considering such a move. What had happened to Delaware since 1655, when it was seen as so graced by nature that it was called a land "flowing with milk and honey?"

Trees

Among the first natural assets to be consumed were Delaware's trees. The Finns, of course, were so extraordinarily efficient and comfortable in clearing virgin forests that turning them loose in Delaware's sylvan setting was like turning loose "Brer Rabbit in a briar patch."[2] In addition to adopting Finnish timbering techniques, other Delaware colonists emulated the Finns in annually burning off underbrush in surrounding woods and marshes. Although the practice of torching the countryside has been commonly credited to Native Americans, setting fires to improve grazing land for livestock and to return a short-lived fertility to the soil was also a centuries-old practice carried on in certain parts of Europe by the Scots, Welsh, and some west-country English farmers as well as by the Finns.[3] In short, the annual burning of Delaware's landscape probably had its roots at least as much in the customs of Europe's northern and western periphery as it did in the Native American example.

Like most routine human activities, the annual burning of Delaware's landscape was rarely recorded in detail during the seventeenth century. In fact, the earliest surviving detailed record of widespread firing of fields and woods by whites only dates back to 1714, when Delaware's colonial legislature restricted legal burning to March 10 and 11. The legislature expected the spring dampness of those two days to diminish the danger of a runaway conflagration. The purpose of human-set fires, as then recorded, was to "burn up all the great quantities of leaves and rubbish which after last fall and winter have, like snow drifts, covered the ground." Human-set fires on those two days in March were so widespread that the entire Delaware countryside was "full of smoke and flame." But for those who watched their homes and outbuildings, "no damage" happened. The annual burning prevented a long-term accumulation of dead branches and other highly combustible detritus so that, most of the time, the fires were not hot enough to seriously harm larger trees. And the results were almost instantly gratifying because, after only a few days, one could "see how gloriously the grass begins from the black ash in the clean woods."[4]

Nevertheless, the torching of the countryside proved to be a double-edged sword because fires were unpredictable and sometimes raged out of control, causing the destruction of valuable timber, fences and buildings. In the years that followed, many Delawareans probably refused to restrict the firing of their fields to March 10 and 11. In 1740, the colonial assembly stretched the legal annual burning days from March 10 to May 1. Those dates were reaffirmed by the state legislature in 1827.[5]

As previously noted, the Finns and their earliest imitators left the stumps of fallen trees in the ground. But unlike the Finns, Delaware's other Europeans had come from an Old World countryside in which stumps had long ago been removed from farm fields.[6] On arriving in the lower Delaware Valley, it soon became apparent to them that corn and tobacco, the two Native American crops that would prove very significant to early Delaware agriculture, could be raised in stump-filled fields. Wheat, however, was another story. By the mid-eighteenth century, Philadelphia had become a major wheat-exporting port and most of Delaware was part of its expanding agricultural hinterland. Successful wheat cultivation demanded that stump-filled fields be cleared so that the land could be first plowed and then, after seed was broadcast, dragged to provide a thin layer of soil to cover the seeds. Increasingly driven by the profit potential of wheat, Delaware farmers acquired oxen to jerk tree stumps out of the ground, to pull iron-tipped plows to prepare the fields for planting and then to drag the fields to provide a thin soil cover over the sowed wheat kernels. Oxen were preferred over horses on most eighteenth century Delaware farms because, "after serving at the plow and cart to a good age, they could be fatted for the table," while horses were perceived by most Anglo-Saxons as "not good to eat." Moreover, unlike horses, oxen did not need to be fed grain and, in general, ate less hay and other food.[7]

Stump-filled fields gradually diminished in number during the second half of the eighteenth century, but they continued to be common in certain parts of the state as late as 1867. That year a reporter from New York City noted that large tracts of "stumpland" could be found in the Blackbird Creek area of southern New Castle County. Further south in Sussex there were "barren tracts with high, burnt pine stumps." Clearly, the seventeenth century Finnish precedent of leaving stumps in fields after cutting down or burning trees continued well into the second half of the nineteenth century. In other areas of the state, however, such as the Hockessin Valley of northern New Castle County, "every foot of land," except in wood lots, was under "careful cultivation" in fields whose stumps had long since been removed.[8]

Increasing pressure was exercised on Delaware's diminishing timber resources by the demands of a rapidly growing population. By 1800, there were about 64,000 people in Delaware, which may have been as many as twenty-five to forty times the number of humans as in the pre-contact era. In each successive decade from the establishment of the first permanent European colony by the Swedes in 1638 until the end of the eighteenth century, increasing numbers of trees were cut down, not only to meet the soaring demand for more cleared farmland, but also for more heating and cooking fuel to feed the "great blazing fires" that were often "halfway up the chimney." During the eighteenth century, approximately ten to twenty cords of wood—a cord measured four feet by four feet by eight feet—may have been burned annually in each of Delaware's rural households for cooking and, most of all, for heating. Although Delaware's fireplaces may not have consumed as much timber as fireplaces further north, they were in use for at least three-fourths of the year to provide heat, and year-round to cook meals.[9]

Because log cabins and other colonial homes were so drafty that an estimated eighty to ninety percent of the heat produced in open fireplaces went up the chimney, enormous amounts of wood were consumed in an often vain attempt to keep Delawareans warm. One example was the George Read house, built in the town of New Castle in the first few years of the nineteenth century. A very large and elegant home, it initially consumed more than fifty cords of wood per year for warmth and cooking. If all of that wood were stacked together it would stand four feet high, four feet deep, and stretch about 448 feet, which is considerably longer than a football field.

Wood-burning stoves were in general use in America's northeastern seacoast cities by the 1840s because they dramatically reduced the loss of heat through chimneys and thus required only about one-fourth to one-sixth as much wood to produce the same amount of warmth as a fireplace. But wood stoves were slow to catch on in most of rural and small-town Delaware until after the Civil War. As late as 1856, for example, there were only a handful of wood-burning stoves in the entire village of Seaford.

Although the Read House turned to coal-burning stoves by the third decade of the nineteenth century, they were even less popular then wood-burning stoves in rural Delaware throughout the nineteenth century. Rural and small-town Delaware resisted the more efficient wood and coal-burning stoves because, despite the fact that old growth forests had disappeared from most of the state by the early nineteenth century, second and third growth timber, springing up from abandoned crop land and cut-over forests, continued to provide relatively cheap firewood for people living in the immediate area. Under these conditions, it is not surprising that "wood choppers" continued to abound in rural Delaware throughout the twentieth century to meet the high demand for wood as fuel. Although the number of wooded acres Delawareans annually consumed to keep hearth fires going is impossible to calculate—one estimate is that an average American farmer cut down one or two acres of fairly mature timber each year to provide fireplace fuel for his family—there is no doubt that the constant demand for firewood helped keep Delaware's new forests from reaching maturity.[10]

At first, hickory was the favored firewood, then oak. As the supply of these and other hardwoods was depleted, many Delawareans turned to the fast-growing pine tree to warm their homes despite the fact that the latter produced less heat than most hardwoods and, when burned, caked chimneys with flammable creosote which, in turn, often caused chimney fires.

Initially, when trees were cut down to clear land for farms, firewood was plentiful because it was a by-product of the clearing process. By the later half of the eighteenth century and early nineteenth century, after much of the landscape had been stripped of most of its trees to make way for fields, the situation changed dramatically. Those people most specifically impacted were residents of

Delaware's larger towns as well as those who lived in more-distant Philadelphia. Both groups were forced to pay higher prices to keep warm because cords of firewood had to be shipped in from forests and wood lots located many miles away. As transportation of bulky and heavy goods by land was so costly, the overland market destination of firewood had to be within five to twenty miles of its source. But if a navigable stream were available, the considerably cheaper cost of water transportation made it economically feasible to ship cords of firewood four or five times that distance. As early as 1745, Benjamin Franklin complained that large coastal towns like Philadelphia had become dependent on firewood which was brought in, at great cost, from, in some cases, nearly 100 miles away. By 1763, some of Philadelphia's fireplaces were supplied by timber cut and shipped by sail up the Delaware River from southern New Castle County's Blackbird Creek Valley.[11]

Iron furnaces were also great consumers of nearby forests. Indeed, as many as four hundred bushels of charcoal were necessary to produce just one ton of wrought iron. This meant that even small furnaces and forges employed an army of wood cutters. Although iron ore existed in most areas of Delaware, workable deposits were found in only a few sites. While some iron mining and smelting sites existed in northern New Castle County, particularly in the Iron Hill area only a few miles south of Newark, it was bog iron mining and smelting at the headwaters of certain streams in Sussex County, after the mid-eighteenth century, that exercised the greatest impact on early Delaware's forested landscape. In 1763, or shortly before, Deep Creek Iron Works was established in western Sussex County on a tributary of the Nanticoke River. This furnace, which was probably the first blast furnace in southern Delaware, was situated some three miles north of the present-day village of Concord. Soon afterwards, a handful of other iron works with furnaces was also established in western Sussex, only to be put out of business by the British naval blockade during the American Revolution.

Most iron works in southern Delaware acquired from 5,000 to 15,000 acres of forest and each year their employees cut approximately five percent of those timbered acres to provide enough charcoal to fire their furnaces. At Deep Creek Iron Works, for example, perhaps as many as 750 acres of virgin forest were cut each year to make charcoal.* Needless to say, on these large company holdings, which were really self-contained plantations that included a large farm to provide food, a store, a gristmill and a blacksmith shop, most employees were involved in felling trees, making charcoal and transporting it, rather than in mining the bog ore or working in the furnace or foundry.[12]

After the American Revolution, the iron industry in Sussex made a modest comeback only to have the last furnace, which was located in Millsboro, shut down in 1836. Just three years earlier, however, the Millsboro furnace

---

* *A similar but more modern example of stripping the landscape of trees to produce charcoal for iron furnaces took place during Mao's "Great Leap Forward" in China from 1958–62.*

demonstrated its impact on the surrounding forests by consuming 180,000 bushels of charcoal. Because it took an acre of woodland to produce 800 bushels of charcoal, in 1833 alone, approximately 225 acres in the Millsboro area were clear-cut to meet the voracious demands of the local iron industry for charcoal. (The mining of bog iron but not the manufacture of it continued in Sussex until the late nineteenth century.) But even while southern Delaware's iron furnaces were gradually shutting down during the early nineteenth century, at least one out-of-state iron manufacturer turned to felling trees in southern Sussex's Cypress Swamp to provide charcoal for his iron furnace in New Jersey.

The overall demand for Delaware charcoal declined in the early nineteenth century only to revive shortly after the Civil War and continued into the early twentieth century. At the time, charcoal was used as a filtering agent for the manufacture of rum, whiskey and applejack. It was also a vital ingredient in DuPont-made gun and blasting powder. Delaware charcoal was also used as a fuel by tool-making foundries in New England and the Midwest.[13] Thus, for almost two centuries, the continuing need for charcoal impacted Delaware's sylvan landscape.

The demands of smaller industries also contributed to the depletion of Delaware's forests. One example, during the late eighteenth and early nineteenth centuries, was salt-making along the Delaware coast from Cape Henlopen south to the Maryland line at Fenwick Island. Shallow wells were dug along the sandy shore and the sea water that naturally filled those wells became even saltier through natural evaporation. Then the concentrated salt water was poured into huge kettles or pans that were heated over wood-burning fires, causing the water to completely evaporate, leaving only salt in the kettle. A significant amount of timber was cut to fuel these salt-making fires. By 1781, "several ill disposed people" had set up salt works on Cape Henlopen, which was public land specifically set aside for residents of Sussex to catch fish, shoot birds, and pasture livestock. But the salt works' employees were "daily destroying" Cape Henlopen's trees so that, in the near future, the resulting unanchored sands would cover fields formerly used to pasture livestock. As angry petitioners pointed out, "in a few years, the shelter afforded to cattle and horses will be taken away and of consequence the feedings thereon destroyed." Subsequent deforestation probably explains the sudden appearance of "the great sand hill," in an 1831 map of Cape Henlopen. By 1888, Cape Henlopen, which was originally "covered with a growth of pines and cedars, but now is almost destitute of trees" presented "the appearance of a sandy waste."[14]

During the colonial era, Delaware found itself on the periphery of a vast Atlantic commercial network that bound together Europe, Africa, the mainland English colonies, and the West Indies. Although it was neither a leading tobacco exporter like Virginia nor an important wheat producer like Pennsylvania, diminutive Delaware made modest contributions to the vibrant coastal trade that stretched from Boston to Charleston with the rapidly growing port of Philadelphia at its very heart.

Moreover, some of Delaware's cornmeal and wheat flour eventually found its way to the West Indies and even into the Mediterranean.[15]

The need to package and transport Delaware's growing agricultural surplus created a strong demand for coopers to construct more wooden barrels and for shipwrights to build more wooden ships. That, combined with an increasing domestic hunger for boards and shingles used to construct houses, outbuildings, bridges, mills, fences and piers increased the pressure to cut down early Delaware's most valuable lumber-producing trees.

As his ship approached Cape Henlopen from the Atlantic in early December, 1632, Captain David de Vries commented on the sandhills that "rise up full of pine trees which would serve as masts on ships." As he sailed north in early 1633 to explore Delaware Bay and River, de Vries noted that coastal Delaware was all "beautifully level, full of groves of oak, hickory, ash and chestnut trees..." In 1764, the Mason-Dixon expedition into Sussex County's Cypress Swamp found beech, holly, cedar and bald cypress (residents of Cypress Swamp often referred to both white cedar and bald cypress as "cedar").

Twenty-four years later, another observer noted that the most common Delaware trees were oak, hickory, poplar, walnut, maple, ash, and, in Sussex, cedar and pine. But these sightings represent only a tiny fraction of the more than sixty tree species that are native to Delaware.[16] It did not take long for farmers, timbermen, and sawmill owners to realize that certain types of trees, if cut down, processed, and then delivered to consumers, were more valuable than others. Unlike charcoal production, where almost every variety of wood could be used—one of the few exceptions concerned the preferred use of dogwood, willow and alder in making charcoal that would become a key ingredient in gunpowder—lumbering called for a selective harvesting of trees.

Lumbering was second only to flour milling as America's most important industry until at least the 1850s. Although specific data is not available, probably the same can be said for lumbering in Delaware for the same period. Indeed, despite the fact that most of the state's most valuable sylvan assets were harvested at an early date, lumbering continued to be a significant industry in Delaware through the twentieth century.

Oak trees were perceived to be the state's most valuable sylvan asset, and Delaware's white oaks, much sought after by shipwrights and barrel stave manufacturers, were subsequently described by one source as "the finest in the United States." Black and Spanish oak were not only used to construct solid sailing vessels and sturdy barrels, their barks were high in tannic acid which was used in tanning hides. In addition, the inner bark of the black oak produced quercitron which was in great demand as a yellow dye for textiles. According to one source, Delaware's black oaks produced "the best quercitron bark in the Union," and brought a higher price in 1870 "than any other for the Liverpool market." Oaks were found in Delaware from the Pennsylvania line in the extreme north to Cypress Swamp on the southern boundary with Maryland.[17]

In early Delaware, pine trees were common in Sussex and southwestern Kent County. Although always an important part of southern Delaware's forests, it wasn't until after the initial harvesting of most of the area's valuable local hardwoods, in the eighteenth and early nineteenth centuries, that the fast-growing pine began to dominate much of the Sussex and southern Kent treescape. By 1867, for example, the forests in the Delmar area were "composed entirely of pines," while along the Nanticoke River almost three decades earlier, the supply of pine trees seemed inexhaustible. When oak and hickory began to get scarce in Delaware, shipwrights turned to loblolly pines to provide ship planking. But it was not until the mid-nineteenth century with the coming of the railroad to southwestern Kent and western Sussex, that many of Delaware's previously inaccessible pine forests were finally opened to commercial lumbering. The loblolly pine, which grows naturally as far north as southern Kent County, has been "first in timber importance ever since." In addition to growing faster and straighter than most other trees, the loblolly pine thrived in abandoned farm fields, and southern Delaware was full of the latter by the second decade of the nineteenth century.[18]

Bald Cypress

The bald cypress, one of early Delaware's most commercially valuable trees, grew naturally no further north than southern Kent County, and was particularly concentrated in Cypress (Cedar) Swamp. In 1764, according to English boundary surveyor Charles Mason, Cypress Swamp stretched thirty miles from north to south and was fourteen miles wide with about one-half of its acreage just across the line in Maryland. Mason reported that the swamp contained "the greatest quantity of timber I ever saw," and that bald cypress towered above all of the other trees, "seeming to reach the clouds." One nineteenth-century Delawarean maintained that some of the bald cypress in the swamp rose 140 feet in height.[19]

Because these giant sentinels were decay resistant, they soon became lucrative targets for timber interests. By the end of the French and Indian War (1754–1763), John Dagworthy and other gentry were sending crews of enslaved blacks and free whites into Cypress Swamp to chop down, process, and then haul to boat landings along the Indian River and its tributaries, bald cypress in the form of shingles, pilings, fence-posts, and lumber.

Testifying to the splendid durability of cypress siding was twentieth century Delaware journalist Anthony Higgins who, during the early 1930s, lived in "Water Chance," an early nineteenth-century house, located in southeastern Sussex and partially mantled by cypress shingles from Cypress Swamp. Although more than 100 years old, the hand-riven shingles, "worn thin by the weather," showed no signs of decay. Cypress Swamp's processed bald cypress found its way into lumber yards as far away as New Castle, Philadelphia, and Trenton. Bald cypress favored land that was seasonally covered by standing water. Set as it was in swamp muck,

The Bald Cypress— A view of bald cypress trees at Trussum Pond in southwestern Sussex County, one of the most northerly cypress ponds in the United States, *photograph by Richard B. Carter, 1976.*

the bald cypress's broad root system provided a firm foundation to support the heavy tree while its root nobs (known locally as "cypress knees"), which could protrude as much as eight to ten feet above ground, may have served to absorb air into the tree's circulatory system. Because these large nobs were hollow, southern Delawareans excised them and turned them into wooden buckets to draw water from wells.[20]

Like bald cypress, white cedar was resistant to rot and was also used for posts, fence rails, pilings, shingles, and lumber. White cedar trees were heavily concentrated in Cypress Swamp but, unlike bald cypress, they ranged much further north and could be found in fresh water swamps and along watercourses in both Kent and New Castle Counties. Perhaps the largest concentration in northern Delaware was in and around a freshwater swamp located a few miles north of the mouth of the Smyrna River. In 1878, however, a major storm surge broke through the low-lying dunes that held back the upper Delaware Bay and poured salt water into the area. Subsequent salt water breaches of the dune line, fires, and lumbering just about wiped out this significant concentration of white cedars. Today part of this wetland area is designated the Cedar Swamp Wildlife Area and is located in south-eastern New Castle County.[21]

The European-rooted perception that forests were obstacles to progress led colonial Delawareans to do their best to drive back wilderness with both ax and fire. For them, transforming a sylvan landscape into cultivated

Worm Fencing— This view of a fenceline at Pemberton Hall Historical Park in Wicomico County, Maryland, near Salisbury is typical of the worm fencing that would have been seen on farmsteads in nearby southern Delaware in the 18th and 19th centuries, *photograph by Richard B. Carter, 2006.*

fields and green meadows was simply "improving" the land. But it was not long before the consequences of broad and indiscriminate forest destruction caused some influential Delawareans to have second thoughts about what was happening. The specific issue that gave a certain focus to the increasing concern about forest depletion was the shortage of wood for building fences.

The amount of timber needed to fence in a farm was enormous. A square plot of land containing 160 acres, for example, required two miles of post-and-rail fencing to enclose it. In addition to hundreds of posts, 8,800 rails, each ten feet long, were needed. Although it is impossible to estimate the amount of land that was fenced in at any one time in early Delaware, the response of Delawareans and their elected representatives to the shortage of fencing timber indicates that a large number of acres were involved. In 1739, the Delaware colonial assembly required that all cultivated fields be enclosed by either post and rail or worm (zig-zag) fences if a farmer wished to collect compensation for crop destruction by free-roaming livestock. However, the worm fence, which originated with the Finns and Lapps on the northern periphery of Europe, was extraordinarily wasteful of timber because its zig-zag nature required almost 50 percent more wood to enclose a field than did the straight post-and-rail fence. In 1783, nine farmers in northeastern Sussex complained that "the want of timber" prevented them from building fences of any sort to confine their livestock. In the 1790s, the timber shortage in southern New Castle County caused one landowner to prohibit his heirs from using timber on his property for anything else but fence construction. In 1803, John Dickinson went one better by demanding that the tenant farmer on his Kent County plantation build only non-worm fences because the worm variety wasted so much wood. Nevertheless, the worm fence was the favorite of seventeenth-, eighteenth-, and early nineteenth-century Delawareans because it was simple to construct and could be easily moved to enclose a second field after the fertility of the first played out. (Post-and-rail fences lacked this mobility.) As late as 1870, the worm variety made up about one half of all Delaware fencing.[22]

Worm Fences

Delaware's colonial assembly first expressed uneasiness about the colony's diminishing forests in 1741, with particular concern about those who, in their enthusiasm to harvest timber, were chopping down trees on other people's property. Moreover, the customary annual firing of the woods destroyed "many timber-trees," and prevented "the growth of young trees and woods." In 1759, Hendrick Hendrickson of Kent County pointed out that timber was a precious economic resource and cautioned against the unnecessary cutting of wood lots. By 1770, the shortage of wood for making fences caused the colonial assembly to encourage Delawareans to dig ditches, construct stone walls and grow hedges as substitutes for wooden fences. But rocks for constructing stone walls were not that common in central and southern Delaware and, because of climatic differences, hedge rows never did grow as well in Delaware as in England.[23]

By the close of the eighteenth century, the state's mature timber resources were seriously depleted. In some areas, deforestation had proceeded so far that fence rails and posts had to be imported from New Jersey. To help preserve the few remaining groves of mature trees, many New Castle County farmers heeded the call of their state legislature and grew hedgerows in place of wooden fences during the late eighteenth and early nineteenth centuries. The clearing of Delaware's forests had another effect: a significant decrease in land values. In the eighteenth century, clearing land and building a house and outbuildings on a specific property usually led to a significant increase in the value of a farm. By 1826, however, the scarcity of timber in central Delaware caused wood lots to sell at twenty to twenty-five percent more per acre than cleared farmland that included a house and some outbuildings.[24]

Only in some of the pine forests along the southwestern border with Maryland and in the dark and mysterious interior of Cypress Swamp, still dominated by a great canopy of bald cypress, white cedar, oak and some pine, did large concentrations of old growth trees survive. But even in isolated Cypress Swamp, determined assaults by timber and agricultural interests had significantly reduced the size of the swamp and systematically cut down many of its most prized trees. A 1782 estimate suggests that Cypress Swamp may have lost more than half of its size since the mid-1760s and measured only ten to twelve miles from north to south and just seven miles in width, with about one half of the entire swamp remaining in Delaware. Maps of Delaware, drawn in 1796 and 1876 respectively, suggest that Cypress Swamp continued to approximate in size the 1782 estimate into the last quarter of the nineteenth century.[25]

By cutting down shade-providing trees, timbermen exposed Cypress Swamp to sunlight. Much of the surface water subsequently evaporated and some of the layer of peat, which coated most of the swamp's floor, dried out and became highly combustible. The potential for disaster was increased when a serious drought struck the Cypress Swamp area in the summer of 1782. In June, a fire began which "burned for many weeks before much notice was taken." On August 20, at approximately two hours before sunset, a strong, hot wind from the southwest whipped up the long-standing fire into a wall of flames that proceeded "with such inconceivable fury, that it mowed down or otherwise destroyed," at least 3,000 acres of prime timber in less than twelve hours. Burning coals from the fire were hurled into the air with such force that some landed fourteen miles away along the Atlantic Coast while, at night, light from the blaze was visible for seventy miles.[26] Unfortunately, this was not the last great conflagration in Cypress Swamp.

Wildlife

The gradual destruction of woodland and the draining of marsh and swampland caused the disappearance of much natural habitat, and this posed a serious threat to Delaware's wildlife. Equally as serious was the relentless hunting and trapping of most of Delaware's fauna. Perhaps no species suffered more than the white-tailed deer.

In 1679, a European traveler reported seeing herds of deer in the Christina River Valley. By the late eighteenth century, however, overhunting throughout Delaware made herds of deer an increasingly rare sight. Because they were such a good source of meat as well as buckskin, Delaware's legislature joined with legislatures in other Anglo-American colonies in attempting to preserve deer for posterity by restricting hunting to a limited number of months and by forbidding particularly efficient hunting techniques such as ring hunting—surrounding a woods with hunters either mounted or on foot and gradually tightening the ring until all game trapped inside were killed—and the use of dogs. In 1740, for example, the colonial legislature established the first deer season in Delaware by passing an act "to prevent the unseasonable killing of deer within this government." Henceforth, it was illegal for all but Native Americans to hunt deer from January 1 to August 5. Hunters were fined forty shillings for each deer killed out of season, and the money was split by the informer and a fund for the poor. If the poacher was a servant or a slave and, therefore, had no money, the punishment was twenty-one stripes on the back. Despite these efforts, Delaware's deer population continued its dramatic decline until even individual deer sightings were rare by the early nineteenth century. But not until 1841 did the General Assembly legislate a total ban on deer hunting.[27]

While hunters and trappers, who were after meat and pelts, were responsible for most of Delaware's wildlife hunting kills, still other mammals, birds and reptiles were gunned down or trapped for bounty money put up by Delaware's colonial and county governments, or even by private citizens. Grackles (the English colonists called them "blackbirds"), woodpeckers, and crows were a threat to grain crops, while most types of hawks ravaged domestic fowl. At different times, each bird had a bounty placed on its head by either the Pennsylvania colonial legislature, whose laws applied to Delaware from 1682–1703, or the Delaware colonial legislature whose laws regulated Delaware from 1704–1776.[28]

A few of Delaware's snakes were deemed dangerous and were added to the bounty list. One mid-eighteenth century observer found poisonous snakes "numerous" in the lower Delaware Valley." As late as 1795, at a new mill site along the banks of the Brandywine not far from Wilmington, "rattlesnakes and copperheads, it was said, flourished...in countless numbers." In the same general area, through the bottom of a ravine of "solid rocks," Rattlesnake Run flowed north into the Brandywine. In 1752, John Crevet of southern New Castle County offered twenty shillings "to any person that destroys the greatest number of rattlesnakes, corn snakes [not poisonous], and vipers [probably copperheads]."

Gray Wolf

Thanks to the clearing of their natural habitat, and to the hostile actions of humans and their free-roaming, aggressive swine, rattlesnakes declined to such a small number that they ceased to be an issue in Delaware during the nineteenth century and are considered extinct in the state today. Copperheads were also significantly fewer in number by the nineteenth century and only a few survive in present-day Delaware.[29]

Bounties were also established on a few particularly troublesome mammals such as squirrels, foxes, bears, and wolves. Corn fields were plundered by squirrels because the bushy-tailed rodents were partial to corn.

They were particularly vexing because they did not "eat all the corn, but only the inner and sweet part." Squirrels also raided storehouses full of harvested corn and made off with the equivalent of bushels of the shelled kernels. As the number of acres planted in corn expanded, easier access to their favorite food caused a dramatic increase in Delaware's squirrel population. By 1712, Delaware's government was directing "each freeholder to present a certain number of squirrel heads" at each sitting of the county court or pay a fine. Subsequently, Kent and Sussex established bounties on these troublesome rodents.[30]

Gray Wolf

While bears and foxes were considered threats to some of colonial Delaware's livestock, wolves were perceived as particularly pernicious predators. During the seventeenth century, "There were excessive numbers of wolves," and "their howling and yelping might be heard all night." They were particularly prone to attacking sheep, pigs, and young cattle which they "frequently tore in pieces." Wolf bounties were probably first established in Delaware by the Dutch, if not by the Swedes before them. By 1676, the going rate for a dead wolf in New Castle County was forty guilders. By 1681, Sussex County was offering its own wolf bounties. Although wolves were gradually driven from much of Delaware over the next sixty years, enough remained in 1742 to cause the Delaware legislature to establish a colony-wide bounty of fifteen shillings for each adult wolf and 10 shillings for each "whelp or puppy." Often the bounty hunters shot wolves after trapping them in deep pits, because this was seen as a particularly effective way to diminish their numbers. In 1678, for example, the New Castle County Court ordered the county's inhabitants to dig fifty wolf pits by the first day in May.[31]

As in the other Anglo-American colonies, most of Delaware's bounty hunters were either financially strapped young white males who had learned to hunt to put food on the table, or male members of the fast-fading Native American population, who brought generations of game-pursuing experience to tracking down outlawed species. One 1698 account reported that in areas of European settlement along the west bank of the lower Delaware River, "wolves are pretty well destroyed by the Indians, for the sake of the reward given them..." For payment, Delaware's county or colonial governments demanded that the bounty hunter present the severed head of the squirrel, wolf, crow or other targeted fauna at one of the county courts. Delaware's Indian bounty hunters would often sell their quarry's head to a white middleman who, in turn, carried the head to court to collect the full bounty price. To prevent repeated use of the same wolf's head, the county courts simply had the ears cut off, a practice also followed in Maryland and Virginia.[32]

Whether for meat, for pelts, or for the cash generated by bounties, hunting became an essential part of the lives of most male Delawareans. But it was particularly central to the lives of Delaware's ordinary people. In most of Europe, by the seventeenth and eighteenth centuries, hunting was a right reserved for the noble and gentry classes and was denied by law to commoners. Eighteenth-century English jurist William Blackstone declared that for one of his countrymen to legally kill a partridge he needed fifty times as much land as he did to qualify to vote for a knight to represent his shire in Parliament. As previously noted, the only European commoners with considerable hunting experience, prior to immigrating to Delaware, were the Finns. Consequently, most of the Swedes, Dutch, English, Welsh, and Scots-Irish who settled in Delaware had to acquire hunting skills before they could expect much success in putting wild meat on the table. Although black slaves—who formed between twenty and twenty-five percent of Delaware's population by 1770—may have brought some hunting skills with them from Africa, their enslaved status and the corresponding reluctance of most of their masters to allow them to carry guns limited their hunting activities.[33]

Early Trappers— Trapping and hunting were central to the life of European settlers in early Delaware. This view of colonial-era trappers is by noted nineteenth-century Delaware illustrator Felix Octavius Carr Darley, *courtesy of the Historical Society of Delaware.*

Once white males other than the Finns learned to use guns and to stalk game, they found in hunting considerable enjoyment as well as a means of putting food on their tables and bounty cash in their pockets. Consequently, protecting the hunting rights of ordinary whites, or even those of Native Americans, was essential to the politically powerful because it preserved an important psychological safety valve that harmlessly vented pent-up energies and emotions that might, otherwise, become socially disruptive. As a result, Delaware's government was reluctant to establish and then enforce really restrictive hunting laws until it was too late to save a particularly endangered species. One example was the unwillingness of the Delaware legislature to enact a long-overdue ban on all deer hunting in the state until the

1841 session. By that date, at least four decades had elapsed since Delaware's deer population had been stalked to the very edge of extinction. The deer-hunting ban of 1841 was too late to spark a recovery in Delaware's white-tailed deer population and thus proved to be an empty gesture rather than a meaningful example of conservation legislation. Just thirty years later, if there were any deer remaining in the state, it was an alleged few who lingered in the deep recesses of the Cypress Swamp.[34]

Fishing rights were also perceived as crucial to both the material and psychological well-being of many "middling" and poorer Delawareans. By the end of winter, the meat and grain stores of the less privileged were either seriously depleted or exhausted. To those just scraping by, the spring runs of shad and herring—most preferred the taste of shad to herring—were a godsend. Thus, they insisted that the Delaware government apply the old English tradition of "common usage" to the state's streams. In 1829, for example, the Delaware General Assembly reaffirmed an early eighteenth century law "that all persons inhabiting in or near any creek or river in this government may enjoy all privileges and advantages that from them are reaped."[35] Clearly those holding this perspective strongly opposed any attempt to control access to fishing grounds or to the catching of fish.

Increasing numbers of Delawareans, however, rejected the principle of common usage. Typically, they would ignore the traditional fishing rights of others by stretching fishing nets from one river bank to the other or by placing devices in streams that funneled huge numbers of anadromous fish into traps. Both tactics not only seriously challenged the concept of common usage by depleting the picine stock available to upstream fishermen, but they so interfered with the efforts of shad, herring and other anadromous fish to reach their spawning grounds every spring that they jeopardized certain future fish populations in Delaware. Between 1726 and 1736, the colonial legislature responded to threats both to the principle of common usage and to Delaware's future fish population by banning bank-to-bank nets and most types of funneled traps in "any creek or river" in Delaware.[36]

But this legislation did not ban the use of other, slightly less efficient types of fish traps, and the latter subsequently took an increasing toll on Delaware's fish population. In 1769, alarmed petitioners demanded that Delaware's colonial legislature clear away "all obstructions" in "every creek within this government occasioned by wears [weirs], racks, baskets, fishing-dams, ponds or other obstruction whatsoever..." Twelve years later petitioners complained that fish weirs were "so thick set" in parts of Indian River that they "totally obstructed" ship traffic. Although there is no evidence that their demands were acted on, the petitioners' complaints and concerns indicate that Delaware's fish population was declining at an alarming rate. Indeed, in 1749, Peter Kalm noted that all along the lower Delaware Valley, "aged people" remembered that "in their youths, the bays, rivers and brooks, had such quantities of fish that at one draught [one drawing in of the fish net] in the morning they caught as many

Wild turkeys, once common in Delaware, were hunted to extinction, probably by the early nineteenth century. They have been reintroduced in southern Delaware in recent decades and their numbers are growing.

as a horse was able to carry home. But at present things are greatly altered, and they often work in vain all night long with their fishing tackle." The reason for the relative scarcity was that now fish were being "caught by a greater variety of contrivances, and in different manners than before." Moreover, the increasing silt in streams, resulting from the destruction of forests and the plowing of the land, caused a decline in piscine numbers by burying fish eggs under silt. No wonder many old timers, when comparing the abundant fish harvests of their youth to the catch in 1749, saw as great a difference as "between day and night."[37]

Fish were not the only form of marine life to experience a significant numerical decline. In the seventeenth and eighteenth centuries, large oyster banks were noted in Delaware Bay, including three within one mile of Cape Henlopen and a number of others just off the Kent County shore. In addition, both Rehoboth and Indian River bays held large concentrations of oysters. In 1732, William Becket, the Anglican rector of St. Peter's in Lewes, found the latter two bays "well stored with excellent fish, cockles and oysters." Later that year he added crabs to the many marine treasures of Rehoboth and Indian River bays. Overharvesting to supply the growing Philadelphia market, however, led to a decline in the supply of oysters and other shellfish throughout the bays in the region. In 1719, New Jersey was first to act to save the oyster banks in Delaware Bay by placing calendar restrictions on harvesting oysters. Initially, the demand for governmental intervention was more muted in Delaware. But by 1796, a petition was sent to the Delaware legislature complaining that in earlier times, "one man in one day might have caught in Rehoboth Bay thirty bushels of oysters..." But now he was successful "if he can take in the same time one sixth part of that number."

Faced with seriously declining shellfish resources but unwilling to bar Delawareans from common access to seafood, the General Assembly decided to strike out at watermen from other states, particularly those from New York, who had given up on their own depleted waters to ply their trade in Delaware's bays. In 1812, the state's General Assembly barred non-residents from harvesting oysters and clams in Delaware's waters. And yet the depletion of the state's coastal shellfish by native Delawareans continued unabated. As historian John A. Munroe points out, "as with its land," early Delaware "was wasteful of its marine sources."[38]

The numbers of birds, particularly edible fowl, also declined dramatically during the seventeenth and eighteenth centuries. From the lower Delaware Valley in 1749, Peter Kalm reported that a once bountiful supply of game birds had been dramatically diminished since the advent of Europeans. Kalm noted that all of the "old Swedes and Englishmen born in America," maintained that "there were not nearly so many edible birds at present as there used to be when they were children..." Whereas, in the late seventeenth century, "a single person could kill eighty ducks in a morning...at present [1749] you frequently wait in vain for a single one."[39]

In addition to ducks, the sharpest population declines in game birds were in geese, cranes, wild turkeys, and partridges. The cause was very clear: by the mid-eighteenth century, Europeans had peopled the countryside, cut down trees, shot or trapped edible birds, and stolen many of their eggs. Those birds not killed were simply "frightened away." One might ask why the Pennsylvania or Delaware colonial legislatures did not introduce strict regulations to end the slaughter. The answer, according to Kalm was that "if any had been made, the spirit of freedom which prevails...would not suffer them to be obeyed."[40]

The wild turkey was a particularly painful example of the extinction in the state of a species that strict hunting regulations could have averted. One estimate maintains that there were approximately 10,000 wild turkeys in Delaware before the advent of Europeans, which works out to about five per square mile. "Uncontrolled hunting, particularly by market harvesters," probably ended the existence of wild turkeys in Delaware by the early nineteenth century. There were a few non-game birds, notably grackles, that actually increased in number as forests were cleared and fields were planted because they adjusted so well to the new habitat and enjoyed the fields of corn so obligingly planted by colonial farmers. No wonder grackles were among the birds for which bounties were paid.[41]

The First Environmental Crisis

Often, it seems, the only theme of interest to environmental historians is the sad story of humans, particularly those of European ancestry, plundering and degrading a pristine natural world. As William Beinart and Peter Coates point out, however, a more thoughtful examination of humans interacting with the natural world would also include "reconstruction and restabilization" of that "natural world in various guises." In short, "transformation should be set alongside that of destruction," in telling the story. But for Delaware in the seventeenth, eighteenth, and nineteenth centuries, degradation of its land and seascape was so unprecedented and of such great consequence to the future economic and social well-being of the state that it deserves to be the dominant theme of Delaware's early environmental history from the arrival of Europeans until at least the end of the Civil War.[42]

Amidst the environmental carnage in the name of "improvement" that marked Delaware's early history, only a few voices were raised in general protest. One was Will Andrew, an eighteenth century Native American from the Cypress Swamp region of southern Sussex County, who pointed out to a member of the local white gentry that the European drive to conquer nature led whites to "make slaves of everything, the wind, the water, the fire, and the earth." But the great irony was that, in Delaware, the frenetic drive to improve nature by clearing the forests, harvesting the fauna, and cultivating the fields, systematically shackled white Delaware to the same environment that it had so recklessly despoiled.[43]

At the dawn of the nineteenth century, most Delawareans were scratching a meager existence from a pillaged land that was wanting in mature valuable timber, game birds and mammals, fish, shellfish, and fertile fields. No wonder itinerant preacher Francis Asbury urged many of his fellow Methodists in southern Delaware to move west to the virgin lands of the Ohio country where "the means of rearing a family and advancing in the world were in reach of the inhabitants."[44] Clearly, Delaware was facing its first major environmental crisis, and that crisis seriously threatened the economic prospects of a people primarily tied to the land.

CHAPTER THREE:

The Impact of the World Beyond

"Land, like your horse or your ox,
if worked, must be fed or it will die."

THE OUTSIDE WORLD EXERTED CONSIDERABLE INFLUENCE ON HOW Delawareans interacted with their natural environment. The best example was the growing commercial port of Philadelphia and its expanding economic hinterland. By 1728, Delawareans as far south as Lewes were routinely sending products of their fields and forests to Philadelphia and bringing home "such European or West Indian commodities as they want for family use."[1] In subsequent years, the economic influence of Philadelphia would become a major force in the reconfiguration of Delaware's rural and urban landscape. Put another way, with the possible exception of the Nanticoke Basin, the footprint of the Quaker City was apparent everywhere in the state. The vital conduit for this Philadelphia-Delaware commercial/ecological connection was the Delaware Estuary.

The Delaware Estuary

The Delaware Estuary includes the lower Delaware River and the entire Delaware Bay. (For the purposes of this study, the Delaware Estuary does not include tidal streams that flow into the Delaware Bay and River.) In the north the Delaware Estuary begins at the rapids of the Delaware River at Trenton, New Jersey, and extends approximately 130 miles south to where the Delaware Bay meets the Atlantic Ocean at Cape Henlopen, Delaware, and Cape May, New Jersey. Both large, ocean-crossing vessels and coastal-sailing craft found enough depth in the estuary's waters to push upstream, at least as far as Philadelphia. From the beginning of European settlement, the Delaware Estuary served as one of America's most important marine superhighways, providing Delawareans who lived on the bay or river with commercial access to other coastal areas of Delaware, to the port of Philadelphia, and to the entire Atlantic world. In describing the estuary in 1653, New Sweden's Governor Johan Rising put it quite simply: "What splendid advantages this...has."[2]

But once sailing ships from the Atlantic Ocean entered the Delaware Estuary, the sail north to New Castle, Wilmington and Philadelphia, could be very challenging. From the beginning of European colonization, the many shoals and sandbars of the Delaware Bay and River presented potential risks which required considerable maritime skill. In 1609, for example, Henry Hudson entered Delaware Bay in search of a northwest passage to the Orient. Probably the first European to round Cape Henlopen, he was discouraged from further exploring the Delaware Bay by the shoals he encountered and soon turned his ship back to the Atlantic Ocean. About eight decades later, a sailing vessel heading up the Delaware Bay to Philadelphia and drawing thirteen feet of water, reported "sometimes" being "stuck upon the sand." In 1657, a Dutch official declared that mariners unacquainted with Delaware Bay would have real difficulties navigating it, "as there are many sand-banks, shoals, and flats." He recommended that money should be appropriated "to render it safer and better for incoming ships."[3]

As early as 1658, there are indications that the Dutch were using marker buoys in the Delaware Estuary to set off safe channels from shoal waters. To make sailing as safe as possible, however, navigators with local maritime experience, and charts and maps based on recent soundings in the Delaware Estuary were essential. Perhaps as early as the mid-seventeenth century, local men were offering their services for hire as pilots to ships rounding Cape Henlopen with the intention of heading north up the Delaware Estuary. Because the best channel up the Delaware Bay lay close to Cape Henlopen, bay and river pilots generally lived in the Lewes area, operated as individual entrepreneurs, and were ferried out in small sailing vessels to board incoming ships at the confluence of bay and ocean. On October 30, 1739, for example, the ship *Savannah* was met by a pilot boat. The pilot boarded the ship to direct its journey north to Philadelphia, while three passengers, including the famed British evangelist George Whitefield, climbed down into the pilot boat and headed for Lewes.[4]

Although Augustine Herrman published "the first reliable chart of the Delaware Estuary" in 1661, the most popular printed early chart was produced by Joshua Fisher in 1756, and it was followed by nine additional editions with the last appearing in 1800. In 1779, a map reflecting the latest technology in measuring the depths of Delaware Bay was produced by J. F. W. Des Barres, a Swiss engineer who was in the British army. In 1802, Pierre C. Varle published a map of Delaware and the Eastern Shore of Maryland which, based on soundings made in 1799, 1800, and 1801, detailed oyster beds, shoals, and ship channels in the Delaware Estuary. Although charts and maps were helpful to sailing the Delaware Bay and River, the constantly shifting shoals made the continued use of local pilots imperative for safe navigation. This caused Joshua Fisher to include a warning on his 1756 map that only the smallest ships should risk sailing the bay and river without a pilot. By the mid-nineteenth century, marine insurance companies required that vessels sailing the Delaware Bay and River Estuary take on a pilot.[5]

Even with the use of local pilots and charts, the entrance to the Delaware Estuary continued to pose a challenge when storms swept through the area or fogs blocked visibility. To improve maritime security, the British government constructed a lighthouse on Cape Henlopen in 1765—tradition has it that a Cape Henlopen lighthouse may have been built as early as 1725. Until the late eighteenth century, Lewes Creek provided anchorage and shelter during storms to all but the largest ocean-going vessels. By the early nineteenth century, however, Lewes Creek was partially silted up and this caused a dramatic change in the economic life of this once sea-faring community. Because now even the modest-sized ships used in the coastal trade could no longer drop anchor in the shelter of Lewes Creek, "numbers of coasting vessels were wrecked along the shore in front of Lewes." Predictably, Lewes' maritime commerce declined dramatically and this caused many local sailors to turn to farming for a living.

The lack of an effective, navigable sheltered port in all of coastal central and southern Delaware, caused Philadelphia merchants to pressure the United States government to take action. In 1828, construction by the federal government began on a 2,500 foot-long stone breakwater and a nearby but separate 1,400 foot stone ice breaker pier, and they were 90 percent completed by 1840. Both stone barriers served as artificial shelters for ships in the Delaware Bay just off Lewes. For a time, the Lewes Breakwater was the second largest in the world. In 1893, one historian called the breakwater near Lewes "the most important improvement of the kind in the United States." Shoaling, however, eventually developed inside the breakwater making this artificial harbor too shallow for large ships. In 1901, with the depth of the Lewes harbor behind the breakwater now measuring less than twelve feet, construction was finished on a one-and-one-half mile long second breakwater further out in Delaware Bay.[6] But when major storms hurled their full fury at Delaware's coast, even with the breakwaters, there was no really effective coastal refuge for either boats or people living in low-lying coastal areas. Threatened residents simply did what they could to protect their boats and then retreated inland until the winds and flood tides subsided.

To some early residents of the Lewes area, turbulent storms out of the Atlantic provided economic opportunities. The happy convergence of overpowering winds and treacherous sandbars just off Cape Henlopen's ocean shore caused ships to founder on nearby shoals. Indeed, some Lewes locals may have emulated a common practice of Englishmen along the Cornwall coast by setting out misleading signals to draw unsuspecting cargo ships to shoal waters and disaster. Once vessels foundered off the cape and the crews and the captains were forced to abandon ship, the locals rowed out and looted the stricken vessels. In 1786, to put an end to such libertine practices, the Delaware legislature made it a felony to "plunder, steal, take away or destroy any goods, merchandise or other effects belonging to any ship or vessel whatsoever which shall be in distress...or shall be wrecked."[7]

The hurricanes and northeasters that periodically pounded Delaware, and the continuing slow rise of water levels in the Atlantic and in the Delaware Estuary, caused significant coastal erosion. One observer maintained that, between the 1630s and 1870, the Delaware coastline retreated one half mile in some places and two miles in others. The inland advance of the sea followed a predictable pattern where water covered sand, sand covered marsh, and marsh covered fast land. In 1870, near the town of New Castle, foundations of houses could be seen nearly a quarter of a mile out in the Delaware River while in Kent County, the Mahon River lighthouse had to be rebuilt three times in the four decades before 1870 because of the encroachments of Delaware Bay.[8]

The Atlantic coast from Cape Henlopen south to the Maryland line, was marked by sand dunes that rose forty to fifty feet above sea level and extended about one half mile inland. The dunes, which were anchored by pine trees and partially covered with grass, were often used to graze cattle and other livestock. In 1832, a great storm that devastated the Atlantic coastline and much of the Delaware Estuary, "swept down all these hills on the ocean side, made them level," and caused the Atlantic to "encroach on our state about half a mile, and made what was before fastland [sic] part of the ocean." Not surprisingly, it killed "many people and a great number of cattle." The storm caused the Atlantic Ocean to break through the barrier dunes into Rehoboth and Indian River bays with such volume that the salinity level of both bays was dramatically increased.[9] Clearly, despite the advent of increasing human technology, nineteenth century Delaware's physical environment—particularly extreme weather fronts and rising sea levels—continued to shape much of the human experience in the First State.

Freshwater Streams

Initially, the many streams that flowed eastward from interior Delaware to the Delaware Estuary provided certain timbermen, farmers and millers with direct commercial access to this marine superhighway. In 1683, William Penn described the land adjoining the Delaware Estuary as having "the advantage of many creeks and rivers...that run into the main river or bay; some navigable for great ships, some for small craft." Indeed, Penn described the Christina, Brandywine, and Shellpot Creek, a tributary of the lower Brandywine, as having a depth of four to eight fathoms—twenty-four to forty-eight feet—and "any one of which have room to lay up the Royal Navy of England." As for Delaware's streams further south, Penn maintained that many rivers and creeks, such as St. Georges, St. Jones, Mispillion, Cedar, and Lewes, were deep enough for "sloops of good burthen," while other streams could be navigated for much of their length by shallops, which were small open boats propelled by sail or oars. Although Penn's comments about the Christina, the Brandywine, and Shellpot Creek were probably rare examples of Quaker hyperbole, most early observers did remark on the depth and clarity of the silt-free streams that crossed the Delaware landscape from west to east at fairly regular intervals before emptying into the Delaware Bay or River. Just one example was the Christina River which, during the colonial period, was navigable by sloops of up to fifty tons from Wilmington to Christiana Bridge, approximately ten miles upstream.[10]

The depth of early Delaware's creeks and rivers was hard for later generations to comprehend. In 1870, Wilmington newspaper publisher and writer Francis Vincent pointed out that many once navigable Delaware streams had entirely disappeared while others "that once floated vessels that bore the original settlers across the ocean" are now so shallow that they will "hardly float the smallest kind of boat without grounding." In fact, according to Vincent, "every creek between the Christina and Cape Henlopen has had its navigation injured" since colonial times.[11]

While marker buoys, local pilots, and charts and maps were making it safer for large ships to ply the waters of the Delaware Bay and River, the streams that emptied into this marine superhighway were developing an aquatic form of atherosclerosis. By the late eighteenth and early nineteenth centuries, most of Delaware's creeks and rivers were at least partially obstructed with silt and other debris. Just one example was the once navigable St. Jones River which, by 1824, was too shallow for six of its winding fifteen-mile course from Dover to the Delaware Bay to carry commercial traffic. The clogging up of formerly navigable streams meant that cargo-bearing ships were no longer able to service large areas of Delaware's interior. The reasons for this maritime hardening of the arteries were many. The first trees to be cut down were along river banks so that strategically located farmland could be cleared, timber sold, and a river landing established. That and subsequent timbering further inland resulted in increased sediment deposits and floating logs in Delaware's waterways. Moreover, the growing presence of fish weirs in creeks and rivers at the end of the eighteenth century also tended to trap sediment in specific locations along stream beds that soon developed into shoals. The situation was compounded at landings when ships at anchor dumped unneeded ballast overboard.[12]

The clogging up of Delaware's streams dramatically accelerated after the mid-eighteenth century thanks to increasing use of ox-pulled plows. Prior to that time, the "hoe and hill" agriculture, which was commonly used to cultivate Delaware's fields, caused only a modest loss of top-soil to rain runoff. The efficiency of plow agriculture, by contrast, dramatically increased the acres of land under cultivation and so deeply turned the soil that it was far more susceptible to being blown away by winds or being washed away in rainstorms and thus ending up as navigation-choking silt in Delaware's waterways.[13]

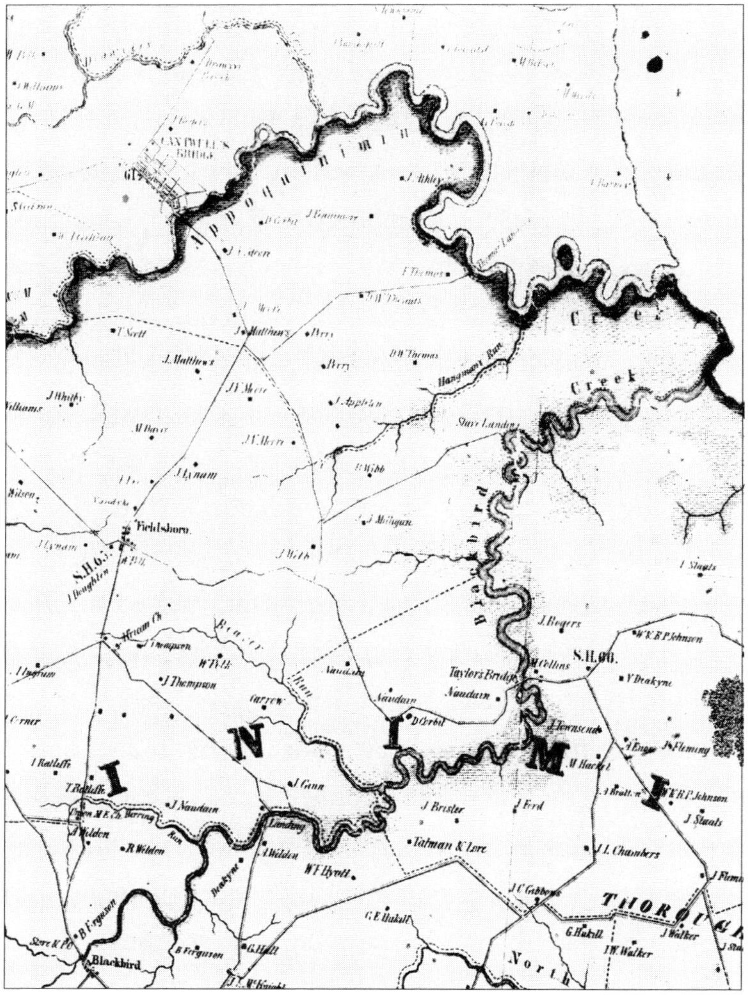

Serpentine Rivers—
The winding and twisting nature of Delaware's tidal estuaries is evident from this detail of a map entitled "New Castle County Delaware From Original Surveys." The area shown is east and southeast of what was then Cantwell's Bridge, now Odessa. The stream at top is Appoquinimink Creek. To the south is the equally circuitous Blackbird Creek. The map was drawn by Samuel M. Rea & Jacob Price and was published by Smith & Wistar of Philadelphia in 1849. *Courtesy of the Delaware Public Archives*

Canals

Water transportation was so crucial to early Delaware's economic prospects that at least some of these silted-up streams had to be reopened. As early as the 1720s, Delaware's legislature was granting permission for certain residents to cut through and dredge out a more navigable mouth to Kent County's Murderkill River so that marine traffic could more easily pass to and from the river to the Delaware Bay.[14] And yet the silting-up of formerly navigable streams was not the only issue that had to be addressed. Of almost equal importance was the fact that, compared to "as the crow flies," Delaware's twisting and looping tidewater streams increased shipping distances from the interior to Delaware Bay and River by three, four, and even five times. If only something could be done to bypass some of the loops and oxbows in creeks and rivers, considerable time and money would be saved.

Even prior to the great canal-building era of the early nineteenth century, some Delawareans seized on the canal as the best way to bypass silted-up stretches of inland waterways and to straighten out some of the serpentine coils in their creeks and rivers in order to both deepen and shorten the water route to the Delaware Estuary. Eastward flowing Duck Creek, was one of the first streams to shoal-up. Near its mouth, it significantly extended its length by dramatically turning south for a number of miles before entering Delaware Bay. To avoid its shoals and to shorten the route to the bay, as early as 1682 a wide ditch was dug eastward from Duck Creek, just above its turn to the south, that directly connected with the bay. Despite the successful construction of this early and primitive canal, by 1786, Duck Creek was "greatly obstructed by [many] shoals, and its course...[was] exceedingly crooked and circuitous, whereby the transportation of produce to market is rendered very difficult and tedious..." In spite of protests by wharf and fishery owners, whose commercial properties along Duck Creek subsequently would be bypassed, Delaware's legislature gave local entrepreneurs permission to dig a series of canals across oxbow necks to produce short cuts for much of Duck Creek's length.*

In general, canal builders expected to make a profit by charging tolls on their newly fashioned waterways. But often the toll rates were regulated by the General Assembly. In a few cases, however, tolls were not allowed, and in at least one case the General Assembly reimbursed the canal builder.[15] Although the exact number of Delaware's streams that were "canalized" is unknown, it is clear that in the seventy years prior to the Civil War, the courses of most of the state's once navigable rivers and creeks were partially altered and straightened in their journey to the sea by legions of shovel-wielding men and dirt-hauling horses and oxen.

Canals were popular in early America because the transportation of bulk cargo was so much cheaper per mile by water than by land. Therefore, it made good sense to invest considerable money in "canalizing" rivers and creeks, and in cutting through miles of soil and rock to connect commercially important bodies of water, such as the Chesapeake Bay/ Susquehanna River, with the Delaware Estuary.

In 1654, Johan Rising, Governor of New Sweden, wrote of the need for a canal across the northern neck of the Delmarva Peninsula. With the subsequent increase in overland transportation of bulk cargoes between the upper Delaware Estuary and the upper Chesapeake Bay, the cost-efficiency of a trans-peninsula canal became increasingly apparent. A 1697 account indicated just how expensive and physically demanding it was to haul cargo the fifteen or twenty miles overland across the neck of the Delmarva Peninsula. Not only bulk products but often boats of up to thirty tons had to be transported on sleds pulled by teams of oxen or horses.[16] The only other viable option for those who wished to ship merchandise and produce from the mouth of the Susquehanna to Philadelphia was to send cargo by sailing vessel the long way, almost 500 miles around the full length of the Delmarva Peninsula.

It was one thing to dig a short ditch to straighten out a couple of coils and oxbows in a river bed or to bypass specific shoals in an otherwise navigable creek, and quite another to make a wide and deep overland dig that would connect the upper Chesapeake with the upper Delaware Estuary. Indeed, the size of

---

* *Today, the partially straightened stream is known as the Smyrna River.*

the proposed transpeninsula canal project demanded far more financial support than Delaware could hope to provide. In any case, most Delawareans were not particularly enthusiastic about a transpeninsula canal, particularly if the eastern terminus was not at or near the port of Wilmington. The real support for the project came from Philadelphia's millers, merchants, and bankers who were increasingly intent on tapping the burgeoning commercial traffic of the Susquehanna River which was currently ending up on the docks of the growing rival port of Baltimore.

Philadelphia's idea was to build the transpeninsula canal so that the western terminus had access to the Chesapeake, just below the mouth of the Susquehanna and significantly north of Baltimore. This would cause some of the Susquehanna-born commerce, which normally headed for Baltimore, to be diverted, via canal, to Philadelphia.[17] Understandably, some Marylanders were less than enthusiastic about the proposed Chesapeake and Delaware Canal because its construction was so clearly an attempt to dramatically expand Philadelphia's economic hinterland at Baltimore's expense.

After considerable debate about the most feasible route to follow, construction on the Chesapeake and Delaware Canal commenced in 1804, only to halt the next year because the canal company faced financial problems. Construction resumed in 1824 and the canal was finally opened in 1829, thanks, in part, to purchases of stock in the canal company by the governments of Pennsylvania, Delaware, Maryland and the United States. Stretching 13.6 miles—about twelve of those miles were in Delaware—the Chesapeake and Delaware Canal's western terminus was on Back Creek, a tributary of Maryland's Elk River, and its eastern terminus was on the Delaware River, about six miles south of the town of New Castle.[18]

While Philadelphians celebrated the Canal's completion, most Delawareans were simply awestruck by this unprecedented physical assault on a very narrow east/west band of their state that stretched across central New Castle County. They were amazed to see, by June, 1824, a work force of 850 men and 150 horses digging and hauling dirt, and leaving behind a deep cut of extraordinary proportions in the land. By 1826, there were 2,600 canal laborers, including some African-Americans and many Irish immigrants, hard at work extending the deep horizontal gash. One contemporary observer remembered that it was "the Irishman and his spade," who dug the Chesapeake and Delaware Canal.

At Dividing Ridge, in western New Castle County, sweat and perseverance produced an enormous wedge-shaped ditch that was 230 feet wide at the top, 80 feet deep, and 36 feet wide at the bottom. This giant excavation project was simply the most astonishing rapid transformation of landscape hitherto experienced in Delaware. The canal site was often visited by tourists and one twentieth century historian proclaimed that, in its day, the finished canal was "one of the greatest works of human skill and ingenuity in the World." In view of its short length, however, it was also a very expensive project. The construction cost per mile of the four-lock Chesapeake and Delaware Canal was eight to nine times that of New York State's Erie Canal, which was completed in 1825, and more than seven times the average cost per mile of the Pennsylvania canals of the same era. The extraordinarily high cost per mile was attributed to the comparatively greater depth and breadth of the cut made into the Delmarva landscape to produce the Chesapeake and Delaware Canal.[19]

In the seventy-five years that preceded the Civil War, the construction of canals to expand and improve Delaware's waterways made economic sense, but at what cost to the natural environment? Although we will never be able to answer that question precisely, the enormous amounts of earth displaced, the numerous wetlands dredged to produce artificial waterways, or filled with dirt to provide tow paths for teams of horses and mules, and the numerous streams that were interrupted, redirected, and even reversed, must have produced a considerable environmental impact. One troubling result of the Chesapeake and Delaware Canal's dramatic cut into Dividing Ridge was that the deep excavation released numerous springs "holding in solution so great a quantity of sulphate of iron... as to kill all fish" in the canal's waters near Summit Bridge.[20]

Philadelphia's growing demands for the products of Delaware's fields and forests, and the increasing desires of Delawareans for the manufactured goods shipped from its docks meant that the Quaker City would continue to be the major outside force in shaping the First State's economy and landscape well into the early twentieth century. In a very real sense, most Delawareans lived and worked in Philadelphia's vast economic hinterland which was held together by the Delaware Estuary and its navigable feeder streams. Only western Sussex County, commercially organized around the Nanticoke River system, which emptied into the Chesapeake Bay, seemed beyond the reach of Philadelphia's economic tentacles.

Land Transportation

Fundamental to Philadelphia's economic and commercial hegemony was an evolving network of roads and bridges that tied the Quaker City to Delaware. While overland transportation was less important than water transportation to cementing ties with Philadelphia's growing markets, by the early nineteenth century an expanding network of roadways had become increasingly indispensable to Delawareans. Unlike other areas of North America, there may have been only a handful of extant Indian trails in Delaware that white colonists could build upon and turn into roads.

During the mid-eighteenth century, when most of Delaware's rivers and creeks were deep and navigable, it was far cheaper and more convenient for farmers and timbermen to ship the products of their fields and forests by water to Philadelphia and to ports beyond than to use overland routes. Moreover, the fact that most of early Delaware's communities were too small to justify connecting them by building expensive roads caused the construction and maintenance of overland routes to be a low priority issue for Delaware's colonial government.[21]

No wonder colonial roads and bridges were hastily engineered, cheaply constructed, and poorly maintained. Narrow, unpaved for the most part, and often laid out in twisting, circuitous paths—"exceedingly crooked and irregular" was the description of one roadway—most of Delaware's early roads had the modest goal of serving men on horseback rather than providing a conduit for the more demanding ox and horse-drawn cart and wagon traffic. Poor roads continued to plague Delaware into the late eighteenth and early nineteenth centuries.

In the late 1790s, a touring Frenchman noted that Delaware was "distinguished by the bad condition" of its roads and bridges. One example was the road through northern New Castle County that formed part of the highway connecting Philadelphia and Baltimore. In 1796, it was characterized as "in a fearful condition," with huge ruts occurring at numerous intervals, causing coaches to be overturned and passengers to be killed. As late as 1811, some public roads in New Castle County were "almost impassible in rainy seasons."

South of the fall line, eighteenth century roads were described as somewhat better, probably because the land was more level and the amount of sandy soil, which was more porous, generally increased as one traveled towards Kent and Sussex. Consequently, in the better-drained sections of central and southern New Castle as well as in much of Kent and Sussex, travelers and their vehicles usually encountered fewer mudholes and fewer deep ruts after rainstorms than further north.[22]

Poor roads, however, were better than no roads at all. Particularly crucial in the seventeenth century was an overland connection to the tobacco-producing farms and plantations of Maryland's Upper Eastern Shore. By 1669, Augustine Herrman completed a "fine broad cart road" about eight miles long from his Maryland plantation on the Bohemia River, which flows westward into the Chesapeake, to the Appoquinimink River, which empties eastward into the Delaware River. Evidently, a second road was constructed about the same time running southwest from the town of New Castle to connect with Herrman's cart road. Other roads were added over the years so that by 1737, an overland route stretched almost the length of the state from Pennsylvania all the way south to Lewes.[23]

By the American Revolution, a network of roads which, in most cases, was really not much more than a web of poorly maintained paths, served a large number of Delawareans who lived some distance from navigable waterways. The best maintained of this sorry lot was a series of linked overland arteries that stretched all the way south from the Pennsylvania line to Milford and from there formed two branches with one leading southeast to Lewes while the other pushed directly south through Dagsboro and on to Snow Hill, Maryland. These roads were used to carry the mail of the British Empire and, in order to separate them from other public roads,

A Sussex County Plantation, circa 1800—
Shown above is a conjectural sketch of a large, coastal Sussex County plantation as it might have appeared about 1800. It includes such typical features as worm fencing, a wharf on a navigable, tidal estuary, an orchard, and a variety of outbuildings, including housing for a small number of slaves, *sketch by Richard B. Carter, 2001.*

were collectively called the King's Road or the King's Highway. Delaware's colonial legislature attempted to set high construction and maintenance standards for the King's Highway, but somewhat lower standards for the other public roads. The constant need to pass additional highway legislation concerning the King's Highway, however, indicates that the legislature's good intentions did not always translate into good results. Despite the problematic condition of Delaware's roads during the eighteenth century, they did offer to many in the overwhelmingly rural, widely scattered population some relief from economic and social isolation by providing access to mail service, river landings, and to "church, mill and market."[24]

But the nature of Delaware's rural economy caused some farmers to object to new roads if they were constructed through their fields. The problem was that by cutting through the wooden fences of already enclosed farmland, new public roads served as both escape routes for a farmer's penned livestock who wished to roam the countryside and mate with less genetically desirable animals, and as entry-ways into a farmer's planted grain-fields for a neighbor's free-roaming and constantly hungry cattle and pigs. In short, a public road through one farmer's fields meant added personal expenses and labor to restrain livestock and protect maturing crops. John Bird, for example, petitioned for relief from a New Castle County Court in 1753 because a new road, about to be cut through his property, would force him to build and maintain a new fence to protect his grain. Seven years later Jacob Grantham of New Castle Hundred—hundreds were subdivisions of counties—asked a court for "the liberty to hang gates" across the new road at the points that it entered and left his land "to prevent the expense of two [new] fences." Grantham's plea was echoed, in 1783, by nine petitioners who lived along the road from the Mispillion River south to Cedar Creek in northeastern Sussex.[25]

Even more challenging than the winding, poorly maintained roads were the many stream crossings used by overland travelers. The ford on White Clay Creek on the road from Christiana Bridge to Newport was described in 1760 as extremely hazardous with travelers "frequently in the most imminent danger of losing their lives..." Much of the lower Brandywine, because of its steep banks and rapid current, was both difficult and dangerous to ford. Other streams, such as the lower reaches of the Christina, Duck Creek, St. Jones, Murderkill, Broadkill, Lewes Creek, and Indian River, were too deep or too marshy along the banks to ford under any condition. A growing problem for travelers was the constantly widening marshes along river banks caused by streams overflowing their sediment-clogged channels. The fact that tidewater rivers were becoming shallower but wider, meant that formerly firm ground along river banks was metamorphosing into narrow ribbons of wetlands. The only option for a horseman or wagoner was to take a long roundabout route upstream to a fording place where the water was shallow at low tide and the river banks still offered firm footing. In 1679, for example, a party of horsemen rode upstream to the future site of Christiana Bridge where, at low tide, they were able to cross the Christina River because the water was not much higher than their horses' knees.[26] The general flow of most Delaware streams from west to east was particularly disruptive to overland travel along a north/south axis. Clearly a safer, more efficient manner of crossing creeks and rivers had to be introduced if both people and goods were to move overland in a systematic fashion.

Some evidence indicates that in the Wilmington area, ferries were carrying people, livestock, and goods across the Christina River as early as 1669 and across the Brandywine as early as 1675. But a fairly reliable ferry service was missing from both rivers until well into the eighteenth century. By the last decades of the eighteenth century there were also ferry crossings at a number of other sites such as Newport, Red Lion, St. Georges, Cantwell's Bridge (Odessa) and Smyrna. Ferries appeared first in northern Delaware because roads in New Castle County were more heavily traveled than those further south. These early ferries were usually pole-driven rafts large enough to hold an ox cart, wagon, or even a coach with the attending horses. But ferries did not work if the creek or river froze over, or if the water level dropped so low that the rafts could no longer float, or if the banks of the stream became so marshy that men and animals could not approach the ferry.[27] Bridges, however, were good for all seasons and for most physical settings.

By the late seventeenth century, crude bridges spanned some of Delaware's smaller streams but not its broader and deeper creeks and rivers. It was an engineering challenge to throw a span across the lower reaches of the Brandywine, Christina, St. Jones, and Indian rivers. Moreover, the construction of a bridge further downstream would act as a tourniquet to the free flow of upriver commerce because, with the exception of a few shallops that had removable masts, sailing ships could not fit under bridge spans. One way out of this dilemma was to construct drawbridges that could be lifted up to let masted vessels through and then lowered so that the overland traffic could resume. But raising the bridge had to begin ahead of time so that sailing vessels would not be held up as they moved up or down the river. In 1813, for example, Delaware's General Assembly gave permission to a company to construct a drawbridge across the Christina River at Newport, provided that the company would begin raising the bridge as soon as it heard a horn-blown signal at a half mile distance from an approaching masted vessel. If the bridge were not up by the time the masted ship reached it, the bridge company was subject to a fine.[28]

During the first two decades of the nineteenth century, a growing population and an increasing desire to move both people and some bulk goods over significant distances placed a new strain on Delaware's mostly unpaved, narrow, and poorly maintained highway system. Just as it turned to private initiative to build badly needed canals, beginning in 1808, Delaware's General Assembly turned to turnpike companies to provide at least some of the new roads that the state needed. The hundreds of New Castle and the county governments of Kent and Sussex also continued to build some new roads with tax dollars.

Delaware's turnpikes were wide, hard-surfaced toll roads, and they were built only in northern New Castle County because that was where the demand for improved roads was greatest. Typically, in adhering to the road construction reforms of the Scottish visionary, Robert McAdam, the turnpike companies were required by the General Assembly to construct highways that were "bedded with wood, stone, gravel, clay and other convenient materials, well compacted together,..." and topped with a layer of "clay, gravel or stone," with "an even surface rising towards the middle, by a gradual arch." Finally, a liquid mix was added to bind together the gravel or stones to give the "macadamized" highways their hard and consistent surfaces. Although the turnpike companies were also required by the General Assembly to keep their "macadam" roads "in perfect order," in 1841 a critic maintained that the macadamized toll roads in northern Delaware were neither well constructed nor well maintained examples "of this kind of road." Governor William Cooper agreed and indicated that one of the problems was that the pieces of stone used to pave the turnpikes were too large.[29]

By 1825, a number of privately constructed toll roads connected the northern Delaware communities of Wilmington, New Castle, Newport, Christiana Bridge, and Newark, with turnpikes beyond Delaware that provided dependable overland access to Philadelphia, to the portion of the very fertile Great Valley that stretched from east to west through southeastern Pennsylvania's Chester and Lancaster counties, to the northern Chesapeake, and to Baltimore.

In 1822, there were nine turnpikes in northern Delaware and they represented a combined total of sixty-five paved miles. That year all but one of the many roads that led out of Wilmington was a turnpike.[30] The newly-built turnpikes, along with canals and older, government-maintained roads, linked much of Delaware to the rapidly expanding market economy of the entire nation during the early nineteenth century. Although water transportation of bulk goods continued, on a per mile basis, to be significantly cheaper than even the horse-pulled wagons of the macadamized turnpikes, the improved efficiency of the privately constructed toll roads made them viable until the appearance of a dramatically new and even more efficient form of overland transportation, the railroad.

Road construction had an immense impact on Delaware's natural environment. Not only were trees cut down, and roots grubbed up, but stone, clay, gravel and even timber were taken from other locations across the landscape and hauled to road construction sites. Moreover, constructing roadbeds across swampy terrain dramatically altered natural drainage patterns and permanently scarred wetlands. But even more important to the natural environment was the impact of new and better highways on Delaware's timbering and farming practices. Woodsmen proceeded to cut down previously inaccessible and therefore untouched stands of trees because now, with new and improved roads, they could get the timber to a navigable stream and thus to an urban market. Those new and improved roads enabled farmers to increase their production of such cash crops as wheat, because now staple crops could be more easily hauled from inland fields to local mills or to river landings for shipment by sail to Wilmington's or Philadelphia's mills and flour markets. All of this led to the clearing of additional acres that were turned into farmland, and to a greater commitment by individual farmers to repeated plantings of the same staple crops simply because the Wilmington and Philadelphia markets demanded it. Not surprisingly, the greater dependence on monoculture combined with the old practice of not fertilizing fields to increasingly drain Delaware's farm land of its nutrients.

Agriculture

By the mid-eighteenth century, most Delaware farmers had committed themselves to at least some commercial wheat production, and this had significant implications for the treatment of the rural landscape. Unlike tobacco and corn, which could be raised successfully on fields prepared by the hoe, productive wheat cultivation required that the entire field be plowed prior to planting. In addition, the search for greater efficiency in the use of ox-pulled plows led to the laying out of fields in a more geometric pattern. But even before plowing, the preparation of commercial wheat fields required a more complete scouring of the landscape of trees, bushes, stumps, roots and other detritus than was a necessary precondition for tobacco and corn.

Because of the subsequent decades of plowing—corn fields, too, were commonly turned by the plow once oxen were widely available in the 1750s—significant amounts of Delaware's top-soil had been lost to wind gusts and to rain runoff by the early nineteenth century. One example was Linden Hill, a farm located in flat Delaware about a half mile north of St. Georges in central New Castle County. By the 1830s, its fields were so scarred by wind and water erosion that gullies were everywhere. This contradicts the perception that, compared to the piedmont, farm fields on the Atlantic Coastal Plain generally suffered only minimal soil erosion.

Although it is impossible to be more specific about the annual loss of topsoil almost 200 years ago, an approximation for the end of the twentieth century is suggestive. Despite following soil conservation practices that were not present two centuries ago, flat Delaware's farmland was annually losing an estimated two tons of topsoil per acre in 2000, while in Delaware's piedmont, the estimated loss was at least twice as much.* Given the wasteful agricultural practices of early Delaware, yearly topsoil losses must have been considerably higher. Because of substantial topsoil loss, the leaching of plant nutrients by rain, repeated monocrop cultivation, and the reluctance of most early Delawareans to fight declining soil fertility by applying livestock manure and marl—soil that included lime—to their fields, by the early nineteenth century most of the state's farms resembled one in central Sussex where the corn and wheat productivity per acre grew "less and less as the unfertilized fields [became] more and more sterile." In 1818, in an address to the Agricultural Society of New Castle County, S. H. Black, a Glasgow physician, pointed out that Delaware's farmland had lost so much of its former fertility that "for the last fifty years [it] has done little more than starve its proprietors or driven them to a sale of it." Black warned that without dramatic steps to improve soil fertility, such as using fertilizers, the state's farmers would be forced to move west.[31]

The unwillingness of most Delaware farmers to abandon the soil-depleting practices of the past may have reflected, in part, their physical isolation from the rest of America. In 1839, William Huffington of Dover observed that, except for northern New Castle County, Delaware was so far off the beaten path that "the progress of improvements" in the state had been "arrested" over the last fifty years. Huffington blamed "the utter want of all cheap and economical means of intercommunication with the rest of our country." He was convinced that the two-thirds of the state's population that lived in Sussex, Kent and southern New Castle Counties was more isolated from the rest of the nation "than any other population of corresponding numbers in the United States."[32]

This isolation factor negatively impacted

* It must be noted that, at the annual loss of five tons of topsoil per acre, it would take 36.3 years to erode evenly an inch of topsoil over one acre.

One-room-plan Houses— An example of two typical southern Delaware one-room-plan houses, joined together into a single structure at some point in their histories, survived into the early 1980s near the southwestern Sussex hamlet of Hearn's Crossroads, *photo by Richard B. Carter, circa 1980.*

---

farming practices well into the second half of the nineteenth century. In 1868, for example, the fertility of most of Delaware's fields continued to be adversely affected by poor crop-rotation practices, the failure to fertilize with animal dung and marl, and the unwillingness to plant such cover crops as timothy and clover. Not only were Delaware's farm fields and meadows "generally destitute of grass" for livestock grazing, there were few "suitable farm buildings for the proper housing" of farm animals. As a result, cattle, sheep and sometimes horses," were seen "running at large, wasting their manure," and simply struggling hard to survive. Because most Delaware livestock was not housed in barns, barnyards, and stables until at least the late nineteenth century, even those farmers who wished to fertilize their crop land found it next to impossible to collect enough manure to do the job. Completing this bleak picture of Delaware's agrarian landscape in 1868 were the typical farmhouses which were either "small frame or log one-story tenements, with seldom a tree or a shrub to make them attractive," and often with "pigs and cattle enjoying themselves, in company with the dog and chickens in the front yard."[33]

Livestock

European livestock joined European guns, axes, shovels, and plows in dramatically altering Delaware's physical world. Indeed, Alfred Crosby's characterization of the environmental impact on North and South America by European farm animals as a "self replicating and world-altering avalanche," eventually applied to Delaware.[34] The first livestock to arrive in Delaware were a mere handful of horses and cattle that were brought to Swanendael (Lewes) in 1631, only to be killed within a year when the entire Dutch settlement was wiped out by local Native Americans. Of more permanent nature were the six horses landed at Fort Christina in 1639, a year after New Sweden was founded. By 1650, a small number of horses, cattle, pigs, sheep and goats were present in the Fort Christina area, but not yet a "world-altering avalanche" of livestock.[35]

Over the next century and a half, however, natural reproduction and almost ideal environmental conditions caused a population explosion among Delaware's domestic animals in general and its swine in particular. As early as 1660, there were probably more than 2,000 cattle and even more pigs in Delaware. Usually turned loose to roam marshes and forests, cattle and horses were drawn to the abundant marsh grass while pigs feasted enthusiastically on the fruits, nuts, roots, seedlings and other edible forest growth that were once the prime food of the increasingly rare white-tailed deer. Particularly significant was the fact that, by the mid-eighteenth century, large farm animals had little to fear from such livestock predators as wolves and cougars because both carnivores had been driven from the state by bounty hunters and by the growing scarcity of their favorite food, white-tailed deer. Smaller farm animals such as lambs, piglets and chickens, however, continued to be vulnerable to foxes, raccoons, opossums, and wild dogs. An additional contributor to the population explosion in livestock was the fact that America's indigenous animal diseases were little or no threat to European farm animals. All of this caused one nineteenth century observer to point out that Delaware's wild animals "have given place to domestic animals, who are under the dominion of men." Predictably, the number of livestock on individual farms dramatically expanded over time. One example was John Curtis of the St. Jones River area in Kent County who owned 300 cattle by 1698.[36]

Under the so-called "dominion of men," free-ranging livestock altered the face of Delaware as they browsed, grazed and rooted their way across the landscape. While considerable study needs to be done on the specifics of the environmental impact of free-ranging livestock, it is clear that their hooves compacted the soil and that their voracious appetites seriously damaged the forest understory. During the brief period of Swedish and Dutch hegemony (1638–1664), some of Delaware's livestock were restricted to barns, barnyards, and fenced-in fields. By the English era, however, most large domestic animals were freely roaming the countryside with ownership signified by patterns or brands cut into the ears of horses, cattle, and pigs. Because horses were the most expensive to purchase, in subsequent years they were the first livestock to have their freedom curtailed. By the mid-eighteenth century, Delaware's colonial legislature had barred horses more than two years old from roaming "the woodlands and marshes," because this led to unsupervised and uncontrolled breeding, which guaranteed a decline in equine blood lines. Indeed, during the free-roaming era, promiscuous breeding by free-ranging farm animals caused Delaware's horses, cattle, and pigs to be physically smaller than their European counterparts.[37]

During the ante-bellum period—1800 to 1860—the days of free-ranging farm animals finally came to an end throughout most of Delaware. Historian Bernard Herman maintains that livestock raising in Delaware was dramatically transformed by a gradual but significant shift in the perceived demarcation line between the traditional common rights of "ordinary" people and the property rights of large land owners. The most visible evidence of this legal shift can be seen in the changing function of fences.

During the colonial era, poor Delawareans customarily assumed that privately owned but "unimproved" land was open to the common use of all. This meant that they had the traditional right to turn loose their livestock to graze, browse and root in forests and marshes owned by others. If horses, cows and pigs wandered out of those forests and devastated the landowner's fields of corn, wheat, or other grain, it was just a regrettable but unavoidable result of common grazing rights. For a farmer whose crops on his own improved land were devoured by wandering hordes of alien farm animals, the only legal recourse was to build fences around his fields, a most expensive thing to do. Roaming livestock were most common in Kent and Sussex Counties. In 1788, Dr. James Tilton, a keen observer of agricultural practices, noted that in central and southern Delaware the state's cattle were "bred in the greatest number on the marshes and [in] the forests" from which they were "driven in large droves" to the lush meadows of New Castle County where they "were grazed and stall fed for the markets of Wilmington and Philadelphia."[38]

By the first three decades of the nineteenth century, several new state laws indicated a shift in the thinking of Delaware's General Assembly on the issue of traditional grazing rights versus the rights of land owners. In short, poor farmers were losing the legal right to turn loose their domestic animals on the privately owned, unimproved land of wealthier farmers and timbermen. The reason was that cattle and pigs were now seen as a significant threat to the well-being of valuable forests and marshes, to improved fields of corn and wheat, and to the aesthetic aspirations of towns and villages. Increasingly, it was the owners of livestock rather than the owners of crop lands who were required, by Delaware law, to build fences so that domestic animals no longer threatened private property rights. In doing so, Delaware predated by a number of decades the actions of the Atlantic coast states further south or the states and territories located on the grasslands of the Great Plains. In all three geographic areas, the end of open-range grazing and the substitution of closed-range grazing was largely dictated by the way fences were used to regulate and control livestock.[39]

Although some cattle continued to roam the forests and marshes of downstate Delaware into the late nineteenth century, most were now confined to free-range grazing on necks of land along the Delaware Bay and River, or on the grass-covered dunes of barrier islands that abutted the Atlantic Ocean. Necks and barrier islands had long been favorite sites for turning loose livestock because they offered good grazing and because their natural borders of water and marsh kept horses, cattle, and pigs from roaming. One example was the neck between the Mispillion River and Cedar Creek in northeastern Sussex. A second was the neck at Plumb Point, just north of the mouth of the Broadkill in eastern Sussex which, in 1818, was "known to be the best shelter for stock of any on the Delaware." A third was on grass-covered dunes of Fenwick Island, in the southeast corner of Sussex. In 1824, the state legislature authorized the area's farmers to erect a gate on the public access road to keep their livestock from wandering westward from Fenwick to the mainland.[40]

A fourth example was the land that extended from the east bank of Lewes Creek to the Delaware Bay and the Atlantic Ocean, and was generally designated "Cape Henlopen." But unlike the other necks and barrier islands, where the rights of individual land ownership were increasingly honored during the first few years of the nineteenth century, this timber, meadow, dune, and marsh-covered landscape east and northeast of Lewes, continued to be open to grazing by the livestock of any Sussex County farmer, thanks to the "commons" status bestowed on the area by William Penn. Prior to the early nineteenth century, the only land use restrictions, other than keeping the Cape Henlopen area wild, concerned the cutting of timber. As with commons elsewhere in history, perceived self-interests of Sussex farmers soon led them to drive ever more cows and pigs across Lewes Creek, and they subsequently over-grazed the public grasslands on and near Cape Henlopen.

In 1814, recognizing that some restrictions on the number of livestock were necessary to halt further degradation, the trustees of the Cape Henlopen commons asked for and received from the Delaware legislature the power to charge for grazing rights on "the land and the marsh of Cape Henlopen." In 1701, William Penn established a second commons of approximately 1,000 acres just north of the town of New Castle. By 1850, however, it too had been turned into rental lands with the income going to town projects.[41]

Of all the free-ranging domestic animals, swine had by far the greatest impact on Delaware's flora and, indirectly, its fauna. Although sheep and goats joined pigs in being more environmentally destructive than cattle and horses, neither sheep nor goats were ever numerous enough in Delaware to compete with the tough, rapidly reproducing porcine hordes. The number of swine roaming the Delaware landscape at any one time in the eighteenth and early nineteenth centuries is impossible to estimate. But judging by the number and vehemence of complaints registered across the

Hogs running wild—
This sketch by artist Jason Burleson-Gibson captures something of the chaos that reigned in southern Delaware towns when hordes of hogs were allowed to run wild in the streets.

state against free-ranging pigs, their numbers were considerable. They attacked the remaining forests, consuming very young trees and other forest understory. One estimate maintains that along the Atlantic Coastal Plain, one free-ranging pig annually needed exclusive grazing, browsing and rooting rights to approximately twenty to thirty acres of "grasses, woody plants and new growth hardwood trees" to remain healthy. A second estimate assigns less acreage per animal, but points out that swine were particularly destructive of smaller pine trees by digging up their roots. A recent environmental history of Delaware and the Eastern Shore of Maryland and Virginia maintains that the "herbaceous layer in many woods has not recovered" to this day from the free-ranging swine onslaught of two centuries ago.[42]

While most early Delawareans seemed to accept the cutting down of forests and the draining of wetlands by humans as clear evidence of progress, the growing porcine degradation of surviving woodlands and marshlands was another matter. In 1803, the General Assembly barred free-ranging swine from a large area of rural southern New Castle County because of "the scarcity of timber." But it was the invasion by pillaging pigs of improved areas of the countryside, such as diked marshes, meadows, wheat and corn fields, and, in particular, the gardens of town and village residents, that ignited anger against these marauders and led to very restrictive state laws. Intense hostility towards marauding swine even led to the "Great Pig Massacre" in Milford in 1809, where "many of them were shot down in the streets." Subsequently, the Delaware General Assembly declared that large stretches of Delaware's rural landscape and most of its towns and villages were off limits to free ranging swine.[43]

But laws that put an end to free-ranging swine were more than simply victories for grain farmers over livestock owners, or triumphs of the aesthetic sensibilities of town and village leaders over the more rustic sensibilities of country folk. These swine-restricting laws, as previously pointed out, can also be seen as victories for the economic interests of the large landowners over the rights, aspirations and needs of the rural poor. In fact, as David J. Grettler points out, many of the rural poor understood the stakes and responded with petitions to the General Assembly that demanded the repeal of restrictive swine legislation. As a 1797 protest petition to the state legislature from northern Kent County insisted, by taking away grazing, rooting and browsing rights to unimproved land, Delaware's government was making life more difficult for the already hard-pressed rural poor. After all, it was only by turning their pigs loose to root on the unimproved property of others that the rural poor could afford to feed the hogs which were so essential for nourishing their families. By forcing the rural poor to restrict their swine to pig-pens or other enclosed areas, the General Assembly made it necessary for them to buy lumber to build pens, and to purchase corn to fatten their swine. Under these conditions, the rural poor of Delaware were in serious economic trouble because "not one poor man in ten could [afford to] raise one hog in the pen."[44]

Until 1816, petitions by the rural poor were usually effective in causing the General Assembly to reverse most swine-restricting legislation. Just as in its hunting and fishing regulations, Delaware's legislature demonstrated at least some sensitivity to the needs of this group as its members struggled to feed their families. All of this changed, however, when the end of the Napoleonic wars in Europe caused a dramatic decline in grain prices in the Delaware Valley. This coincided with the onset of the series of remarkably cold summers, beginning in 1816—the result of a cataclysmic volcanic eruption in the East Indies—that made for almost impossible growing conditions.

The resulting economic hard times sparked a drive by the state government to rid the agrarian landscape of inefficient practices and to support the necessary agricultural reforms that would produce a rural economic recovery. Tolerating free-ranging swine was both wasteful and archaic, and had to stop. What followed was a series of piecemeal laws aimed at protecting individual towns and specific rural areas from marauding pigs. This time, however, the pleas of the rural poor went unheeded. In 1829, free-ranging swine were banned from most of the state. By this date, according to David J. Grettler, the message was clear: "either pen your hogs or lose them to a well placed bullet." A Superior Court ruling in 1845 and further legislation in the 1850s pretty much ended the era of free-ranging hogs in the First State.

In 1893, the Delaware General Assembly drove the last nail into the coffin by making it illegal "for any livestock to run at large" on any "unenclosed lands within this state."[45] In outlawing marauding pigs, Delaware not only protected some of its physical landscape from environmentally destructive porcine hordes, it also protected the property rights of landowners against the perceived traditional grazing rights of the rural poor. Or at least that is what the latest conventional wisdom would have us believe.

Moving, however, from an open-range grazing pattern, where pigs and other livestock roamed freely through the countryside, to a closed-range grazing system, where free-ranging pigs and other livestock that foraged on another's property were no longer legally tolerated, may have been a bit more complicated than simply a land-rights grab by the rich and powerful. Sam Glasscock, Master in Chancery of the Chancery Court of Delaware, argues that from the very beginning of the jurisdiction of English law in Delaware, in the late seventeenth century, "private property rights in real property were fully developed." Thus, open range grazing was not based on "a public right to use private property." Rather, at any time a land owner had the legal right to seek redress from the courts to protect his or her property from trespassers. Given this option, why did so many property holders, initially at least, allow alien livestock to be turned loose on their undeveloped land? After all, under open range grazing practices, those same property owners were expected to build, at their own expense, fences to protect their crops from trespassing swine and cattle.[46]

Glasscock is persuaded that there are probably a number of reasons for legal support of open range grazing, and most of them generally benefitted the large holders of real estate who included a considerable number of absentee owners. As one example, many large landowners in central and southern Delaware simply turned loose their own considerable numbers of livestock to wonder across the countryside and, in some cases, into the garden plots of the rural poor who, if they had not already constructed costly fences, had no legal recourse for relief from the havoc wrought by pigs and cattle to their crops. In examining the Delaware situation, Glasscock draws an analogy to the American West, where powerful ranchers strongly championed open range grazing so that they might let their cattle herds ride roughshod over the crops and therefore the land rights of homesteaders. The popular movie *Shane* brought these questionable open range grazing practices of powerful cattlemen to the big screen.[47]

A second possibility worth considering, according to Glasscock, concerns the English legal doctrine of "adverse possession," which was based on the tendency of English property law to always encourage "the economic development of real property." Adverse possession was achieved through occupancy of another's land in an "open, notorious and hostile" manner for a consecutive number of years—today in Delaware the occupancy must be for twenty consecutive years. The purpose, of course, was to put someone else on the land to make it productive if the owner failed at that task. "Open, notorious and hostile" occupancy presupposed that the original owner knew about, but did not give permission for occupancy. But if a large landowner allowed his poorer neighbors to use that property for livestock grazing, that kind of use could "never ripen into a claim of adverse possession." In short, for many years the rich and powerful protected their Delaware holdings from seizure under law by allowing their land to be foraged by the pigs and cattle of their less fortunate neighbors.

When certain areas of Delaware began to increasingly rely on the production of staple crops, however, an open range grazing system lost its appeal. Large landowners now insisted that neighboring livestock be fenced in so that their own valuable cash crops would not be destroyed by wandering horses, cattle and pigs. By the late eighteenth and early nineteenth centuries, it was only natural that members of Delaware's landed gentry abandon support for open-range grazing and push to enact statewide legislation that would make a closed-range grazing system the law of the land. They did so because it was in their economic best interests.[48]

Draining the Land

Once the closed-grazing system was in place, farming in Delaware became more profitable. The economic situation was further enhanced by the continued growth in the demands by Philadelphia, Wilmington and other growing urban centers for the products of Delaware's fields. But many of those fields were so worn out by the beginning of the nineteenth century that they were no longer productive. To one observer, the reason was quite simple: "land, like your horse or your ox, if worked must be fed, or it will die."[49]

In the colonial era, when Delawareans wanted to expand their cropland or to replace worn-out fields, they simply cleared adjacent forests. By the late eighteenth century, however, there was greater hesitancy about cutting down the remaining groves of mature trees because, increasingly, good timber was in short supply. Instead, many Delawareans turned to diking and draining wetlands to produce new fields for pasturing livestock, growing fodder, and cultivating staple crops.

Although there may have been a few attempts at draining Delaware's marshes during the Swedish period (1638–1655), the real impetus occurred during the Dutch era (1655–1664). Under Dutch and subsequent English rule, throwing up earthen dikes and digging ditches across wetlands transformed marshes into pastures, hay meadows, and fields for growing staple crops. Diking and ditching began in the spring, but was most intense in August because then—despite hordes of mosquitoes—there was a brief lull in farm work and water levels were low.

Maintaining dikes and ditches against the tunneling of muskrats and fiddler crabs, the rooting of free-roaming swine, and the pressures of rising tides, sedimentation, and clogging plant growth was both difficult and costly. In 1782, for example, at least one half of Caesar Rodney's dike, thrown up to hold out what was probably a tributary of the St. Jones River, was "destroyed with the [rising] tide and muskrats." Initially, the draining of wetlands was most pronounced in saltwater marshes, but by the late eighteenth century, inland freshwater marshes became the site of most of the new dikes and drainage ditches.[50]

Historic records give us an occasional glimpse at the transformation of Delaware's wetlands into grazing meadows and farm fields. In 1695, for example, the county court gave permission "for a ditch to be cut through the most convenient place of the town of Lewes into [Lewes] Creek for the convenience of draining the savanna..." Throughout the eighteenth century, the Delaware legislature was constantly approving projects to bank and drain wetlands. In 1766 alone, of the nine acts passed by that body, five dealt with drainage projects. At first, because the original land grants were so large, individual property holders handled the construction of dikes and ditches. After the subdivision of the large land grants, however, it became apparent that ditching organizations or companies were necessary because most draining projects were too substantial in size to be efficiently handled by individual landowners. Only groups of farmers, willing to pool their money and able to make appropriate plans for the ditching of thousands of acres of marsh and swamp, could efficiently drain the land. By the early nineteenth century, much, if not most, of the ditching and diking work was probably contracted out to hired teams of men who specialized in this kind of work. To better facilitate the organization of ditching companies, the General Assembly set out detailed rules in 1822.[51]

All three Delaware counties experienced dramatic landscape changes as wetlands were transformed into fertile fields and pastureland. At first, transformations were most extensive in New Castle County south of the Fall Line, and in northern and central Kent. Not until the mid-and late eighteenth century did large areas of southern and western Kent, and interior regions of Sussex begin to be drained. The rewards for sweat equity and monetary investment were quickly evident. Because, over the centuries, a concentration of organic materials had accumulated in Delaware's marshes, the drained wetlands immediately became some of the most fertile fields in Delaware. Particularly fertile were those located on the west side of Dividing Ridge in western Kent. A notable example was Heronton Bog—called Herrington Marsh by Delaware's state geologist—whose waters eventually drained southwestward into the Chesapeake via the Choptank River. Covered with forest, marsh grass and shrubs as late as 1790, Heronton Bog was transformed by 1838 into an immense area of "arable land of the finest quality," thanks to judiciously located drainage ditches. The fertility was understandable because the bog's soil was made up of "rich black mold containing the accumulated vegetable matter of ages, averaging three feet in depth, but frequently upwards of six feet in depth." Most of Delaware's farms probably averaged between ten and twenty bushels of corn per acre during much of the ante-bellum era. By contrast, the fertile fields carved out of Heronton Bog, initially at least, may have produced as high as ninety to one hundred bushels of corn per acre.[52]

Transforming wetlands into productive farmland made good sense according to the ethos of the Delaware Valley's growing and expanding market economy, fueled as it was by capitalist investments from both in and out of state. The drainage ditches that crisscrossed Delaware became symbols of progressive farming and enlightened land management. One ditch was described in 1838 as increasing "from twelve to twenty-four feet in width from its source to its mouth, a distance of nine miles," and as having enough water in it after spring rains, "to float a moderately sized vessel."[53] Draining wetlands, however, had unexpected environmental consequences. Fields adjacent to the recently dug ditches, but formerly high enough to be prime farmland, now shed much of their rainwater into the newly-constructed drainage systems. Thus, while former wetlands were turned into extraordinarily fertile fields, adjoining but elevated croplands experienced a decline in productivity because of diminished moisture retention.

Delaware's freshwater streams were dramatically altered by the disappearance of marshes and swamps as well as by the destruction of forests. Lost was the sponge effect of wetlands and woodlands on rainwater delivered by thunderstorms or more moderate showers. Because a land lacking in old-growth trees and extensive marshes and swamps simply could not absorb significant amounts of falling rain, the resulting tumultuous runoff increased soil erosion and produced wildly erratic, muddy streams that overflowed their banks one day, only to become almost bone-dry the next. By contrast, the sponge effect of extensive wetlands, swamps, and mature forests meant that excess water was released gradually, limiting soil erosion and causing stream flow to be clearer and more consistent throughout the year. In general, wetland drainage projects also seriously diminished local wildlife populations. Marshes and swamps host the tiny organisms that form the base of the food chain that supports higher forms of life. Without that protein base, the local populations of fish, reptiles, birds and mammals declined. An example was the thousands of herons that once made Heronton Bog their roosting place before the bog was drained. By the late ante-bellum period, they were gone.[54]

The Disease Factor

Ironically, by draining many of their marshes, Delawareans inadvertently improved the health of those living south of New Castle County's Fall Line. A number of disease vectors, such as mice, mosquitoes, and certain types of flies, are indigenous to Delaware. But other disease carriers such as rats—more specifically the black and Norway (brown) varieties—and black flies were accidentally introduced to the Delaware Valley during colonial times after crossing the Atlantic from Europe as stowaways in the holds of ships. A threat to stored grain supplies, rats were even more menacing as reservoirs for pathogens that caused typhus, rabies, spirochetal jaundice, trichinosis and, potentially at least, bubonic plague. But more significant to early Delaware than diseases spread by flies and rats, was the introduction of malaria.

English sailors may have introduced vivax malaria, a relatively mild form of the disease which existed as far north in Europe as England's Yorkshire, to Virginia during Jamestown's first few years. Subsequently, falciparum malaria, the most deadly form of the disease, was probably first carried to the English and Dutch colonies in the bloodstream of enslaved Africans and then spread by the bite of the indigenous anopheles mosquito to both Native Americans and whites. Falciparum malaria probably reached Delaware by the mid-seventeenth century.[55]

Only those Delawareans living in the rolling piedmont, which had a less favorable mosquito habitat than the central and southern sections of the colony, seemed to escape the late summer and fall malarial fevers (then known as "ague") which caused bodies to shiver and teeth to chatter. In 1739, Benjamin Franklin's *Poor Richard's Almanac* highlighted the prevalence of malaria when it observed that

> Ships sailing down Delaware Bay this month [August] shall hear at ten leagues distance, a confused rattling noise like a shower of hail upon a cake of ice. Don't be frighted good passengers! The sailors can inform you, that it is nothing but lower county [Delaware] teeth in the ague.[56]

Throughout flat Delaware, people routinely suffered from malaria. The work crews who cut the Chesapeake and Delaware Canal through the marshes of Red Lion Hundred in central New Castle County in the 1820s, for example, were "killed in great numbers" by the disease. But it was the residents of southern Delaware who were particularly victimized by malaria. In 1782, the Anglican rector in Lewes pointed out that Sussex Countians "are remarkably subject to ague and fever, always in the fall of the year." The worst months for malarial fevers began in August, and extended through autumn until a deep frost killed all the mosquito vectors. Although malaria usually led to the direct death of only a small percentage of its victims, it left others so physically debilitated that they were unable to work in the fields during the fall harvest. In addition, malaria so weakened people that they became highly susceptible to other serious diseases. Visitors to Kent and Sussex, during the summer or early fall months, were also subject to malaria. One example was the English-born botanist, Thomas Nuttell, who contracted the disease after making a plant collecting trip to Sussex in the summer of 1809. For the rest of his life he experienced periodic recurrences of malarial fever.[57]

Understandably, residents from further north were often unwilling to move to Delaware for fear of contracting ague or, once established in the colony, wished to leave during the fever season. In 1706, the Anglican missionary sent to Dover and Appoquinimink was scheduled to be removed from Delaware "for three months in the heat of the summer" to some place free of mosquitoes so that he "would be much more healthful and able to do more service in the Church." By the late eighteenth century, Methodist preachers considered Delaware a high risk area and, if not already seasoned to ague, were reluctant to locate in the state. From time to time, other Methodist preachers were moved out of Delaware by their clerical superiors to recover from lowland fevers.[58]

Centuries of exposure to malaria in their ancestral homeland caused some, but not all, Africans to develop a genetic resistance to symptoms of the disease through the same gene that causes sickle-cell anemia. It was not long before whites in the New World noticed that many enslaved Africans seemed less susceptible than Europeans to the lowland fevers of late summer and early fall, and thus very desirable as field hands. The great irony was that the first waves of enslaved Africans to arrive on this side of the Atlantic probably introduced the falciparum form of malaria which, because of its devastating impact on whites, helped fuel the demand for even more African slaves to work in the fever-infested fields of the New World's coastal lowlands. This inherited resistance combined with a number of other factors, made African slaves a pragmatic solution to early Delaware's need for field laborers.[59]

The threatening presence of malaria helped increase the popularity of slavery to the point that, on the eve of the American Revolution, between twenty and twenty-five percent of Delaware's total population was made up of enslaved blacks. This was a far higher percentage than found in any of the less malaria-prone English colonies to the north, but a significantly lower percentage than in the more malaria-prone English colonies to the south. Even in little Delaware, the level of slavery's popularity coincided with malaria's geographic frontier lines. In New Castle County's rolling hill country, where ague was not a significant problem, the piedmont hundreds of Mill Creek, Christiana and Brandywine reported only four, three, and two percent of their respective populations were enslaved African-Americans in 1800. That same year, only a few miles further south in flat New Castle County's malaria country, slaves in Red Lion, Pencader and St. Georges hundreds were, respectively, sixteen, twelve, and eleven percent of the total population. In fact, according to one source, the low number of African-Americans in Delaware's piedmont caused Eleuthère Irénée du Pont to choose the banks of the Brandywine over the Mattaponi River in Virginia as the site of the DuPont Company powder works because he did not like blacks and Virginia's Mattaponi Valley was full of them. The fact that E. I. du Pont never owned slaves may have further reflected his desire to live in a neighborhood free of African-Americans.[60]

The problem for white Delawareans was that they, unlike large numbers of African-Americans, had little or no inherited resistance to malaria, particularly the falciparum strain. When a Dutchman, Englishman or a Swede crossed the Atlantic to Delaware it was simply expected that within a year or two of arrival, he or she would be stricken by the disease. In August, 1712, for example, Andreas Hesselius, a Swedish cleric who arrived in Delaware the year before, was put to bed for six weeks with "a heavy fever," a "sickness no stranger escapes." As a result, malaria was a serious problem in Delaware from almost the onset of European settlement. During the Dutch era (1655–1664), which coincided with the first significant importation of enslaved Africans into Delaware, malarial fevers were very common, particularly in the area of New Amstel (the town of New Castle). According to a Dutch official, during one year 100 people "perished in consequence," and, in another, the fever so afflicted the Dutch that "house-building for the commencement of a city, and the tillage of the land for a suitable harvest of grain, went forward but poorly."[61]

Luckily for most white Delawareans, exposure to malaria led, over time, to a conditioned resistance to the symptoms of the disease. Using the misguided medical theories of his day, Hesselius explained, in 1712, that this development was due to the fact that "European blood becomes completely changed and, so to say, naturalized with the climate of the country and the common air." Almost four decades later, Peter Kalm noted that once whites had an initial bout with malaria, they tended to develop a resistance to its symptoms.[62]

Although Peter Kalm associated ague, as malaria was then called, with the inhalation of vapors from stagnant waters found in wetlands, and an 1838 article in a Delaware farmers' magazine noted that "the exhalations from swamps and freshwater marshes" produced agues, there is no evidence that Delawareans understood the relationship between draining mosquito habitat and preventing malaria. What is clear, however, is that draining Delaware's wetlands reduced the threat of malaria. By 1868, writer Henry T. Williams noted that in Delaware, "chills and fevers rarely occur except in the southern [Sussex County] portions of the state."[63] Malaria was able to hang on in some areas of Sussex because Delaware's largest county continued to host the largest acreage of the state's undrained swamps and marshes.

Yellow fever, another serious mosquito-spread disease of probable African origin—in this case the vector was the indigenous Aedes aegypti mosquito—occasionally visited Delaware. Unlike the case of some Africans inheriting a resistance to malaria, both blacks and whites acquired immunity to yellow fever only through surviving a bout with the disease. (Throughout the lower Delaware Valley it was incorrectly assumed by many whites that blacks had an inherited immunity to yellow fever and were, therefore, the logical choices to nurse yellow fever victims.) Yellow fever outbreaks occurred in Kent County, Delaware, in 1773, and in Wilmington in 1797, 1798 and 1802. To escape, some Wilmingtonians fled to the countryside until the disease spent itself after the first severe frost. In 1802, for example, "in consequence of increased alarm of the yellow fever and its approaching nearer to our houses," James Hemphill's family "moved out of town." But on the whole, Wilmington and the rest of Delaware escaped the high mortality and the almost annual visits of the disease that plagued Philadelphia during the 1790s. It was for that reason that a number of French planters and merchants, who fled the revolution in Haiti, chose to temporarily settle in Wilmington rather than in Philadelphia during the 1790s. Thanks to political revolution and disease, Wilmington briefly became the center of "one of the most interesting emigré groups in the history of the United States," with French Street in Wilmington named for them.[64]

Cold Weather

Just when yellow fever ceased being a major concern and malaria was placed on the defensive, Delaware was reminded again how vulnerable it was to natural developments and events originating thousands of miles away. In April 1815, with the two great African-originated afflictions now fading, Delawareans took little note of the eruption of Mount Tambora on the obscure island of Sumbawa in the Dutch East Indies (Indonesia). Perhaps the most explosive volcanic eruption in the last 10,000 years, it and an accompanying earthquake caused approximately 90,000 deaths on Sumbawa and nearby islands. To get a better sense of the enormity of the 1815 eruption, it is important to note that the sun-obscuring volcanic ash produced by Mount Tambora was three to ten times the volume of the more famous 1883 Mount Krakatoa eruption, also in the Dutch East Indies, and one hundred times that of the 1980 Mount Saint Helens eruption in the state of Washington. So much volcanic ash was spewed into the atmosphere by the erruption of Mount Tambora that the climate of much of the world was significantly altered for several years.[65]

Beginning in 1816, which was known on both sides of the Atlantic as "the year without summer," Delaware experienced six successive growing seasons of unusually cold and wet weather. In the late spring of 1816, brown, bluish and red snowflakes fell on neighboring Maryland, an ill omen of things to come. That summer and in subsequent growing seasons, crops in Delaware's fields were stunted, frozen, and then began to rot. Corn, which flourished best in warm weather, was particularly affected. In January, 1817, Governor Daniel Rodney complained of "the failure of our crops... especially Indian corn." Winters, too, were unusually cold and lasted well into spring. By early March, 1817, for example, the ground was still covered with snow, and the ice on the Brandywine was fifteen to eighteen inches thick. The Christina was frozen all the way out to the Delaware. By June, a Delaware miller noted that the growing season was "cold and backward." The summer of 1818 must have been somewhat warmer because, although the wheat crop was poor, the corn crop "looked good." But for the next three years the summers turned cold again and the frigid winters, particularly the harsh winter of 1819–1820, caused a number of trees and bushes to die.[66]

The prolonged cold spell not only reduced the productivity of Delaware's rural economy, it also contributed, along with the ongoing decline in soil fertility, to a deep pessimism about the economic future of Delaware's countryside. This, in turn, caused a significant flight of rural Delawareans to the more promising Ohio and Mississippi valleys, and to such expanding urban centers as Wilmington, Philadelphia and Baltimore. Despite continuing high birth rates, Kent and Sussex, Delaware's most rural counties, lost about nine percent of their combined populations from 1810–1840. Even southern New Castle County was having demographic problems, with the number of residents in rural Appoquinimink Hundred declining by approximately eleven percent during the same period.[67]

Delaware's economically blighted rural landscape, and the resulting flight of thousands from the land, cast an ominous shadow over the state's future. Draining marshlands may have reduced the threat of malaria and provided new, fertile fields for staple crops, but soil-depleting agricultural practices soon robbed even the new farmlands of their nutrients. At this critical juncture, some Delawareans discovered economic hope in a natural asset concentrated in the piedmont region of Northern New Castle County. Here streams that began in southeastern Pennsylvania cut their way through hill country, and then plunged towards Delaware's Atlantic Coastal Plain. For those entrepreneurs with capital, engineering skills, energy and imagination, these fast-flowing streams could be harnessed by technology to produce enough water power to drive machines in mills and factories. Delaware's natural environment would once again play a dramatic role in shaping the next act in the state's economic development.

CHAPTER FOUR:

Water Power, Steam Power and the Urban Environment, 1730–1920

"[Wilmington's] steam industry of the later nineteenth century smeared every natural beauty it touched with smoke, dirt and waste."

RELIGIOUS VISIONS AND REVELATIONS HAVE OFTEN SHAPED BOTH human and natural history. They played a particularly crucial role in carving out of wilderness a number of American communities such as John Winthrop's Boston, Roger Williams's Providence, and Brigham Young's Salt Lake City. Wilmington, too, may have been carved out of wilderness thanks, in part, to a religious vision.

Elizabeth Shipley was a Quaker preacher and the second wife of Quaker William Shipley, who owned a farm southwest of Philadelphia near Ridley, Pennsylvania. In 1730, Elizabeth allegedly dreamed that, while riding horseback, she came to "a wild and turbulent stream which she forded with difficulty." She then ascended a hill that offered a fine view of "a broad, treeless savannah in the distance." Crossing the crest of the hill she saw below, on the other side, a second river that twisted and turned in a "snake-like" fashion. At that moment Elizabeth discovered that she was not alone. Turning to her new-found guide she asked, "Friend, what country is this that thou has taken me to?" Her guide responded that she was standing in the midst of a new and fertile land, and that God directed her to move to this land where "ye shall be enriched even unto the seventh generation."

Then the guide vanished and Elizabeth awakened from her dream.[1]

In 1731, with her dream gradually fading from memory, Elizabeth Shipley rode south on horseback, via the King's Highway, to speak at Quaker meetings on the Delmarva Peninsula. After crossing into Delaware and fording the turbulent Brandywine, she ascended a hill from which she saw level meadows—probably marsh grass—stretching east to the Delaware River while, to the south, she observed the "snake-like" Christina River winding its way eastward to junction with the Brandywine and then to union with the Delaware. All of this brought back the dream of the previous year and, more specifically, her guide's admonition to move from her Pennsylvania home and "enter into and take possession of this land, even as the children of Israel took possession of Canaan." After processing all of this, Elizabeth is said to have cried out, "Behold it is the land of my vision, and here I will pitch my tent." Included under that "tent" was her husband, William Shipley, who later would be called "the virtual founder" of Wilmington.[2]

Although a Quaker, William Shipley found business considerations more persuasive than mystical visions. But he was looking for an opportunity to become a merchant. Possibly for that reason, or perhaps because he wished to please his wife, he visited the lower Brandywine and Christina Valleys in 1735, where he encountered a tiny hamlet, populated by descendants of Swedish pioneers, near the site where Fort Christina had once stood. Looking west from the small settlement, William Shipley saw only a few farm houses and an occasional cultivated field amidst a predominately forested landscape that ran up the slopes of the same large hill which, four years earlier, had supposedly afforded his wife, Elizabeth, such a good view of the surrounding countryside. But certain recent developments a number of miles to the north and northwest, in Chester and Lancaster counties, Pennsylvania, caused Shipley to look closely at some plots for sale on the north bank of the Christina, about a mile upstream from the tiny hamlet.[3]

Grain

During the early eighteenth century, Pennsylvania's farm frontier pushed westward from Philadelphia through Chester and Lancaster counties to the Susquehanna River Valley. (Delaware County, Pennsylvania, was subsequently carved out of southeastern Chester County in the nineteenth century.) The limestone rich soil of this region was very fertile and the newly-cleared fields produced bountiful harvests of wheat. So prolific were its grain harvests that southeastern Pennsylvania soon become known as the breadbasket of the Anglo-American colonies. Because shipping wheat east by wagon to Philadelphia's grain market was costly, a more economical route was needed that included transportation by water.

William Shipley was aware that most farms in Lancaster and Chester counties were significantly closer by overland routes to Delaware ports along the Christina River than they were to Philadelphia. Strasburg in western Lancaster County, for example, was almost twenty miles closer to Wilmington than to Philadelphia. Clearly, much of southeastern

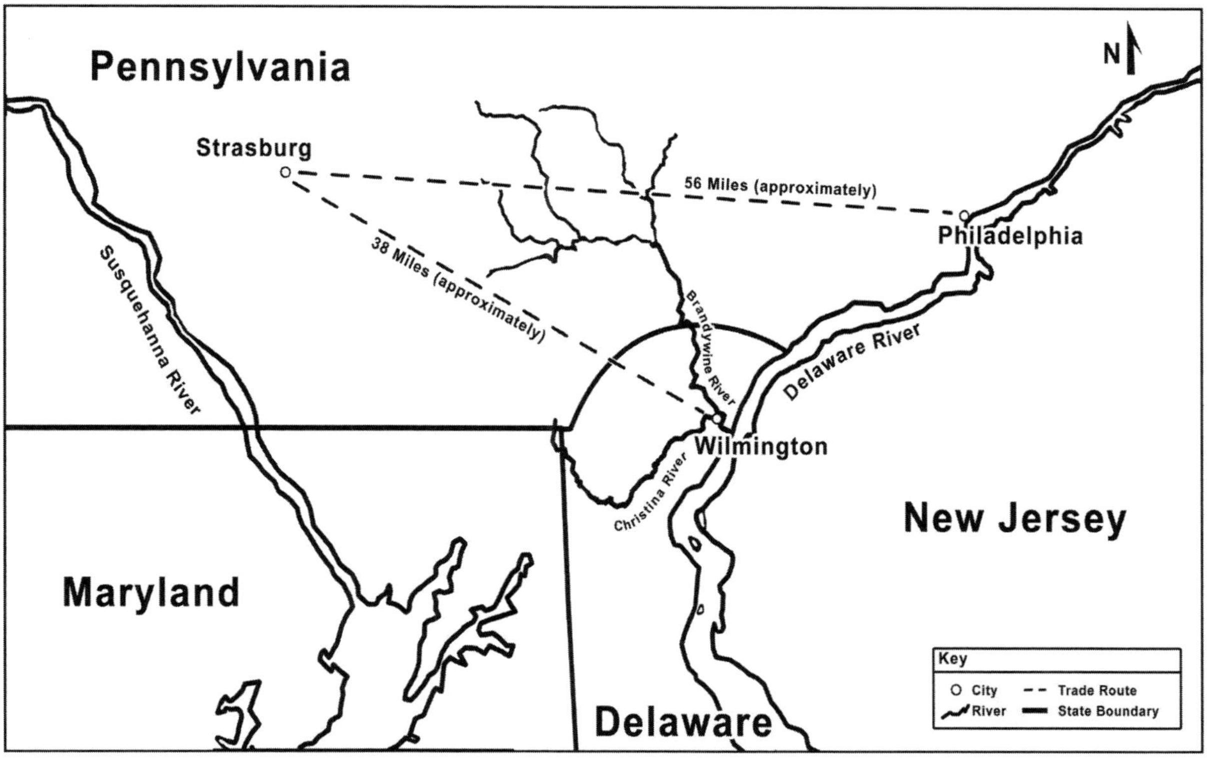

Figure 5—A map of the Brandywine Valley of northern Delaware and southeastern Pennsylvania showing the relative distances from Strasburg, Pennsylvania, to Wilmington, Delaware, and Philadelphia. The shorter distance to Wilmington was the basis for the city's growth in the mid-eighteenth century, *map drawn by Jason L. Burleson-Gibson.*

Pennsylvania's grain harvest would reach Philadelphia more inexpensively if it was first hauled overland by wagon to ports on the north bank of the Christina such as Newport and Wilmington, and then shipped by water to Philadelphia. By the mid-eighteenth century, long lines of huge, grain-laden conestoga wagons, pulled by six-horse teams, rumbled down the sometimes problematic roads from southeastern Pennsylvania, bound for the Christina. Also a common sight were smaller ox-drawn carts and, later, horse-drawn wagons loaded with grain from northern Delaware and Maryland's upper Eastern Shore, delivering their cargo over equally challenging roads to the same destinations.[4]

William Shipley was well aware of the future commercial advantages of locating along the north bank of the Christina. In addition to its direct overland connection with southeastern Pennsylvania's wheat farms, Shipley preferred locating on the north bank because, unlike the marshy south bank, it offered solid ground upon which wharfs could be easily built. Moreover, most of his expected grain supply would be coming overland from Pennsylvania to the Christina's north bank. In 1735, he purchased plots of land between the Christina and the hill to the north from real estate developer Thomas Willing, and proceeded to construct a home, a commercial building, a warehouse, and a wharf on his newly acquired property. In 1739, the small but growing community of 610 people that was clustered around Shipley's commercial facilities was officially named Wilmington. At this point, Wilmington's geographic location along the north bank of the Christina assured its growth as a grain port. After all, ocean-

Sussex County Gristmill— Warren's Mill at Betts Pond near Millsboro, built in 1924 after an earlier mill on the site burned, is at least the third gristmill in this same general location. This mill was operated until 1945, *photograph by Richard B. Carter, 1978.*

going sailing ships drawing up to fourteen feet of water could make it upriver to Wilmington at high tide. That meant that Wilmington's harbor would be navigable for all but the largest cargo ships of the eighteenth century.[5]

Waterpower

At first, Wilmington shipped primarily unprocessed grain because it took both time and a new vision to set up commercial grain mills. When Europeans arrived in Delaware in the seventeenth century, they brought with them the practice of using a variety of power sources to grind wheat, rye and barley into flour, and corn into meal. In 1644, the Swedes built a windmill near Fort Christina. By the 1680s, a windmill was also in operation in the town of New Castle. In 1661, a horse-powered mill was in operation at Dutch-controlled Fort Christina and a year later a second horsemill was grinding grain at the Dutch town of New Amstel (New Castle). The reason for using horse and wind power rather than waterpower was that the Christina and the streams in the New Amstel area were tidal and, initially at least, were considered unsuitable for powering gristmills. But windmills only worked when there was a strong wind, and horsepower was expensive because the animals had to be fed and looked after. It was only a matter of time before waterpower rather than wind or animal power would supply almost all the muscle required for early Delaware's industrial development.[6]

Probably the first water-powered mill in Delaware was built in 1646 in northeastern New Castle County, at Cobb's Creek, under the direction of Swedish Governor Johan Printz. By 1679, a second water-powered mill was in operation on Shellpot Creek—a tributary of the lower Brandywine—which, at the time, was a fast-flowing stream with some rapids and even a waterfall. By the 1680s, it was clear that water-powered mills were much preferred by Delawareans to horsemills and windmills. Early Delaware millers found congenial sites along a number of fast-flowing streams in northern New Castle County such as White Clay Creek, Red Clay Creek, Mill Creek and, eventually, the Brandywine.[7]

Although the best mill sites were usually located along the often turbulent streams rushing through Delaware's piedmont, a large number of mills were also built further south, along the banks of less promising tidewater creeks and rivers. Dams were indispensable to mills throughout the colony, but were particularly crucial in transforming parts of the sluggish tidal streams of flat Delaware into millponds. With the exception of whale wallows, central and southern Delaware had no natural lakes or ponds. Thus, the introduction of millponds produced a very significant environ-mental change in downstate Delaware. When rising water in a millpond was released down a millrace, it was pulled forward by the powerful force of gravity until it reached a velocity sufficient to turn a water wheel. By the end of the colonial period, almost every tidal creek in Delaware had a dam with a millpond behind it. By 1867, Sussex County alone had over 100 millponds of which 95 were still providing waterpower for operating mills.[8]

Gristmilling machinery— a view of the interior of the Hearn & Rawlins Gristmill north of Seaford, the last operating gristmill in Sussex County, which once had more than one hundred mills, *photograph by Richard B. Carter, circa 1980.*

But it was only along the fast-flowing streams of Delaware's piedmont, and more particularly along the Brandywine, that water moved with enough natural force to drive the wheels of large commercial mills. In fact, the extraordinary amount of waterpower that was harnessed to turn mill wheels in northern New Castle County, and particularly along the lower Brandywine, caused one outside observer, in 1795, to label the entire county as "the greatest seat of manufacturing in the United States."[9] It is to the Brandywine, the very center of early manufacturing in Delaware, that we turn to examine, in a more specific manner, the interplay of the forces of nature and human agency in the state's history.

In the early eighteenth century, the Brandywine River was an untamed stream that rose in the Welsh Hills of southeastern Pennsylvania and wound its way through the fertile hill country of Chester County towards the Delaware line and, eventually, juncture with the Christina. After crossing into Delaware, the Brandywine cut a gorge through ridges of hard granite while dropping 120 feet over the next four miles. Driven by gravity, the Brandywine turned tumultuous as it rushed over rapids and minor falls. The river banks along this four-mile stretch, particularly at the very end where the Brandywine's Great Falls met tidewater, offered ideal locations for water-powered mills. (Today the specific place where "Great Falls" met tidewater is just upstream from Wilmington's Market Street Bridge.) For the next two miles of tidewater to where it joined the Christina, the Brandywine turned placid and was at least seven feet deep at high tide, which meant that large grain-filled shallops could move up to the head-of-tidewater, deposit their cargoes and then, filled with barrels of wheat flour and cornmeal, make their way downstream to the Christina and then either upstream to the port of Wilmington or downstream to the Delaware.[10]

Entrepreneurs with vision understood that the two miles of placid tidal water below the Great Falls was just as critical to the future profitability of flour milling on the Brandywine as the four miles of surging, turbulent water further upstream.[11] In fact few rivers in America could match the lower Brandywine's happy proximity of waterpower and navigable stream with an easy outlet to a major transportation estuary and to the Atlantic trade network beyond. All of this would cause the Brandywine to join with the very different Christina in dictating that early Wilmington's economic development would rest on the twin pillars of the grain trade, and the manufacture of wheat flour and cornmeal.

During much of the seventeenth century, however, the turbulent Brandywine struck early Delawareans as too large and too unruly to control with the dam-building and grain-milling technology of the day. They turned, instead, to building mill sites along its smaller tributaries such as Shellpot Creek, because those streams were more manageable. By 1687, however, there was a barley mill in operation on the south bank of the Brandywine near the fording place just above the Great Falls. In 1727, Samuel Kirk and associates purchased the barley mill and superimposed their own gristmill on the same site. Kirk and his partners also constructed the first dam on the lower Brandywine, a very crude structure of logs and rocks which were laid across the river to divert just enough water into a short millrace to power an undershot wheel, a relatively inefficient type of waterwheel.

Like other Delaware gristmills of the time and most of those to come, the two early Brandywine mills were "custom mills," because they simply ground grain from nearby farms for a fee. After flour was manufactured at the mill from their grain, these local farmers set aside some barrels for their own use and sold the rest. The miller neither paid for the wheat nor got involved in the marketing and sale of the flour.[12]

In 1741, the arrival of Quaker Thomas Canby, a 73-year-old Bucks County, Pennsylvania, farmer and miller, marked the beginning of a new era in Brandywine milling. Canby and his two sons, Thomas, Jr., and Oliver, were drawn to the Wilmington area because of the availability of large supplies of grain, because of the navigable nature of the Christina and the tidewater sector of the Brandywine, and because of the lower Brandywine's potential for turning the huge grinding stones used in large, commercial mills. In 1742, Oliver Canby built a dam, a millrace and a gristmill just downstream from Samuel Kirk's earlier mill. Canby's mill stood "just below the northeasterly termination of Orange Street" at the south bank of the Brandywine. But unlike previous mills along the Brandywine or elsewhere in Delaware, this new facility was operated as a commercial mill where the owners purchased grain from farmers, ground it into flour, and then shipped it to Philadelphia or to other, more distant markets on ships owned by Wilmington merchants or eventually, in some cases, by the commercial millers themselves. As early as 1742, for example, in a very modest but precedent-setting step, a brig owned by merchant William Shipley carried a large barrel of flour from the Canby mill, via Wilmington, to the West Indies.[13]

But the Canby mill was built on the south bank of the Brandywine, upstream from the Great Falls, which meant that grain-loaded shallops from the Christina and grain-bearing wagons from southeastern Pennsylvania had to be unloaded at Wilmington's waterfront. From there the wheat was trucked by wagon northwest, over the hill, to the south bank of the lower Brandywine where it was ground into flour. Most of the barrels of flour produced by the mill then retraced the trip southeast over the hill to Wilmington, which was both time-consuming and costly, to be stacked in the holds of waiting ships. The only way to eliminate this costly, two-way wagon trip was to build new mills on the Brandywine further downstream, beyond the Great Falls, at the head-of-tidewater where, at high tide, shallops could dock next to the mills where their grain cargo could be directly unloaded and barrels of flour directly loaded. But the head-of-tidewater lacked waterpower.

In the early 1760s, a new dam across the Brandywine was constructed upstream from the Canby mill and a long millrace—subsequently called the south race—was dug on the south bank, from just above the new dam all the way downstream, beyond the Great Falls, to the head-of-tidewater. Now, rushing water powerful enough to drive overshot wheels, the most efficient type, was available at navigable sites along the Brandywine's south bank, downstream from the Great Falls. There remained, however, a man-made problem concerning access by shallops. Some years earlier, a crude bridge had been constructed across the Brandywine below the head-of-tidewater, making it impossible for masted vessels to squeeze under the span on their way upstream to the new mill sites. In 1760, in an effort to open up potential mill sites to marine traffic, New Castle County officials decided to remove the old bridge and replace it with a new one that would cross the Brandywine at the head-of-tidewater—where King Street meets the Brandywine today. In 1765, the new bridge was ready for use, and two years later the old bridge was gone.[14]

At first, development of mill sites on the lower Brandywine's north bank was delayed by the presence of rocky terrain which made it exceedingly difficult to dig out a duplicate of the south bank's long millrace. In the early 1770s, with Quaker miller Joseph Tatnall providing the leadership, this formidable natural obstacle was penetrated by a long millrace which, henceforth, would be "the north race." By 1772, the surge of water down the newly completed millrace delivered power to four mills on the Brandywine's north bank, at the head-of-tidewater.

Now, on a quarter-mile stretch of the lower Brandywine there were eight commercial gristmills: four on each bank. At the end of the American Revolution, a German traveler found the mills at the head of the Brandywine's tidewater "so conveniently placed," that "large shallops can lie close to them and load and unload wheat and flour with great ease." A few years earlier, another observer called this concentration of mills "the most extraordinary and valuable in the world."[15]

By 1791, there were at least twelve water-powered gristmills on the banks of the lower Brandywine at the head-of-tidewater. Along the length of the Brandywine that same year, from its origin at the juncture of two small streams in the Welsh Hills of southeastern Pennsylvania to where it ended in merger with the Christina, there were approximately 131 water-powered mills of various types.[16] The entire river had been dramatically transformed by the destruction of nearby forests, the drainage of the basin's wetlands, and the construction of milldams, millraces and mill-ponds. What was once a turbulent and challenging river for at least part of its journey to the sea had been putatively reshaped to meet the predictable rhythms demanded by water-powered mills. In short, the Brandywine now seemed to some to be no more than a domesticated stream whose flow could be turned on and off to suit the power needs of mill owners.

The commercial gristmills at the Brandywine's head-of-tidewater probably reached their height of productivity and prosperity about 1815. Fourteen mills, at times employing up to 600 men in mill-related jobs, produced enough annual profit to help bankroll the future transformation of nearby Wilming-ton's waterfront, along the north bank of the Christina, from a provincial shipping facility to a specialized manufacturing center. Eventually, however, a shrinking agricultural hinterland and increasing competition from other mills led to a decline in the number and economic health of the gristmills along the lower Brandywine.

The problem was that America's westward-moving frontier shipped increasing amounts of grain to such older milling centers as Richmond and Baltimore. To the north and northwest, the Erie Canal and the Great Lakes opened up vast new agricultural hinterlands for recently founded milling centers such as Rochester and Buffalo. Even Philadelphia expanded its immediate source of grain westward by completing a turnpike to Lancaster in 1796, and, in the late 1820s, a railroad that provided cheap bulk transportation from the town of Columbia on the lower Susquehanna to the Quaker City. The latter transportation developments redirected to Philadelphia the flow of southeastern Pennsylvania's wheat, which had formerly gone to the ports and mills along the lower Christina and Brandywine.[17]

Faced with the loss of its most important source of inexpensive wheat and battered by increasing competition from rapidly growing milling communities elsewhere, Brandywine millers were forced to import new supplies of wheat from a considerable distance, including such far away European ports as Danzig on the Baltic. In 1836, Brandywine miller Edmund Canby noted that it was "a very singular state of things when old and populous Europe" would find it profitable to ship grain all the way across the Atlantic to the lower Brandywine. By that time the handwriting was on the wall; strong competition from better located mills across the nation made the decline of the Brandywine mills inevitable. Edmund Canby put it quite simply: "every day seems to illustrate the fact that millering as a regular business is done for here." A few of the Brandywine's head-of-tidewater gristmills continued to manufacture wheat flour and cornmeal into the twentieth century, but no longer did the lower Brandywine offer "perhaps the best situation for [grist] mills in the world."[18] Although other mills that produced lumber, textiles, paper and gunpowder were located further upstream, it was clear by the mid-nineteenth century that the industrial center of Delaware had moved from the lower Brandywine Valley to the north bank of the Christina, only a mile or two away.

The dam construction that physically transformed the Brandywine Valley serves to remind us what can happen to ecological relationships when humans choose to dramatically alter the natural rhythm of waterways. Anadromous fish, such as shad and herring, were turned back by stone masonry when they attempted their annual spring spawning runs up the Brandywine. By halting these essential migrations, dam-building millers threatened both the river's short term and long-term fish populations. Of course, all of this was in direct conflict with English common law, which protected public fishing rights on streams and rivers by granting government the power to license individuals to remove dams, weirs and other obstructions that threatened the present and future supply of fish.[19]

In 1727, when a crude structure of stones and logs was built on the lower Brandywine, upstream Lenape protested to the Delaware colonial government that the dam threatened one of their most important food sources. Delaware's colonial legislature responded that year with an act for "the preservation of peace and promoting harmony," between whites and those "Indians" who lived on the banks of the Brandywine. Based on English common law, the Act of 1727 empowered Delaware's magistrates or other designated individuals to remove all obstructions from the river, including dams, that interfered with the upstream migration of fish. By the mid-eighteenth century, however, only a few Native Americans still lived along Delaware's stretch of the Brandywine or just over the line in Pennsylvania, with the last, Indian Hannah, dying in 1803. Now it was the turn of whites, who were also dependent on the spring run of spawning fish, to step forward and oppose the dam-building proclivities of the Brandywine's millers.[20]

Dams were more acceptable to fishermen if they included "fishways" or fishruns that allowed at least some shad, herring and other anadromous species to fight their way upstream to spawning grounds beyond the dams. But often these fishways were not kept open by the millers. In 1756, petitions from white residents of the Brandywine Valley living in both Delaware and Pennsylvania caused a commission to be appointed to police the river and to break through dams that did not have fishways or to open those fishways that were now closed up. In 1760, four Brandywine dams—their exact locations are unclear—were breached by government-appointed officials. After that date, however, "there were no more dam wars, but also no more shad."

The shad and other anadromous fish stopped making their annual spring spawning runs because Delaware's colonial government decided that it was good public policy to support Brandywine millers in their ongoing dispute with the river's fishermen. Encouraging entrepreneurs to construct water-powered mills and their related mill-dams along the Brandywine made good sense because the mills provided "a public benefit and ought therefore to be encouraged." Thus, in 1760, the same year that four dams on the Brandywine were breached, Delaware's colonial legislature annulled the law of 1727 that provided official support for the breaching of those dams.[21]

To a colonial assembly intent on encouraging the construction of water-powered mills, abolishing the 1727 law made good sense because that piece of legislation discouraged the building of mill-dams and was, therefore, a barrier to progress. As Henry Seidel Canby later observed, after 1760 "[water] power was now dominant on the lower Brandywine" and the "fish were somewhere turned back." Canby remembered that during his childhood in Wilmington, in the late nineteenth century, the migratory instinct of a certain type of herring—sometimes called alewives—continued to be so strong that every spring they fought their way, by the thousands, through the Brandywine's Great Falls, "vainly trying to get past the bulwarks of dams and races that blocked" their way upstream.[22] The action taken by the colonial legislature to favor the Brandywine's dam-builders indicated that, in the future, the public's traditional fishing rights along streams and rivers would be protected by Delaware's government only where and when those traditional rights did not conflict with the legislature's more modern commitment to industrial development.

Despite the construction of numerous milldams along its length, the supposed domestication of the Brandywine was far from complete during the nineteenth century. The problem was that extremes in the river's flow were accentuated by man-induced changes in the watershed's landscape. As previously mentioned, forests, marshes and swamps absorb water during heavy rainfalls and then, during dry spells, gradually release the water into nearby streams and rivers. This moderating influence on the Brandywine's flow had been dramatically decreased by the cutting down of trees and the draining of wetlands during the seventeenth and eighteenth centuries.

All of this meant that, despite the construction of numerous dams, the flow of the Brandywine was often either too strong or too weak to meet the industrial needs of the river's millers. Days when its volume reached flood proportions and caused cascading water to thunder over its dams and surge across its rapids, were more than matched by long periods when there was so little water flowing downstream from the dams that much of the Brandywine's riverbed looked like "a rocky ridge in a thirsty land." During the mid-nineteenth century, long-time Wilmington resident Elizabeth Montgomery correctly characterized this sometimes unpredictable river as simply a "whimsical stream."[23]

When major flooding did occur, the entire milling community was in jeopardy. On February 21, 1822, the Brandywine rose more than twenty feet after a hard rain. About 9:30 p.m., miller Edmund Canby was sitting in his parlor when he heard "a tremendous crash...a considerable part of Samuel Shipley's mill next to the bridge fell in." By the next day, the bridge over the Brandywine that connected Wilmington to the Philadelphia Pike was in the river and a number of other mills, mill dams and mill races had been destroyed or badly damaged by the devastating surge of water. Even more destructive was the flood of January 26, 1839. Again, surging water smashed into a series of mills, mill dams and mill races, and knocked out the bridge over the Brandywine that connected Wilmington with the road north to Philadelphia. Much of the flood destruction was directly attributable to ice accumulating behind milldams which, initially at least, held back part of the river's surge. But when the ice broke, huge chunks rode downstream on the crest of a cascading avalanche of water, scouring out the river's banks and any structures built on those banks. Obversely, droughts caused water levels in millponds to drop so low that it was impossible to channel enough water into millraces to turn mill wheels. Perhaps the most significant example occurred in the summer of 1838, when there had not been a real rain for three months and Brandywine millers were frustrated by the lack of available waterpower.[24]

Steam Power

The mid-century emergence of the north bank of the Christina as the new industrial heartland of Delaware was predicated on the introduction of steam power. As late as 1832, water continued to provide power for the overwhelming majority of the state's mills and factories. Steam, by contrast, provided power for only about six percent of Delaware's manufacturing sites. Industrial dependence on waterpower, however, had its drawbacks.[25] Factories and mills had to be constructed along fast-flowing streams that often were many miles away from significant pools of skilled laborers, important transportation routes, essential sources of raw materials, and major market centers. By contrast, steam-driven machines could be located just about anywhere that suited the manufacturer's needs. If steam power were to become available, Wilmington along the Christina was an even better location for industry than the river banks of the lower Brandywine.

While lacking the waterpower found just over the hill in the lower Brandywine Valley, Wilmington had a number of natural advantages that were attractive to manufacturers. Although its harbor on the Christina was not as deep as Philadelphia's, nature did give Wilmington a slightly warmer winter climate, which meant that its harbor was ice-free for a number of days each year when the Quaker City to the northeast was ice-bound. How serious a shipping problem this could be is demonstrated by the fact that Philadelphia's port was ice-bound for fifty days in 1817 and fifty-two in 1822. In 1792, a Wilmington merchant-trader observed that "many times in winter, we can send our vessels out from hence when it is not practicable from Philadelphia, and always at the breaking up of the winter our vessels can depart several days sooner than from..." the Quaker City. The commercial advantages "derived from these circumstances" were "too obvious to need recital." This slight climatic edge contributed to the growing prosperity of Wilmington's port which, by 1794, was home to numerous sloops, schooners and approximately thirty square-rigged ships. Philadelphia finally minimized Wilmington's advantage in ice-free days when it obtained an ice cutter in 1837.[26]

By the first decades of the nineteenth century, however, there was a growing problem with the port of Wilmington. While many of the newly constructed Atlantic-crossing sailing vessels were larger and drew more water than ever before—by 1870, an average ocean-going ship drew approximately twenty to twenty-four feet of water—the Christina was silting up. Soil runoff from plowed fields, the dumping of ballast from ships, and the general practice of throwing city refuse into the Christina all contributed to diminishing the depth of its ship channel. The diminished depth limited the flushing effects of tidal flow. As historian Carol Hoffecker points out, "the silting process not only retarded trade, it also imposed limitations on the draft of ships built in Wilmington." Although the federal government occasionally dredged the river during the nineteenth century, the harbor never was deepened enough to really compete for overseas trade until the Marine Terminal, located further downstream near the confluence of the Christina and Delaware, was built in 1920.[27]

While the port of Wilmington struggled with the problem of silt, its only nearby rival port in Delaware was dying. The harbor at New Castle, located on a bend of the Delaware River, faced an even greater threat from accumulating silt. In 1827, "a gradual formation of a deposit of mud" threatened to seriously restrict ships from landing at the town. The Delaware

General Assembly called on the U.S. Congress to respond to the crisis by appropriating money, as allowed by an 1802 federal statute, for the improvement of New Castle's harbor. The United States government responded with a plan to construct piers further out into the Delaware River only to have local residents, who were experienced in the "peculiar action of the current and set of the tides," predict that such construction, "if carried into effect, would destroy the said harbor."[28]

Despite protests from a delegation of Delawareans who traveled to Washington, the federal government persevered in its plans and the resulting erection of new wharves rapidly increased the deposits of mud, "whereby the said harbor is now [1835] completely choked up and destroyed, and scarcely affords adequate safety and protection for a single vessel."[29] This brief examination of New Castle's silting issue demonstrates that, in some cases, human and government attempts to provide solutions to perplexing natural problems can lead to unintended consequences. In this case, the federal government's new wharves sealed the port of New Castle's fate.

Ironically, failure to expand Wilmington's overseas shipping potential during the nineteenth century encouraged the construction of steam-powered factories along the Christina's north bank. Because of the limited growth in maritime commerce, real estate values along the river remained moderate compared to the more commercially active waterfront areas of Baltimore, Philadelphia, and New York. This meant that manufacturers found land along the north bank of the Christina to be reasonably priced and therefore a good place to build their steam-powered mills and factories. Another prerequisite, a large pool of skilled mechanics—trained primarily in the Brandywine mills—was also locally available. A third prerequisite, investment capital, was generated from the profits made by the mills along the Brandywine and the ripple effect of those profits as they moved through Wilmington's commercial community.[30]

Easy access to raw materials and markets via superior transportation links was a fourth prerequisite. But here the case for Wilmington was less clear because silting had caused its main link to the outside world, the Christina, to become somewhat problematic by the early 1830s for anything more than the coastal trade. At this crucial juncture, talk of a railroad that would link Wilmington with Philadelphia and Baltimore stirred the imaginations of Wilmingtonians and attracted their support. It became clear to Wilmington's business community that, unlike the Chesapeake and Delaware Canal, the proposed railroad would directly link Wilmington's future industrial sites with raw materials and markets.

The completion of the Philadelphia, Wilmington and Baltimore Railroad in 1837 was a crucial step in changing the economic and social character of Wilmington from a grain-and-flour port to a specialized manufacturing center. Running south from Philadelphia, the rail line crossed the Brandywine into east Wilmington and then swung westward, paralleling the north bank of the Christina. The narrow strip of land between the railroad and the Christina became

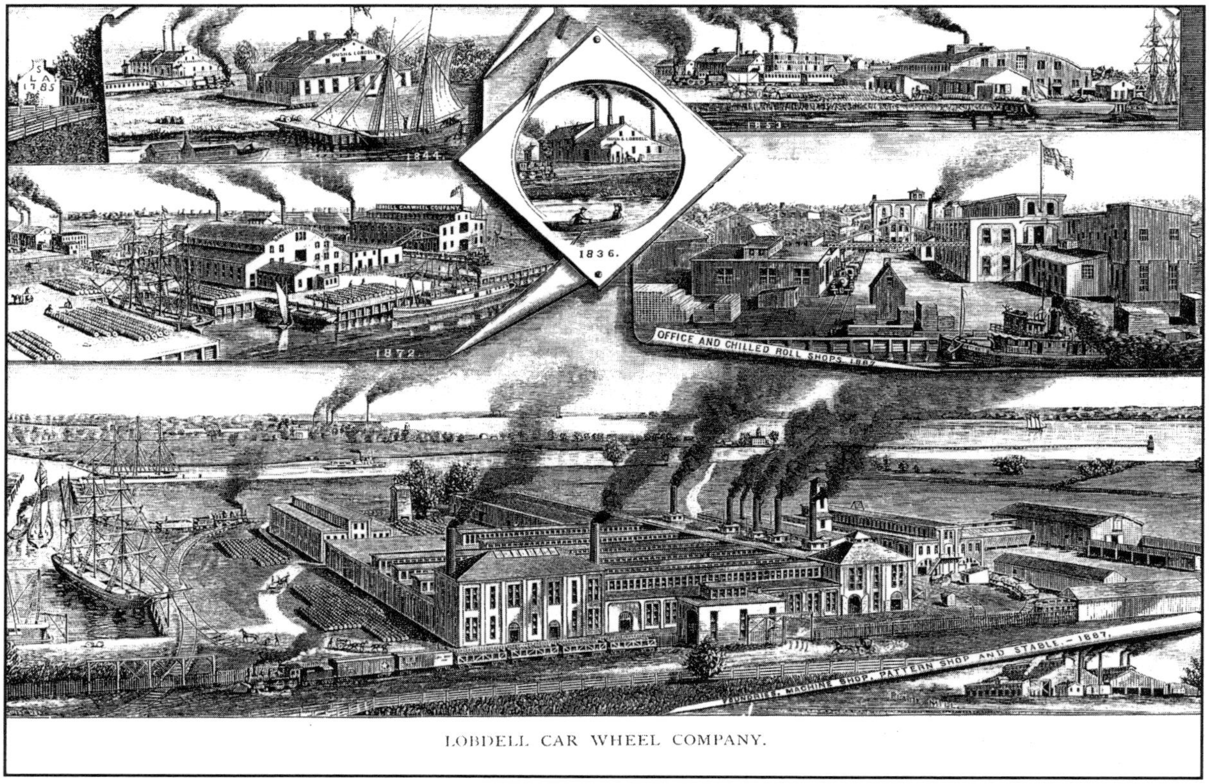

Industrial Smoke— An illustration from volume two of *History of Delaware, 1609–1888*, by J. Thomas Scharf illustrating the industrial growth of Wilmington in the years after the Civil War, with the accompanying clouds of industrial smoke, *reproduced courtesy of the Delaware Public Archives.*

increasingly valuable for factory sites because, in addition to water transport, now there was a direct rail link to markets and to specific sources of coal and iron throughout urban eastern America.

By 1870, coal-fueled factories along the north bank of the Christina made Wilmington the national leader in turning out railroad cars and iron ships, while foundry work, tanning, and carriage-making also prospered there or in nearby neighborhoods. Meanwhile, the lower Brandywine Valley, much of which would be annexed by Wilmington, continued to manufacture significant quantities of flour, textiles, paper, and gunpowder. Reflecting its industrial expansion, the city's population increased from 8,452 in 1830, to 30,841 in 1870, and to 110,000 in 1920.[31] All of this caused the once clean air over Wilmington to become increasingly fouled by smoke from factory and hearth, the city's streets to smell of mounting quantities of garbage and animal excrement, and the nearby Christina and its proximate tributary streams to show signs of rising levels of pollution.

Why should a study of this nature concern itself with Wilmington? After all, some would argue, environmental history is better suited for examining the "wild and rural" rather than the "industrial and urban." Indeed, the 1869 report of the Wilmington Board of Trade maintained that the physical world was far less important than human agency in shaping the growth of successful urban areas because "men, and not natural advantages, make great and prosperous cities." The evidence already presented concerning Wilmington's development, however, indicates that the specifics of physical setting played a major role in the founding and development of Delaware's largest city. In fact, Wilmington's history was at least as much shaped by its navigable waterways, its nearby fast-flowing streams, its tree-covered central hill, the lime-laced soil of its grain-producing hinterland in southeastern Pennsylvania, and its moderate climate, as it was by human initiative. Therefore, examining the impact of the natural environment is essential to understanding the rise of Wilmington. Equally as clear is the impact of Wilmington's growth and expansion on its air, water, soil, flora, fauna, disease environment, population concentration, and increasing noise level. In fact, this brief examination of how nature shaped Wilmington's past and, conversely, how Wilmington shaped and reshaped its natural setting, is in keeping with the view that at the dawn of the twenty-first century, urban environmental history has become "arguably the most flourishing part of environmental history."[32]

The Air

Because they were dependent on energy generated by fast-flowing water rather than by the burning of wood or coal, the early mills along the lower Brandywine did not pollute the air. As writer Henry Seidel Canby observed, compared to what was to follow along the lower Christina, "the capture of the swift currents of the Brandywine for work and profit was more like seduction than rape." By contrast, the coal-fueled, steam-powered manufacturing plants, constructed along the north bank of the Christina in the mid- and late-nineteenth century, blotted the blue sky with black smoke and covered the city with grimy soot. The factory fuel of choice was bituminous (soft) coal, of which, when it was burned in furnaces, only a small portion was consumed. The remainder escaped through smokestacks to coat the homes and lungs of Wilmington's residents. As one Wilmingtonian recalled, unlike waterpower, the coal-fueled "steam industry of the later nineteenth century smeared every natural beauty it touched with smoke, dirt and waste..."[33]

Wilmington's air was also fouled by domestic smoke. During the nineteenth century, wood stoves and then coal stoves replaced fireplaces in providing heat for homes in urban Delaware. The demand for firewood stripped the rural areas surrounding Wilmington of most of their trees. By the mid-nineteenth century, barges were pushing up the Christina from the Delaware carrying bituminous coal from western Pennsylvania and anthracite (hard) coal from northeastern Pennsylvania, while additional shipments of coal were coming in to Wilmington by rail from Philadelphia. It was not until well after World War II that oil finally replaced coal as the primary heating fuel for Wilmington's homes, causing the air above the lower Christina and Brandywine Valleys to take on a distinctly different odor.[34]

Rustic Ways in an Urban Setting

In the midst of its industrial expansion and rapid urban growth, nineteenth-century Wilmington, like most American cities, continued to reflect rural ways and values. When people from the countryside moved into the city, they not only established vegetable gardens on their small plots, but also brought along their poultry, goats, cattle, horses, and, most perplexing, their swine. While a variety of domestic animals and fowl roamed Wilmington's streets, additional livestock and fowl were confined to fenced-in back yards or enclosed empty lots. In 1848, for example, Edmund Canby, kept horses, two cows, and several pigs on his property on the north edge of the city.

As late as the end of the nineteenth century, most Wilmington families had flocks of chickens, which meant that, "in every part of town the crowing roosters could be heard." During the same period, vacant lots in the poorer sections of Wilmington "swarmed with goats." Occasionally adding to the variety of animal smells and the cacophony of animal sounds were the cattle and pig drives through the city's streets that brought protein on the hoof from Wilmington's agricultural hinterland to the city's slaughterhouses.[35]

Despite dropping their manure in yards and streets, some farm stock—particularly swine—were relied on to help keep streets clean by eating the garbage that littered urban thoroughfares in Wilmington and in other American cities. It was not then known that there was a link between the feeding habits of free-roaming pigs, particularly those that ate rat and swine carcasses, and serious human disease. Not until the 1930s did medical research prove that humans who butchered and consumed free roaming pigs often contracted trichinosis.[36]

Swine roaming Wilmington's streets presented other problems: they were often unruly animals with the potential to injure and even kill small children. One source described urban swine in general as "a nasty and brutish lot" that "copulated in public and had the annoying habit of defecating on people." By the 1830s, Wilmington's civic leaders were no longer able to ignore the filth, disorder, and destruction associated with free-roaming pigs. In 1833, swine were banned from Wilmington's streets. But this hurt many of the city's poor, because it meant that urban swine were denied access to street garbage which was their main source of food—America's urban poor of the eighteenth and early nineteenth centuries generally considered city streets to be urban commons and therefore open feeding ground for their livestock. Perhaps because closing the streets to pigs dealt a serious economic blow to many low-income residents, in 1857 Wilmington found a compromise that allowed swine back in the streets, but only if they bore yokes on their necks and rings through their noses so they could be better controlled.

In 1867, free-roaming horses, cattle, donkeys, mules, goats, and sheep were banned from the city's streets, but no action was taken to control pigs. Finally, in 1871, the city council passed an ordinance that required every swine "found at large in any of the public streets, lanes, or alleys," to be seized by police or other appointed government officials. The owner was ordered to pay a fine or have the offending pig sold by the city with the proceeds going to Wilmington's coffers. As to the trash and livestock carcasses left in the streets, for decades men had been hired to remove dead animals and, by the early 1870s, Wilmington had in place a crude, citywide garbage collection program.[37]

Horses posed an even greater threat to the health and aesthetic sensitivities of Wilmington's residents than free-roaming pigs. Because they were so much larger, horses produced far more manure and urine per animal. Moreover, their shod hooves struck the cobblestone streets with such force that the resulting staccato sounds added significantly to the perpetual din that enveloped the city. During the second half of the nineteenth century, increasing numbers of horses were harnessed to the growing numbers of wagons, carts and other vehicles that were so essential to the city's ascending economic vitality. Playing a key role in transporting urban workers and shoppers was the horse car, which was pulled over iron rails laid in Wilmington's major streets. Because each horse produced somewhere between fifteen and thirty pounds of dung every day, the magnitude of horse manure in Wilmington and in other American cities was astonishing.

By 1900, Wilmington probably had more than 6,000 horses which annually left on the streets and in their stalls enough dung to cover a full acre of land with a pile seventy-two feet high. All of this manure attracted and supported approximately six billion disease-spreading flies. Fortunately for the city, neighboring farmers collected some of the manure to fertilize fields that produced hay for Wilmington's horses and fruits and vegetables for the tables of Wilmington's residents.[38]

The air, particularly on windy days, contained tiny particles of dried horse manure that irritated eyes, lungs, and bronchial passages. On the streets, the smell and sight of horse dung was offensive to increasing numbers of residents who sensed the connection between poor sanitation and disease, or who were simply angry that they had to pay taxes to clean up the unsightly mess. One resident remembered that, during the last two decades of the nineteenth century, flies, mosquitoes and other insects "were plentiful because of the presence of ... manure piles and the number of horses using the streets." Particularly troublesome, in view of the large numbers of insects present, was the fact that then "it was not customary to protect the houses by means of window screens."[39]

Horse manure continued to be a major source of Wilmington's air and street contamination until the first decades of the twentieth century, when electric trolley cars and then automobiles gradually replaced horses on the city's roads. The rise and decline of horse dung also had a dramatic impact on the size of Wilmington's English sparrow population. By the late nineteenth century, English sparrows were numerous because they found an ample food supply in the droppings left by "the number of horses using the streets and the numerous stables, each with a manure pile." With the replacement of the horse by the automobile, the number of English sparrows plummeted to the point that, in 1938, "only an occasional English sparrow is to be seen in the town." The reason was obvious: it was "due to the present scarcity of food [i.e. horse manure] to be found upon our clean streets."[40]

Supporters of the "horseless carriage" predicted that the automobile would be a vast improvement over the horse because it would purify the air and purge the streets of contaminants. The irony, of course, was that the dramatically increasing numbers of gas-powered vehicles added significantly to the urban noise level and fouled the air and streets of Wilmington in a different way. As the constant "clip clop" of horses' hooves and the piles of equine manure gradually faded into history, they were replaced by the falsetto beat of gasoline engines, the constant honking of car horns, the acrid smell of automobile exhaust, and by oil-stained streets littered with trash thrown from car windows.

Disease Affects the Hinterland

Other sources that joined in fouling nineteenth century Wilmington's air and, often, its waterways included slaughterhouses, soap factories, and tanneries. The odor and the unsanitary liquid runoff from slaughterhouses was a significant problem in certain parts of the city during the first half of the nineteenth century. In 1833, Wilmington placed restrictions on slaughterhouses and other industries that created strong odors. In 1848, the city banned the construction of new slaughterhouses unless they first obtained approval from the city council by promising to meet certain sanitary standards. The floor of a proposed meat-processing plant, for example, was to "have a gradual descent to a point where the blood and garbage will be received in some suitable vessel," and the vessel was to be emptied of its contents "at some point without the city."

In the process of turning cow and pig fat into soap, soap factories emitted a particularly obnoxious smell. After a Wilmington soap plant burned down in 1828, Edmund Canby hoped that no more would be built in the city "to stifle the people with its stench." Tanning, which relied on an "adequate supply of water, a quantity of tan bark, and the hides and skins to be processed," had an early start in Wilmington and rivaled flour milling and ship building as the city's principal industry during the eighteenth century.[41] Unfortunately, tanneries could also spread disease carried in the hair and skin of the hides being processed to locations far from Delaware's largest city.

One Wilmington tannery became the source of an anthrax epidemic that spread through the cattle herds of coastal New Castle and coastal Kent during the early 1890s and lingered for a number of years thereafter. Arthur T. Neale, who was appointed the second director of the Delaware College (University of Delaware) agricultural experimental station in Newark in 1889, played a key role in limiting the ravages of the disease. Although he was terminated for insubordination in 1907, Neale was an "energetic director" who noticed that cattle infected with anthrax were from farms located south of Wilmington along marshes and tidal streams bordering the Delaware River. In 1893, when a tannery worker in Wilmington was discovered to have an arm infection caused by anthrax, Neale and some others were able to pinpoint Wilmington's tanneries as the source of the bovine outbreak. Clearly, tannery hides were infected with anthrax microbes and the disease was spread by the traditional industrial practice of dumping the hair and other waste byproducts from leather manufacturing into waterways leading to the Christina or the Brandywine. Eventually the infected materials were transported downstream into the Delaware and from there as far south as southern Kent County. At that point, high tides moving up the Delaware Bay and River pushed the infected waste material inland to contaminate the salt marshes and the meadows along river-banks that were commonly grazed by cattle. To address this serious threat to Delaware's dairy and beef cattle industry, Neale quickly introduced the Pasteur anthrax vaccine and promoted other steps which gradually reduced the anthrax epidemic. Not until 1915, however, did the epidemic in Delaware finally end.[42] The anthrax experience taught downstate farmers a valuable lesson: those who live and work downstream will always find the condition of their physical world subject to the polluting practices of the industries and people who live upstream.

Oh, for a drink of clean water

Early in the nineteenth century, it was clear that Wilmington needed a safer, more dependable source of water than was provided by shallow, individual backyard wells that too often were sunk in close proximity to overflowing privies. A reasonably clean water supply was more important than any other government-provided sanitary service in insuring the health of city residents and the future growth of the city. The then-common method of testing water was by checking color, taste and smell. Even by this primitive method, Wilmington's water supply was found wanting. To improve the quality of drinking water a private water company was chartered in 1804, and it proceeded to tap a spring part way up the hill at Third and Tatnall. From there, gravity carried the water downhill through center-bored logs to a number of cisterns scattered around Wilmington. Subscribers then hand-pumped the water into buckets and other containers and carried it home. In 1810, the city of Wilmington purchased the water company.[43]

In the decade after Wilmington acquired the water company, the first recorded clash took place between environmentalists and the city's government. A large number of willow and Lombardy poplar trees grew in the neighborhood of the natural spring and the city's waterworks at Third and Tatnall, as well as along the city's streets, lanes and alleys. Concerned that the roots of these trees were drawing off a significant amount of water from the natural spring that supplied the city, and convinced that the tree roots also clogged up the center-bored logs buried just beneath the city's streets that served as water conduits, Wilmington's government ordered most of the city's willow and Lombardy poplar trees cut down. The decision sparked considerable protest from homeowners who argued that these trees provided Wilmington's streets with both beauty and shade. As city woodsmen proceeded with their work, one old man wrapped himself around a tree in front of his house and declared that the ax must first strike him before it touched "a single bough." As the woodsman moved "from one tree to the other, the owner would follow and clasp it lovingly."

Finally the tree owner was worn out and "retired to his house."—Delaware's first recorded "tree hugger," perhaps? Evidently there was a brief pause in tree removal, but it ended when an appeal by tree preservationists was turned down by a Delaware Supreme Court Justice because "trees on the streets of Wilmington are a public nuisance." Subsequently, all of the offending willows and poplars were cut down.[44]

As Wilmington's growth pushed up the hill from the Christina and the spring at Third and Tatnall could no longer produce enough water to meet the needs of its growing population, the city's government decided to follow Philadelphia's lead and construct gravity-flow reservoirs. In 1827, Wilmington purchased a mill located on the south race along the Brandywine, from where water was pumped uphill through iron pipes, by power generated by a waterwheel, to two adjacent reservoirs located on the present site of Rodney Square. Then, as Delaware historian Carol Hoffecker explains, gravity "moved the water down [South] Market Street through iron pipes of decreasing size until it reached Water Street along the Christina." In subsequent years the city would continue to depend on Brandywine water, diverted through the south race and then pumped to the top of the hill. In 1855, the city's first steam-powered pump was purchased and, in 1873, Wilmington built its first pump house along the Brandywine. The city's growing population led to the need for more reservoirs so, in 1859, a second site was established on the high ground in the city's northwest. In 1877, Cool Spring Reservoir was also constructed in the city's northwest sector and was so much larger than the earlier reservoirs that it quadrupled Wilmington's water storage capacity. To fill these reservoirs, the city purchased the water rights of additional mills located along the Brandywine's south race.[45] On the surface this may seem to be an engineering success story with Wilmington taking, albeit slowly, certain sequential steps to meet the expanding water needs of its growing population. But supplying volume must not be confused with providing quality.

Despite the imprecise nature of the medical science of the day, as early as the late eighteenth century there was a general sense that impure water contributed to poor health by spreading diseases. But not until 1863 did an uneasy Wilmington Board of Health admit that the city's drinking water contained too much mineral and organic matter. Seventeen years later Wilmington physician L. P. Bush told the Delaware State Medical Society that, because deaths from typhoid were so evenly distributed throughout the city, he suspected that Wilmington's water supply system played a key role in the spread of this disease. If anything, the quality of Wilmington's drinking water seemed to be deteriorating. In 1882, a Brooklyn-based scientist was of the opinion that the city's water was more impure than ever before.[46] What was the source of this pollution?

After 1827, Wilmington's water supply system depended exclusively on Brandywine waters that were diverted down the south race and then pumped uphill to one of the city's reservoirs. This was all right because, to the naked eye at least, the Brandywine seemed relatively clean. One observer claimed that, in the late nineteenth century, the Brandywine contained "no oil or other pollution." But others, including local physicians, suspected that the river was contaminated and urged an end to the dumping of pollutants taking place upriver. From the farms, villages and mills further up the Brandywine Valley, "all sorts of infected material was constantly being poured into the stream to become a source of danger to anyone lower down who drank the water." Although approximately four-fifths of the river's basin was north of the Delaware line, Delaware mills located just upstream from the south race entrance were particularly egregious sources of contamination. According to Dr. L. P. Bush in 1877, the factory pollutants poured into the river were "quadrupled by those which are swept into the Brandywine by every heavy rainfall from the hillsides and valleys, consisting of earthy and excrementious substances."[47]

An 1881 letter sent by the Wilmington Board of Health complained to the DuPont Company of the many privies for workers on its property that overflowed and spilled sewage into the Brandywine from "which our people are obliged to drink."[48] On a more positive note that year, a textile mill and a paper mill owned by the Bancroft family agreed to empty their industrial wastes into the north race, thus freeing the water heading down the Brandywine into the south race from two important, site-specific sources of industrial pollution.

But it may have been Wilmington that was the single most important polluter of its own drinking water. Although, during the nineteenth century, increasing numbers of the city's households were directly connected by iron pipes to Wilmington's water system, other city households remained unconnected. In the 1880s and 1890s, for example, many city residents continued to drink from backyard and kitchen pumps that drew from wells that were too close to a privy. Even more troubling for those who drank from the city's water system was the household wastewater that was commonly emptied into street gutters which, on Wilmington's north slope, washed downhill towards the Brandywine's south race. On reaching the south race, the gutters crossed over the race and emptied their contaminated liquid contents into the Brandywine beyond. The gutters over the south race, however, "were of only primitive construction and not at all water tight so that a little dirty water was constantly leaking...into the drinking water below."[49]

To make matters worse, by the late nineteenth century a number of houses and stores with privies "for which no wells had been provided," extended from the top of Wilmington's north slope, all the way downhill to the vicinity of the Brandywine's south race. Moreover, the Wilmington and Brandywine Cemetery, with its diseased and decomposing corpses, also extended down the north slope to the south race, just upstream from the pumping station. After a heavy rain, "all manner of pollution" from these two sources washed downhill into the same water that would be then pumped uphill to the city's reservoirs to become the drinking water of most of the city's residents. The presence of typhoid, "the worst" of the water-born diseases to threaten urban

America in the nineteenth century, was not surprising. After all, it was spread primarily by drinking water contaminated by human feces.[50]

Of course, the production of human waste was even greater on Wilmington's more heavily populated east and south slopes as well as in the adjoining lowlands, and much of this human waste, along with factory waste, washed directly into the Christina. Ironically, surface contamination throughout Wilmington by human feces was increased by the construction of the city's central water system. The resulting accessibility of piped-in water made possible the introduction of water closets, which were seen as an improvement over the old privy system. Although specific figures for Wilmington are unavailable, approximately one-third of all American urban homes had water closets by 1880. The problem with water closets was that human wastes were flushed away, but no further than the old privy vaults or cesspools that were used prior to the advent of the city's central sewerage system. In Wilmington and across urban America, the greater volume of water, "flooded cesspools and privy vaults, inundated yards and lots, and posed not just a nuisance but a major health hazard."[51]

Particularly at risk were the poorly drained lowlands along the Christina where most of the families of Wilmington's blue-collar laborers were crowded together in slum housing not far from the factories where they worked. Here, amid the stench of sewage, "congestion was painful, dirty water ran over broken pavements and the yards behind [the tenements] were reduced to a dump heap." The nearby Christina was heavily polluted, partly by rain water that washed down the two sides of the hill and across the adjoining lowlands and over the river bank. As contaminated as were the waters of the Brandywine's south race, Wilmington was fortunate to be drawing its drinking supply from the Brandywine rather than from the even more contaminated Christina. The periodic removal of human feces and urine from the city's backyard privies and other depositories was so inefficient that, in 1882, the Wilmington Board of Health pointed out that the soil was saturated with human waste which "renders the whole substructure on which the city of Wilmington is built, a pestilential mass." That same year Wilmington's Board of Trade declared that the entire city was in danger of becoming a huge cesspool.[52]

Getting Rid of Human Waste

The story of the construction of Wilmington's sewer system and the deposit of its contents some distance from the city is integrally connected with the general theory of that era concerning the spread of disease. By the mid-nineteenth century, the English idea that filth was the medium that transmitted diseases was beginning to have an impact on American thinking. By the 1870s, "The primacy of the filth theory was reaching its zenith," and considerable energies were turned to developing new environmental sanitation programs with civil engineers playing an increasingly vital role.[53]

By the 1880s, there was a growing consensus among Wilmington's business elite and its health professionals that a centralized system of sewers was desperately needed to dispose of the city's human waste. Older ways of cleaning out individual privies and cesspools were simply too inadequate to meet rising sanitary standards. In 1881, Wilmington's Board of Health reported "that many attacks of serious disease, and in some cases deaths have resulted in defective plumbing in our city." It went on to say that Wilmington's disposal of human waste "is not only defective but scandalous." Unless some "proper system" of sewers was adopted, "a terrible penalty of disease and death" awaited Wilmington's residents.[54]

In 1883, urged on by both the Wilmington Board of Health and the Wilmington Board of Trade, the city council hired Rudolph Hering, an outside sanitary engineer—a civil engineer specializing in sewers—to study Wilmington's needs and to make recommendations. Driven more by certain political priorities than by long-term health and sanitation concerns, the city council ignored Hering's subsequent recommendation that Wilmington immediately construct a relatively inexpensive system of connected sewers throughout the city. A very frustrated group of residents, however, went over the head of the city council to plead with the Delaware General Assembly to set up an independent commission to deal with the increasingly serious sewage problem in Wilmington. In 1877, a similar case of city-council intransigence caused Wilmington's civic leaders to turn to the Delaware General Assembly to appoint a bipartisan commission to more honestly and efficiently run Wilmington's waterworks.[55]

In 1887, the new, state-appointed commission to deal with Wilmington's sewage and street problems hired T. Chalkley Hatton as chief engineer. After a sanitary inspection of parts of the city that clearly proved how badly sewers were needed, Hatton developed a plan for a sewer system that dumped Wilmington's raw sewage, via a conduit that led eastward to Cherry Island, into the Delaware River. Believing in the aphorism that "dilution is the solution to pollution," Hatton justified his plan by insisting that the tidal waters of the Delaware River would easily disperse the sewage. Sewer construction began in the 1890s, and by 1899, approximately thirty percent of Wilmington's houses were connected by pipe to the city's sewerage system. But unlike Hatton's original plan, Wilmington's sewers emptied directly into the Christina and the Brandywine rather than into the Delaware at Cherry Island. Although the lowlands along the Christina and the fast-growing, newly annexed area north of the Brandywine were not yet served by sewers, Wilmington was on its way to becoming a sanitary city. But that is not to say that nineteenth century Wilmington was a pioneering American city in providing a healthier physical environment for its residents. Rather, the historical record affirms Carol Hoffecker's judgement that in constructing a central water system and a central sewer system, Wilmington was never a leader among the nation's municipalities.[56]

CHAPTER FIVE:

Protecting Some Biotic Life, 1825–1945

Arch McLane "killed over 100 ducks with one shot."

LANDSCAPE OFFERS US A USEFUL LENS FOR VIEWING THE INTERACTIONS between nature and people. The first Europeans to round Cape Henlopen and sail north up the Delaware Bay and River found the future state of Delaware almost completely covered by a mantle of old-growth forests through which freshwater streams, once they cleared the beaver dams constructed near their headwaters, wound their way unfettered to the sea. Saltwater wetlands were sandwiched between dry land and the coast, and fresh water wetlands dotted the interior. From north to south, from east to west, except for a few, primarily seasonal Native American settlements and some recently burned-over woodlands, there was little evidence of human presence.

By 1910, almost three centuries later, Delaware's landscape was significantly different. The old-growth forests and free flowing streams were gone, and the heavy imprint of man was everywhere. Only about 25 percent of the original forest cover remained and most of its trees were less than sixty years old.[1] Almost every freshwater river and creek was reshaped and, in some cases, redirected by silt deposits and by canal, dike, and bulkhead construction, while stream flow was interrupted by dams and then backed up into mill ponds. Not only had rising sea levels pushed Delaware's coastline inland—in some areas by more than two miles since the first Europeans arrived—but the state's fresh and saltwater wetlands had been significantly reduced in size.

Registering an indelible imprint on the Delaware landscape was a growing population that reached 202,322 by 1910, or probably eighty to one hundred times the number of humans present when the first European explorers rounded Cape Henlopen approximately three centuries earlier. Expressed in another way, there were probably far fewer than 1.4 people per square mile in Delaware in 1610 compared to 103 per square mile three hundred years later. By 1910, approximately 43 percent of the state's population lived in Wilmington, which meant that the sequential transformation of the countryside from forest to farm, farm to town, and town to city was more complete in and around Delaware's largest urban center than elsewhere.[2] The human imprint was apparent almost everywhere across a landscape increasingly characterized by fields and meadows, smoke-belching factories, and a spreading network of roads and railroads connecting growing villages and towns to urban markets. But the rate and the specifics of landscape change varied a great deal depending on where you were in the state.

Since the colonial era, there has been a consensus that Delaware is divided into two quite distinct geographic entities that differ markedly in the nature of their respective populations. The two Delawares are usually labeled "northern" and "southern," or "upstate" and "downstate." There is disagreement, however, over the exact location of the line that separates the two. Some argue that the dividing line is the Smyrna River, the boundary between Kent and New Castle County, while others point to the Chesapeake and Delaware Canal in central New Castle County ("Below the Canal" or "above the Canal" are commonly heard geographic expressions based on the latter boundary line). A third possible dividing line between the two Delawares is the south bank of the Christina River, where Wilmington ends and the rest of the state begins. In 1879, Delaware writer and illustrator Howard Pyle maintained that just south of Wilmington, "the vim and progress of modern utilitarianism" quickly merges "into the indolence peculiar to southern life." In 1934, Henry Seidel Canby noted that Wilmington "was austere and northern," while the land further south was "an alien state." A fourth possible dividing line is the Mispillion River, which flows east through Milford into the Delaware Bay and separates much of Sussex County from the rest of the state. In the mid-nineteenth century, Sussex was said to be asleep like Rip Van Winkle, and its people were viewed as significantly less educated and more impoverished than other Delawareans.[3]

No matter which dividing line was chosen to separate the two Delawares, like the directional flow of most of the state's rivers, it cut across the state from west to east and was based on distinctions rooted in such basic environmental factors as specific geographic location, climate, level of soil fertility, presence of malaria, and the availability of water power. Inevitably, despite the state's diminutive size, these and other natural factors helped shape Delawareans into two distinctly different populations, divided not only by how they made a living but by how they viewed the world. In the second half of the twentieth century, the philosophical differences that separated the two Delawares would express themselves in disagreements on a number of specific issues. Perhaps the most important one was whether or not Delaware really had an environmental crisis that needed strong government intervention.

Although Delawareans continued to be partially shaped by their natural surroundings during the nineteenth century and first half of the twentieth century, their desire to earn a living, to jettison their waste materials, and to participate in recreational activities placed increasing pressures on the quality of Delaware's natural environment. Draining the soil of nutrition, denuding and trashing the landscape, destroying much of the wildlife, and polluting the air and water, as previously pointed out, were patterns of behavior partially based on some wasteful Old World practices brought to colonial Delaware from the northern periphery of Europe and then strengthened and expanded by human interaction with a natural world that was, initially "a land of plenty." Seemingly unlimited acres of fertile soil, extensive forests of old-growth timber, and numerous flocks of birds, herds of animals, and schools of fish, made the destructive exploitation of nature seem a small price to pay for the resulting improvement of the landscape. But there may have been some other behavior patterns at work concerning the treatment of the natural world that cannot be explained by simply pointing to certain cultural perspectives that were rooted in the peripheral forests of northern Europe or in the subsequent Delaware experience.

To former Delaware wildlife manager Tony Florio, some of the increasing pressure on the state's natural environment in general and its wildlife in particular may have its roots in an instinctive trait present in humans since time out of mind that drives them to hunt and fish for more than just the elementary need to put food on the table. Indeed, according to this view, stalking and killing game and hooking fish provides solace for the soul as well as protein for the empty stomach. To Florio, it is "ludicrous" to think "that our ancestors did not hunt for pleasure."[4] Whatever the motivation, by the late nineteenth century it was clear that hunters, trappers, and fishermen had so stripped Delaware of most of its game birds, mammals, fish, and shellfish that stronger state intervention was essential for the protection of the little that was left of Delaware's "desirable" wildlife.

Birds

Particularly problematic were commercial hunters who killed game and then shipped the meat, skins, and feathers to markets beyond Delaware. After commercial hunters played a key role in the disappearance of deer and turkeys from Delaware's forests and fields, they turned to killing as many of the remaining birds as they could to produce marketable meat. The advent of the steamboat and then the railroad made it possible to rush dead ducks to food markets in Wilmington, Philadelphia, and other urban centers. Soon, nonresident commercial

Great Blue Heron,
photo courtesy of Dolores Michels.

hunters were considered a threat to the survival of the state's most desirable game birds and this caused the General Assembly, in 1839, to ban out-of-state hunters from exporting across the state's borders any dead fowl shot in Delaware. Subsequent developments, however, indicate that this ban was either ineffectively written or simply ignored. By the 1880s, a rising demand by women for feather-adorned hats became another reason for shooting birds. Up and down the Atlantic coast, enormous numbers of egrets, herons, ibises, swans, as well as terns and seagulls, were slaughtered just to acquire their colorful plumage. Commercial hunters turned to giant punt guns and even to light cannon to more

A female mallard at Bombay Hook National Wildlife Refuge near Leipsic, Delaware, *photo courtesy of Dolores Michels.*

efficiently kill, for either meat or feathers, as many birds as possible. An entire flock might be brought down with only one blast of birdshot from these extraordinarily lethal weapons. In Delaware, Arch McLane, a commercial hunter from the Smyrna area, "killed over 100 ducks with one shot," while the aptly named Punky Slaughter from the same area routinely knocked a boat full of ducks from the sky and then loaded them up at Fleming's Landing, on the north bank of the Smyrna River, for shipment by steamboat to Philadelphia.[5]

At the end of the nineteenth century, most commercial hunters were generally scorned across the nation as "lower class layabouts who eschewed steady work." They had few if any friends in high places and were woefully overmatched in the political arena once middle- and upper-class sportsmen began to sense that commercial hunters were a serious threat to the very game birds that sportsmen hunted. Moreover, by killing nongame birds for their feathers, commercial hunters aroused the ire of middle and upper-class women who were horrified that the millenary tastes of at least some members of their gender led to the ongoing slaughter of so many beautiful winged creatures. While the Victorian doctrine of separate spheres argued that women were morally superior to men, the predilection of some women for hats decorated with feathers from slaughtered birds challenged this perspective. Indeed, the middle- and upper-class concerns of both genders were first codified into law in 1885, when the Delaware General Assembly made illegal the use of a "swivel or punt gun other than such as habitually raised at arms length and fired from the shoulder," for hunting ducks, geese, and other wild fowl.

Subsequent state legislation would be supported by the Delaware Audubon Society, which was founded in 1900 at the home of Mrs. William S. Hilles of Wilmington, a member of Delaware's prominent Bayard family, and would become Delaware's first voluntary conservation organization. Modeled after the National Audubon Society, the Delaware Audubon Society sought to protect birds from slaughter by commercial hunters and to discourage their use "in wearing apparel and for the purposes of ornament." The first Delaware Audubon Society eventually disbanded and the modern Delaware Audubon Society was not founded until 1976.

In 1911, while hat-producing New York state banned the sale of the plumes of all native birds, the Delaware General Assembly banned shipping and selling for profit any birds and animals killed in Delaware. That same year Delaware dealt a more specific blow to commercial hunters who shot fowl just for their feathers by specifically outlawing the killing of non-game birds and the possession of "any part of the plumage."[6]

By the first decade of the twentieth century, Delaware's wildlife had been selectively decimated by commercial, recreational, and subsistence hunters, as well as by the continued destruction of its natural habitat. The brant, a small herbivorous goose, disappeared from the state by 1840, but did reappear in very modest numbers four decades later. The ruffed grouse, which was so dependent on extensive old forest habitats rather than fragmented woodlots, was gone from northern Delaware by the 1870s and would subsequently disappear permanently from the entire state by the twentieth century. The migrating American passenger pigeon, once probably the most numerous bird in North America, was no longer seen in the mid-Atlantic area by the mid-nineteenth century. The American passenger pigeon's rapid decline to extinction in Delaware and, eventually, everywhere else in North America, was caused by extensive commercial hunting, by loss of natural habitat, and by a number of other factors. Turkeys, deer and beavers continued to be absent from the Delaware landscape, while the last bear in the state was supposedly shot in Sussex County's Cypress Swamp in 1906.[7]

Comments by interested observers and the passage of additional hunting restrictions indicate that many other birds and mammals were struggling hard to survive in the animal-unfriendly environment of the late nineteenth and the first two decades of the twentieth century. As early as 1870, cranes and curlews were no longer very visible in the state and, by the beginning of World War I, the bobwhite was just about extinct in New Castle County. Prior to 1950, hunting game birds in Delaware primarily meant hunting ducks because there

A member of Bombay Hook's resident turtle population,
photo courtesy of Dolores Michels

was only a modest number of geese present in the state to compete for the attention of hunters. Until well after World War II, most geese that flew south via the Atlantic flyway, wintered in the Carolinas rather than on Delmarva or on the western shore of Maryland. The dramatic increase in the post-World War II goose population came about when more corn was planted to feed the growing flocks of Delaware's poultry industry. Mechanical pickers, increasingly used to harvest corn, worked with less-than-perfect efficiency so that each fall the state's farm fields were strewn with shattered grain, and that led to huge flocks of geese abandoning the Carolinas for Delaware and the rest of Delmarva as their new winter home. But ducks were a different matter. One old progger from northeastern Kent County—proggers hunt, trap and fish the wetlands and bays for a living—recalled that in the first decade of the twentieth century "there was lots of ducks—my God, the sky was black with them." There are other indications, however, that the duck population was under considerable pressure, causing the General Assembly, in 1885, to restrict duck hunting to the five months stretching from September 1 to January 31. But staying ahead of hunters was difficult because ever more effective new tactics were constantly being adopted to take advantage of the latest technological advances. Particularly egregious was the practice of shooting ducks and other birds from the fast moving sailing vessels and motorboats that

Canada geese in a wintertime Sussex County marsh,
courtesy of Richard B. Carter.

plied the waters of Rehoboth Bay and the Indian River. The General Assembly legislated a halt to this style of hunting in 1907.[8]

Turtles, Muskrats, Foxes and Rabbits

According to one source, turtles in general and diamond-back terrapins in particular were so numerous in eighteenth-century Delaware that they were commonly fed, almost daily, to both slaves and servants. Despite the fact that terrapin would one day be considered a delicacy, a constant diet of turtle lost its appeal in a hurry. Slaves and servants protested and that led to an ordinance forbidding masters and employers from feeding terrapin meat to their slaves and servants more than three times a week. In 1812, however, a declining turtle population led to the banning of out-of-state terrapin hunters, only to have the ban lifted in 1851 for those non-Delaware hunters who paid for a license. By 1869, no in-state or out-of-state terrapin hunter was allowed to take a turtle that measured less than six inches long, and the use of dredges to hunt turtles was made illegal on Rehoboth and Indian River Bays. Towards the end of the nineteenth century, turtle meat was increasingly viewed in urban centers as a gourmet food, and this placed particular pressure on the state's diamond-back terrapin population, which was particularly concentrated

Muskrat homes in the Delaware Bay marshes of Kent County,
courtesy of the Delaware Public Archives.

in and near the bays and the brackish creeks of eastern Sussex. By 1907, proggers and others had so diminished the numbers of this once-numerous turtle that the General Assembly outlawed the capturing and killing of diamond-backs in Sussex from March 1 to August 1. Despite a shortened season in southern Delaware, diamond-backs continued to provide a significant source of income for proggers and for a number of farmers with holdings adjoining Delaware's brackish creeks and inland bays. Extensive hunting, however, so depleted the numbers of diamond-back terrapins that by 1938 they were rarely seen in Delaware.[9]

With the disappearance of deer, bears and beavers from Delaware, many hunters and trappers turned to killing muskrats. Living in haystack-shaped homes in brackish and fresh water wetlands, muskrats were primarily targeted for their pelts, which were once sold as "Hudson seal." Their meat was also popular and was served in homes and eating places as "marsh rabbit." Muskrats were also killed by irate millers and angry farmers because the rodents liked to tunnel through earthen dams, dikes, and sea walls. In addition, whenever the number of foxes or wild dogs increased in a specific area, the threat to the local muskrat population was considerable. In 1941, for example, a "great increase in foxes" on Long Neck in eastern Sussex, led to a significant decline in the number of muskrats. But just as

A red fox kit at Bombay Hook,
photo courtesy of Dolores Michels

important in depleting the number of muskrats in Delaware as human and animal hunters, was the continuous reduction of wetlands. As a Kent County progger pointed out, the impact of digging drainage ditches to dry up the land was predictable: the population of muskrats would decline because "muskrats wants water."[10]

For most of Delaware's early history, muskrats remained relatively numerous because the early elimination of many of their most serious predators helped counteract the effect of declining acres of wetlands and the toll taken by hunters, trappers, irate millers, and angry farmers. But as the twentieth century approached, there was growing concern among some Delawareans about an apparent decline in the number of muskrats. In 1899, the General Assembly outlawed the hunting and trapping of muskrats in the Smyrna River, which wound through prime muskrat habitat before emptying into the upper Delaware Bay. Four years later it became illegal to kill or capture muskrats anywhere in Delaware "by any means whatever," during any flood or high tide that caused them to leave their protective marshlands. In 1909, hunting and trapping muskrats anywhere in Delaware "by any means whatever," was banned from early spring to late fall except to protect "any mill dam, ditch bank or river bank from being undermined, injured or destroyed by muskrats." Four years later, hunting muskrats with dogs was prohibited.[11] The growing concern about muskrats had little to do with the modern conception that their presence was beneficial to the ecology of Delaware's wetlands. Rather, the concern was primarily rooted in the fact that any significant decline in the number of muskrats threatened a valuable source of income for a number of Delawareans.

During the first half of the twentieth century, many Delaware farmers who owned wetlands along the Delaware Estuary or in more interior areas derived their principal income from trapping muskrats or by turning over their marshes to proggers and other trappers for about 50 percent of the sale price for each muskrat pelt. At one time muskrat colonies flourished in the wetlands that stretched south from the mouth of the Christina to Fenwick Island. So numerous were muskrats in the marshes southeast of Smyrna that, in 1814, a small pelt-trading center in the area changed its name from Fast Landing to Leipsic in hope of emulating some of the success of that important German fur center.

According to one source, in some years of the nineteenth century, as many as 100,000 muskrat skins annually were shipped out of Leipsic. But the whimsical nature of Delaware's weather caused a number of storm-driven salt-water incursions that breached the natural sand and earthen barriers protecting central and southern Delaware's coastal wetlands from the sea, changing the character of the vegetation in these marshes in a way that had the effect of reducing the number of muskrats. The shifting geographic source of Delaware's annual catch reflected this decline in southern Delaware's muskrat population. Although a majority of the state's muskrat harvest probably came from Kent and Sussex in the late nineteenth century, by 1940 about 61 percent of the year's catch of 150,000 pelts came from New Castle County, where the best muskrat habitat was in the Delaware City area. Only 39 percent came from Kent and Sussex combined. By 1978, excessive trapping had so diminished the numbers of muskrats that the annual state-wide catch had declined to somewhere between 40,000 and 50,000 pelts. By 2000, the annual catch was down to 19,314 pelts.[12]

The drive to preserve the red fox, like the attempt to protect the muskrat, had little to do with what we would label today as ecological concerns. In the case of the muskrat it was because of the income produced by its pelt, in the case of the fox it was to preserve an old Anglo-Saxon sport. Although foxes threatened young lambs and barnyard fowl, their continued existence was key to the fox hunt, an activity that was central to the cultural life of an eighteenth and nineteenth century Delaware gentry determined to emulate on this side of the Atlantic many of the traditions of landed Englishmen. One example was Anthony Higgins (1763–1824) of Red Lion Hundred in central New Castle County, who was described by his grandson as a very close mirror of an "English fox-hunting squire." Probably, the gray fox was the only indigenous fox in Delaware, but the deliberate introduction of Great Britain's red fox, the traditional quarry of hounds and riders, into the Anglo-American colonies during the mid-eighteenth century may have contributed to the dramatic decline in the numbers of gray foxes. The imported red fox seemed superior to the gray fox in competing for the same food in a changing Delaware landscape that increasingly featured open fields and scraggily second- and third-growth forests.[13]

Throughout the nineteenth century, the Delaware fox hunt continued to follow the English tradition of hounds chasing their red-fox quarry across Delaware's fields and through its woods with mounted members of the landed class in hot pursuit. By the beginning of the twentieth century, however, the state's red fox population was in decline and a venerable Anglo-Saxon tradition was in jeopardy. In 1901, the General Assembly responded by banning fox hunting from May 1 to August 31, and made it illegal to kill a fox while it was pursued by hounds. Ten years later the sale of fox furs was outlawed. As Delaware's countryside became more densely populated and the automobile replaced the horse as a means of transportation, fox hunting experienced a dramatic change. In the wealthy estate country of northern New Castle County, fox hunting was gradually abandoned and, by the 1970s, traditionally mounted fox hunters from Delaware's piedmont had moved their activities southwest to the Middletown area and to the adjoining sections of Maryland's Eastern Shore. Further south in Delaware, a very different type of fox hunting evolved that used automobiles rather than horses.[14]

In the early 1930s, journalist and writer Tony Higgins recorded a typical downstate fox hunt amidst its physical setting:

> It was a clear and frosty winter day. All morning we had followed the chase—not on horses but in cars, that being the custom hereabouts—through the flat, heavily-wooded country of Indian River Hundred [located in eastern Sussex]. The woods were arrow straight loblolly pines of deep green, and the roads were sandy trails that often ducked under streams and black swamp water. Farms were little more than clearings in the woods; unpainted houses, small fields of yellow fodder, all in the shadow of the forest. The red fox we had jumped near Zoar Chapel and chased for three hours. Now he vanished for good in the swamp called "Hoss Goin' Over" at the head of Lingo's Creek.[15] *

It was the growing scarcity of certain wildlife important for commercial hunting and trapping or for sport, rather than contemporary ecological concerns, that also explains attempts by Delaware to preserve some of its other fauna during the nineteenth and the first half of the twentieth century. Wild rabbits, for example, in addition to representing potential food for the local population, were the quarry in an important rural sport. But

* Note: Zoar Chapel is located on the east side of Delaware Route 30, about four miles north of Millsboro. The swamp referred to as "Hoss Goin' Over," along the headwaters of the creek dividing Long Neck from the Oak Orchard area, is some seven miles away "as the fox runs."

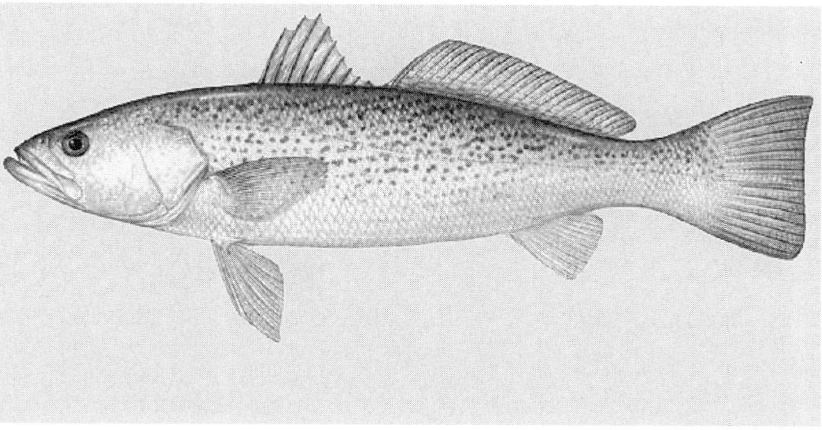

Sea Trout or Weakfish

their declining number led the state, in 1885, to restrict rabbit hunting to November and December. The use of ferrets for driving rabbits into the open where they could be easily shot placed additional pressure on the rabbit population and, in 1909, caused the General Assembly to ban "having ferrets in one's possession while hunting." Nevertheless, Delaware's rabbit population continued to decline, forcing the state's game-and-fish commissioners to import new stock from Kansas in 1921.

Other wildlife partially protected by legislation from 1800 to 1950 to preserve their numbers for hunting, included opossums, raccoons, skunks, otters, minks, black and gray squirrels, snapping turtles, frogs, partridges, quail and pheasants. However, there was no legal protection for blackbirds, crows, English sparrows, starlings, turkey buzzards, and all hawks except the osprey (fish hawk), because they were a threat to barnyard fowl and field crops, or were simply perceived to be public nuisances in other ways. For a number of years the state established a bounty for killing many of these birds.

Even wildlife specifically shielded by law could be shot or trapped under certain circumstances. In 1920, the General Assembly codified what was already common practice by granting Delawareans the right to kill in order to protect their "property and premises from the ravage and depredations of any wild birds and animals protected by law."[16]

Oysters, Sea Trout and Shad

Declining numbers also caused the General Assembly to establish restrictions on the oyster harvest. In some coastal states north of Delaware, the oyster population was seriously depleted by the late eighteenth century. By the beginning of the nineteenth century, watermen from Connecticut and then New York were sailing south to test their skill and luck in Delaware waters. Their presence posed a particularly serious threat to Delaware's natural oyster beds because the northern watermen used an iron-toothed bar dredger that so effectively scraped the bottom that one haul brought up bushels of bivalves. The increasing presence of out-of-state watermen and their superior dredging equipment guaranteed a serious decline in the number of oysters in the Delaware, Rehoboth, and Indian River bays, and therefore threatened the livelihood of Delaware watermen. In 1812, as previously indicated, an alarmed General Assembly made

Shad

this state, for planting oysters." The beds were to be marked by stakes and trespassing oyster pirates were to be arrested. Although, by 1851, there were legal limits on the daily harvest of wild oysters by Delawareans, there were no legal limits on the number that could be taken each day by Delaware watermen from their own planted-oyster beds. Despite all of these efforts, in 1871, Governor Gove Saulsbury warned that "the laws of the state have not been adequate for the protection of our oyster beds." [18]

it illegal for out-of-staters to take oysters and clams from Delaware's waters unless the newly harvested shellfish were loaded on board a vessel "wholly" owned by a Delawarean. In 1835, both the use of an oyster dredge at any time of the year and the raking of oysters from May 15 to August 15 were declared illegal in Delaware waters.[17] Later in the nineteenth century, the use of modified oyster dredges was allowed.

Then as now, state governments were always looking for new sources of income that would sit well with constituents. In 1849, to provide an additional, although quite modest source of funds that would be painless to Delaware taxpayers, the General Assembly reopened the state's natural oyster and clam beds to out-of-staters, provided that the latter pay an annual license fee of $100 per boat. This was decreased to $50 two years later. But these and other restrictions on the harvesting of oysters did little to halt the gradual depletion of Delaware's natural supply. In 1851, the General Assembly tried a new approach to the issue by offering, but only to Delaware residents, the right to claim not more than "an acre, of the bottom of any streams or waters of

The General Assembly responded to Governor Saulsbury's concern by immediately establishing new restrictions on oyster harvesting. By the end of 1871, oyster harvesting was easily "the most regulated and controlled" economic activity in Delaware. Indeed, so many oyster regulations were on the books by 1893 that, to avoid confusion, the General Assembly repealed all of them and substituted a new, detailed set of regulations. But many out-of-state watermen were unwilling to recognize Delaware's restrictions on their raking and dredging rights. Their continued harvesting of oysters in Delaware waters was one of the factors that led to oyster wars featuring at least one exchange of gunfire in the late 1880s.[19]

In 1925, a typhoid fever epidemic blamed on contaminated oysters taken from polluted waters caused considerable anxiety in Chicago and elsewhere. Across the nation, state health commissions and boards nervously began examining the contamination level in both locally-produced and imported oysters. In Delaware there was great concern about what impact this might have on the state's oyster industry. Finally, in 1927, the Delaware State Board of Health decided to halt the harvesting of oysters from the polluted waters of the St. Jones, Murderkill, Mispillion, and upper Broadkill Rivers because at least some oysters from those rivers were contaminated. Fifty years later, scientists from the newly-established University of Delaware College of Marine Studies attempted to transplant young oysters, grown in water tanks, to the Delaware Bay. The project was soon abandoned because, it was claimed, the bay was so polluted that few of these oysters grew to maturity.[20]

Because of increasing water pollution and over-fishing by two-masted sailing schooners—motor-powered vessels were not made legal for oystering until 1949—the Delaware oyster industry reached its zenith about 1920 and then went into decline. Nevertheless, watermen on both sides of the Delaware Bay continued to make at least part of their living from oystering into the second half of the twentieth century. Of particular help was the settlement of an issue that historically had created considerable tension between Delaware and New Jersey watermen. In 1934, the United States Supreme Court ruled that the marine jurisdictional boundary between the two states was a line drawn roughly down the middle of the Delaware Bay.

Although the annual harvest continued to be considerably above sustainable levels and the annual catch throughout the Delaware Estuary had declined from almost 30 million pounds of oysters in 1890 to between 10 and 15 million pounds in 1950, the catastrophe to come was totally unexpected. It started on the New Jersey side of the Delaware Bay in 1956, where planted beds began yielding only dead or dying oysters, and then spread to the Delaware side the next year. A protozoan parasite caused an oyster disease called MSX that so devastated the Delaware Estuary's oyster stock that, in 1959, watermen from both sides of the Delaware Bay were asked to stop tonging and dredging until the disease had run its course. A partial recovery followed. By 1973, oyster harvesting in Delaware's waters was dramatically up over previous years, but it was still only about one-sixth of the 1957 figure.[21]

Fish had always been an important food for Delawareans. To the economically hard-pressed people of Kent and Sussex, the word that sea trout (weakfish) were running in the Delaware Bay, usually from late April to the end of May, was exciting news. Farmers and others who lived only a few miles from the coast quickly loaded their nets and other fishing gear into wagons, hitched up their horses or oxen, and then headed towards the nearest beach to do some serious fishing. They might serve together in a large or a small net-hauling crew, depending on the size of both the boat and the nets that were available. Although some professional fishermen were present, they were a distinct minority of those pulling in nets full of sea trout. Other farmers and mechanics who lived further inland, in some cases almost thirty miles from the nearest salt water, also hitched up their wagons on news that the sea trout were running. But instead of fishing gear they threw in only what was needed to gut, split, and salt down the fish. On arriving at a fishing site along the coast, they purchased from the net-hauling crews enough fresh sea trout to last their individual families an entire year, and then processed the fish on the spot, loaded them into their wagons and headed home.

As late as the 1880s, most of the netted sea trout from Delaware waters were used to feed in-state residents. The little that was left over was sent on to Philadelphia or to other out-of-state sites. As one student of fishing noted in 1887, "it is said that this species of fish [sea trout] constitutes an important part of the food supply of the rural population of Delaware." But as the last decade of the nineteenth century approached, many of Delaware's fishermen were becoming increasingly pessimistic about the future supply of sea trout and of most other edible fish.[22]

Of major concern was the number of shad remaining in the Delaware Estuary. When the first Europeans arrived in Delaware, "every stream...abounded in shad." For much of Delaware's history, the shad was its most important subsistence and commercial fish, as well as a convenient litmus test for the environmental health of the Delaware Estuary and its tributaries. Unlike the sea trout, shad are anadromous. Although they had the potential to reach thirty inches in length and to weigh up to ten or twelve pounds, a five pounder was more typical. Shad were particularly important to the less affluent families of early Delaware, because they were among the first fish to make the spawning run upstream at the end of winter when food larders were low. In fact, the fish is named for the shadbush whose white blossoms are a harbinger of spring in the Delaware Valley. But because it is oily and boney, the shad was never very popular further north, particularly in New England, where cod was the favorite. A native of Rehoboth Beach remembered that, at the beginning of the twentieth century, her father would net shad during the spring run and then complain loudly, while eating them, of "all those damn bones." Nevertheless, in the mid-Atlantic states and particularly in the Delaware Valley, shad were increasingly appreciated during the nineteenth century as a regional delicacy. The subsequent rise in demand, however, led to over-fishing and a resulting significant decline in the number of shad.[23]

The problem was that the Delaware Estuary and its tributaries were seen by fishermen as a marine public commons where market forces rewarded those who caught as many fish as they could as quickly as they could, and thus made individual restraint a formula for failure. As in the case of the land-based public commons on the Cape Henlopen side of Lewes Creek, where market forces led to livestock overgrazing and tree destruction, market-driven fishing practices led inevitably to serious environmental degradation of Delaware's marine resources by significantly reducing the population of shad and other fish. As Governor Ebe Tunnell pointed out in 1901, unrestrained market forces made Delaware's fishermen their own worst enemies because they were driven by the lust for immediate profit to kill "young and unmarketable fish," and therefore destroy the future "source of their supply."[24]

In addition to being overfished from almost the beginning of European colonization, shad as well as other anadromous fish, such as herring and sturgeon, suffered enormously from the erection of milldams that blocked their spring runs to spawning grounds. Prior to 1760, as previously mentioned, there was a government-backed provision that each dam must contain a fishway to provide both upstream and downstream passage for migrat-ing fish. After 1760, however, the Delaware colonial legislature chose to support the interests of industrial millers over the well-being of shad-dependent residents living upstream, by permitting new milldams to be constructed without fishways.

The subsequent decline in shad and other anadromous fish finally caused the General Assembly, in 1839, to take action. But rather than demanding fishways through dams, it banned the use of gill nets within a mile of the mouth of any river or creek that emptied into the Delaware Bay or River. In addition, nonresident commercial fishermen were no longer allowed to export fish caught in Delaware waters to other states or nations. But the decline in shad and other fish continued, reaching epic proportions by the post-Civil War era. In 1871, Governor Gove Saulsbury lamented, with probably some exaggeration, that where "formerly our supplies of fish were caught from our own waters and preserved by our own people, now our entire supply is imported from other states." One specific explanation for this untenable situation, according to Governor Saulsbury, was that state fishing regulations were not adequate to protect Delaware's fish "from depredation by nonresidents." The General Assembly responded by requiring that out-of-staters buy a license prior to fishing for any species of fish in Delaware waters. Even Delawareans who fished specifically for shad were required to pay for a license. During the late nineteenth century, additional legislation was enacted to protect the shad.[25]

Shad made an impressive recovery in the late nineteenth century and reached peak numbers in the Delaware Estuary and its tributary streams at the beginning of the twentieth century, when more than 14 million pounds were sold out of the fishing ports of Delaware City and New Castle. The increase was probably due to more stringent state regulations on commercial fishermen and the massive restocking of area streams with shad and other fish by Delaware and other states that was well underway by the 1880s. And yet by 1940, only forty years after reaching its peak, the shad fishery in Delaware "had almost been eliminated."[26] This time, overfishing had to share the blame with a precipitous rise in water pollution.

Shad begin life in freshwater streams during the spring and remain in fresh or brackish water until fall. Then commences a slow journey that usually finds them in the Atlantic Ocean by the time they are one year old. After maturing over the next three to five years, shad return in the spring to the freshwater stream of their birth to spawn and thus give rise to a new generation. During the colonial period, most shad that returned to the Delaware Estuary and its freshwater tributaries to spawn initially favored streams in central and southern Delaware and in southern New Jersey that emptied into the Delaware Bay. The early despoiling of tidewater streams by dams and siltation, and the presence of so many fishing weirs, however, caused the shad to turn increasingly to tributaries of the upper Delaware River for spawning grounds. But as the Delaware River grew more polluted, anadromous fish found the upstream spring migration an increasingly difficult journey.[27]

Pivotal to the environmental health of the Delaware River was the impact of Philadelphia. By the early twentieth century, Philadelphia's sewers were annually dumping 200,000 tons of raw sewage into the Delaware. Ships plying the lower Delaware also added to the river's pollution. A resident of Chester, Pennsylvania, during the 1930s and 1940s remembered that, "if a ship had anything to discharge—wastewater, oil or fuel—the captain would dump it into the river." The resulting stench was so strong that when boats docked in the Quaker City, individual crew members often jumped ship rather than try to sleep in their berths surrounded by the nauseating odors. It was said that pilots flying as high as 5,000 feet above the Delaware noticed the river's foul smell. During World War II, increased defense manufacturing in the Philadelphia area added substantially to the already significant amount of industrial waste pouring into the Delaware. In 1944, one government official declared that Philadelphia was the foulest freshwater port in the world. A second pointed out that the enormous amounts of organic wastes annually pouring into the Delaware in the Philadelphia/Camden area significantly depleted the river's dissolved oxygen, which was essential for the survival of fish and any other living organisms. The resulting belt of dead water extended from river bank to river bank in the Philadelphia/Camden area and, during the heat of the summer and that season's diminished stream flow, as far as thirty miles downstream. Environmental activist June MacArtor pointed out in 1959, that because "Delaware was at the end of the line," it was aware of what was going on.[28]

Despite this heightened awareness, Wilmington also dumped raw sewage and industrial wastes into the Delaware via the Brandywine and the Christina. But because it was dwarfed in size by the two river cities to the northeast—in 1950, Philadelphia's population was approximately twenty times larger and Camden's was almost three times larger—Wilmington's contribution to polluting the Delaware was, by comparison, far less substantial.

Delaware's commercial fishermen, accustomed in the spring to netting large numbers of migrating shad as they moved upstream along the west bank of the Delaware, noticed by the 1940s that there were far fewer shad than even a few decades earlier and that north of the mouth of the Christina, shad had almost disappeared. Intent on spawning upstream, they could no longer penetrate the Delaware River's dead zone. For the small shad population that survived in the Delaware

Estuary south of this pollution barrier, spawning continued in some of the less despoiled tidewater streams. This, in turn, preserved a breeding stock that became the foundation of a shad revival once the pollution barrier in the Delaware was dislodged and the flow of subsequent pollutants into the river was significantly diminished. That occurred soon after World War II when improvements and expansion in sewage systems and the construction of sewage-treatment plants began to modify the toxic nature and the physical size of the open cesspool that, in warm weather during the 1940s, stretched almost as far south as the mouth of the Christina.[29]

Nature also played a dramatic role in breaking up this riverine pollution barrier in the Delaware. In 1955, flooding caused by two hurricanes "scoured out pollution-saturated sediments" that were in the riverbed or were caked several feet deep on its banks, and then flushed them out to sea. About 1960, five years after this natural cleansing, shad began to reappear in the upper Delaware Basin. But abnormally low rainfall in 1961, 1963, 1965 and in the early spring of 1966, decreased the river flow and delayed the recovery. Nevertheless, in the decades immediately following World War II, the future looked more promising for shad, in part because sportsmen joined with ecologists in pushing hard for a cleaner Delaware River and more accessible spawning grounds in its tributaries.

Many streams, once favored spawning sites for migrating fish but subsequently blocked by dams, were reopened. One example was the Brandywine, where dams had so completely blocked spring runs that a 1952 sampling found no anadromous fish above the Great Falls in downtown Wilmington. In order to attract anglers to the Brandywine, Delaware built a series of fish ladders in the river's dams in 1969 and 1970 so shad and other anadromous species could make it upstream to spawn.[30]

Public support for measures essential for the resurgence of the shad population in the late nineteenth century and again in the late twentieth century certainly reflected economic concerns. But it was also rooted in changing cultural currents and evolving environmental attitudes. During the late nineteenth century, for example, rising enthusiasm for catching and eating shad was fueled in part by the need of urban Anglo-Americans to "reconnect with the natural world and with their rural, pre-industrial ancestors."

Unlike fox hunting, which was driven by the desire to preserve upper-class English customs, the renewed interest in shad fishing was partially driven by the desire to imitate the vocational and eating activities of an earlier, lower class fishing population of English ancestry who lived along the river in Delaware, New Jersey and Pennsylvania. The desire to reconnect with this local ethnic past helped rally support for restocking streams and for restrictive legislation that produced a rebound in the shad population at the beginning of the twentieth century. Predictably, however, this enthusiasm for shad also led to over-fishing and to a crash in the number of shad in the Delaware River and its tributaries after 1905.[31]

When shad numbers plummeted further

with the approach of World War II, increasing pollution of the Delaware River received some of the blame. Subsequently, the presence or absence of shad in the Delaware and its tributaries was seen by many as a dependable barometer that registered the essential environmental health of the river system. Unfortunately, the increasing numbers of shad in the Delaware over the last three or four decades of the twentieth century gave the residents of Delaware, New Jersey and Pennsylvania a false sense of security concerning the quality of their river's water. No longer were sailors nauseated by odors rising from the cesspool that once extended from Philadelphia to Wilmington, but years of accumulated chemical and industrial waste still lay relatively undisturbed in the river's bed. The Delaware River of 2000 could not claim to be an environmentally safe body of water.[32]

CHAPTER SIX:

SAVING FIELD AND FOREST, 1818–2000

"Neglected fields are now reinvigorated by the use of these new fertilizers and made to yield abundant crops."

IN WRITING ABOUT THE INTERACTION OF HUMANS WITH THE NATURAL environment, historians usually are drawn to one of two themes or narratives: the first focuses on the happy story of humans improving on nature, while the second focuses on the sad story of humans destroying nature. The truth, however, is that environmental history is too complicated to be captured by either narrative. In Delaware's case, the trend over time has featured wildlife destruction, soil exhaustion, and increasing pollution of both air and water. But, like the swirling eddies that counter the general flow of a river's current, human effort modified or, in a few cases, even reversed the downward spiral in the quality of Delaware's natural environment. Clearly, increased government regulation of hunting and fishing as well a few initial attempts to improve the water quality of some of Delaware's streams are examples of these efforts. Even more to the point were the innovative actions of some farmers, scientists and civic leaders to bring new, productive life to Delaware's worn-out farmlands. Reinvigorating the soil was crucial to Delaware's economic future because, as former state geologist James Booth pointed out in 1841, within Delaware "agriculture must be pursued and encouraged as the principal source of wealth."[1]

Soil Depletion and Recovery

More than any other human activity in Delaware's history, the pursuit of agriculture was responsible for reshaping the landscape, and it did so by simply erasing raw nature from the land. This included, as previously pointed out, extirpating soil fertility at an early date. All across the United States during the nineteenth century most farmers were poor stewards of the land, and consequently were labeled "earth butchers" and "predatory agriculturalists." And yet, as one mid-century observer pointed out, agricultural practices in much of Delaware seemed "far behind" even those in most of "her sister states."[2]

No matter how Delaware's farming practices may have compared to those elsewhere, one thing was clear by the early nineteenth century: Delaware agriculture was in deep trouble. In 1807, Governor Nathaniel Mitchell stated the obvious when he declared Delaware "is agricultural," and "upon this source depends, in a very great measure, the wealth of our citizens." By 1821, Governor Jacob Stout noted that "agriculture in this state is in a languishing condition, for want of a proper system of husbandry." In 1837, Governor Charles Polk described agriculture in Delaware as "truly in a prostrate and suffering condition." Strong evidence of soil declension was visible almost everywhere. One observer estimated, in 1828, that seven-tenths of the state's arable land, with the exception of drained meadows along the Delaware River, produced "scarcely twenty bushels of corn and ten bushels of wheat per acre," while in the region west of the Appalachians, the figures were, respectively, thirty and twenty-five bushels per acre.[3] Some of the poor harvest figures for Delaware during the early and mid-nineteenth century, however, may have been due to the presence of increasingly active crop diseases that were encouraged by repeated plantings of wheat and, particularly, corn.

In nineteenth-century Delaware, farm tenants were very numerous. In 1890, for example, at least 50 percent of the state's farm operators were either renters or sharecroppers, and this ranked Delaware sixth among all of the states in farm tenantry. Only Louisiana, Mississippi, Alabama, Georgia, and South Carolina had a higher percentage of farmers who were either renters or sharecroppers. Farm tenants continued to be numerous in Delaware throughout the first half of the twentieth century, with sharecroppers outnumbering renters. Although declining to approximately one-third of farm operators by the mid-1930s, tenants climbed back to 49 percent in 1941 only to fall to 10 percent by 1978.[4]

Predictably, because they had no long-term commitment to the land, renters and, more particularly, sharecroppers were often blamed for much of the state's soil exhaustion. As early as 1818, physician and agricultural reformer Dr. Samuel H. Black from Glasgow, New Castle County, complained of the loss of soil fertility and called for "the cultivation of land by owners rather than tenants." In addition to encouraging broad-ranging, soil-depleting agricultural practices by renters and sharecroppers, Delaware's farm tenant system was specifically blamed for the extensive and repeated plantings of corn, a practice that significantly drained nutrients from the soil. Traveling through Sussex County in 1838, Dover attorney and agricultural reformer William Huffington was told that most of the farmland was tenanted out to sharecroppers, and that the landlords "required their tenants to till half of every farm in corn each year." Worse yet, sharecroppers made no effort to counteract soil exhaustion by manuring fields. "Under such a system," Huffington observed, the "destruction of the soil is rapid and certain." In northwestern Sussex, fields that had been fertile only a few years earlier were now, by 1838, "either abandoned altogether or covered with meager corn stalks, many of which had not acquired sufficient strength to bear even an ear of corn." In 1941, more than a century later, Delaware's State Forester William S. Taber continued the chorus of criticism when he insisted that tenant farmers "are not greatly interested in making" long-term improvements "on land that they may possibly not occupy another year."[5]

By 1840, Delaware farmers had become so committed to overplanting corn that the state was annually producing seven times as much corn as wheat, which was the state's second most important crop. Most of the corn was fed to livestock, while most of the wheat was sent to mills and then to the regional and international flour market. No wonder Huffington found a general pattern of soil exhaustion as far north as Odessa in southern New Castle County.[6] What was happening in most of Delaware was common throughout much of the American South, particularly after the Civil War. The only difference was that landlords in the former Confederate states usually pressured their sharecroppers to overplant soil-depleting cotton rather than soil-depleting corn.

Slavery, according to its Delaware critics, also contributed to soil exhaustion. In the 1830s, a farm near St. Georges in central New Castle County was described as desperately poor because "a century and a half of slave labor had extracted its plant food." In 1847, a committee of Delaware's House of Representatives unsuccessfully recommended the abolition of slavery in the state, not because of what slavery did to African-Americans, but because of what slavery did to the agricultural economy and, by inference, to the soil that was the very basis of that economy. The House committee maintained that

> the careless, slovenly and unproductive husbandry visible in some parts of our state, undoubtedly results from the habit of depending on slave labor. It is no longer a question that slave labor impoverishes, while free labor enriches people.

In 1849, Wilmington abolitionist Thomas Garrett wrote that "had it not been for the curse of slavery, Delaware would this day have been a garden instead of a barren wilderness in the middle and lower part of it." In 1868, only three years after slavery was abolished in the state, a writer examining Delaware agriculture noted that in Sussex, where "the peculiar institution" had a more significant impact than

James Curtis Booth, Delaware's first state geologist (1837–1838), *courtesy of Morris Library, University of Delaware.*

in Kent and New Castle, slavery had been "a shackle on land improvement."[7] Just as with the owners of tenanted farms, Delaware's slave masters were driven by market forces to favor repeated plantings of corn or some other cash crop (monoculture) over the less soil-destructive practice of crop diversity.

As with other aspects of Delaware's environmental history, the level of soil exhaustion varied according to location. The lands that were initially most fertile, such as the section of piedmont in northwestern Delaware, the lands at the juncture of piedmont and Atlantic Coastal Plain, the rich soil west of Middletown, and the flat lands in much of the immediate Dover area, continued to be the most productive. The mixed farming prevalent in northern Delaware was probably less soil depleting than the tendency to emphasize one or, at most, two cash crops that was popular with many farmers further south. Understandably, almost all of the potential cropland of northern New Castle County was under cultivation by the mid-nineteenth century while, in the more soil-depleted counties of Kent and Sussex, the proportion of potential cropland under cultivation was only one-half and one-fourth respectively.[8]

In the face of an extended agricultural crisis during most of the first half of the nineteenth century, a reform movement that focused on returning competitive fertility to Delaware's exhausted farmlands slowly gained momentum. Early reform voices, such as Dr. Samuel H. Black's, received some attention in northern New Castle County but were generally ignored further south. By the mid-nineteenth century, however, proper crop rotation, field manuring, and the addition to the soil of lime and guano—the latter a sea bird excrement that was imported primarily from Peru—were being urged by reformers, and a small number of farmers in Kent and even a few in Sussex were beginning to listen. Because the most avid proponents were from northern Delaware, however, it is not surprising that it was the Agricultural Society of New Castle County that, in 1836, first urged the Delaware General Assembly to produce a geological survey for the entire state. In 1837, in part to deal with the problem of exhausted soil as well as to assess the economic potential of the state's rocks and minerals, the General Assembly hired James Curtis Booth of Philadelphia, an exceptionally well-trained geologist and chemist—he would later be elected president of the American Chemical Society—with paternal roots in New Castle County, to serve as state geologist and to spend two years conducting a survey of Delaware's "exploitable resources." One of Booth's more specific functions was to educate farmers about the potential of minerals to restore productivity to fields suffering from soil exhaustion.[9]

For two years (1837–38) Booth intermittently roamed the state and witnessed firsthand the desperate plight of many in Delaware's agricultural community. To survive economically, a number of impoverished farmers were wandering through the countryside, systematically cutting great swaths of bark from the limited number of mature surviving oaks, and then hauling the bark to commercial dealers for use in tanning and dyeing. A horrified Booth noted that the bark peeling caused "the indiscriminate massacre of thousands of black oak trees." Of course, the fundamental problem that had to be addressed was: "How could Delaware's worn-out soil be made more productive so that scenes such as these would not be repeated?" Booth was convinced that the application of livestock manure and lime—the latter provides calcium which is an essential plant nutrient in cell walls and membranes—would dramatically improve the land. But it would be a hard struggle to convince farmers to change their ways. Indeed, some thirty years after Booth traveled through Delaware's countryside and issued reports, the continued backwardness of agriculture in Sussex County was blamed on its people "remaining loyal to the patterns of farming followed by their ancestors." Even to the north, in Kent and southern New Castle, there existed a deep-rooted preference for the old farming ways over the changes dictated by the new, progressive farming theories.[10]

Joining Dr. Samuel Black and James Curtis Booth in proposing reforms was William Huffington who, in addition to practicing law in Dover, was the editor and the primary writer for the short-lived (1838–1839) but first-rate Dover publication, *The Delaware Register and Farmer's Magazine*. To offer hope for the future, Huffington reminded his readers that England's Kent and Sussex counties were also once characterized by their "unproductive farm lands," but now were "among the first agricultural districts" in all of Great Britain. Like Booth, he maintained that the conversion of exhausted farmland to productive fields depended on the extensive use of manure and lime. To Huffington, "nothing" was better for the rehabilitation of exhausted soil than manure. As for lime, all that was certain was that spreading it on fields produced "luxuriant crops" from land that, prior to application, was "comparatively of little value." Huffington also pointed out that applying wood ashes and composted organic materials was helpful in increasing soil fertility.[11]

Where to get manure and lime? Manure was difficult to collect because, as previously mentioned, many of the state's farm animals, particularly its cattle and pigs, were still running free across the rural landscape in 1838. Huffington's solution was that Delaware farmers should construct stables, sheds and barnyards for livestock in order "to collect manure." Even if farmers could be convinced to limit the movement of livestock so that manure could be collected and spread on fields, where could they get lime? Fortunately there were a number of unrefined deposits in Delaware and in nearby areas of Pennsylvania. In the extreme north, quarries in Delaware's piedmont as well as those in Pennsylvania's upper Schuylkill Valley provided blocks of limestone that were then fired in kilns or crushed and ground by machines to produce field dressing that could be shipped and applied. Oyster shells, including those from ancient Native American middens, were burned to produce a lime-intensive residue. At first, however, the most economic source of lime for some farmers was marl, which was made up of layers of dirt or sand imbedded with ancient sea shells. Marl was found concentrated in dozens of sites in Delaware by the perambulating state geologist, James Curtis Booth.

By far the largest concentration was the "green marl" that had been uncovered during the excavation of the Chesapeake and Delaware Canal in the 1820s.[12] This proved particularly fortuitous for St. Georges and Red Lion Hundreds of New Castle County because both hundreds were suffering from soil exhaustion by the early nineteenth century, and both were in close proximity to the canal's supply of marl.

The excavation of the eastern part of the Chesapeake and Delaware Canal was subcontracted to, among others, five Delaware farmers who owned land adjoining the canal route. While the eastern portion was being dug by legions of laborers (1824–27), William Hurlock, one of the five subcontractors, loaded up his wagons at the end of each day with some of the strange looking soil that had been excavated to produce the canal's deep-cut ravine. Marl, as the strangely colored soil was called, was spread, along with manure, on a thirty-acre field he owned, which was then plowed and planted with corn. The results were dramatic enough to convince nearby farmers to spread marl on their own fields where, in some cases, yields were increased tenfold. In 1833, when President Andrew Jackson made his important tour through the northeastern states, he was towed through the Chesapeake and Delaware Canal. While seated on the upper deck of a horse-pulled barge, Jackson observed the Delaware landscape and acknowledged the shouts and waves of his supporters who were standing on shore. When his attention was drawn to a marl-fertilized field of corn next to the canal bank, President Jackson reportedly exclaimed: "By the eternal, it is equal to any that I have seen in Tennessee!"[13]

The discovery of marl deposits, just when Delaware's agricultural economy was facing such a bleak future, was seen by some as providential. William Huffington was convinced that it was all part of God's plan to hide

> for countless ages within the bowels of the earth, a substance likely to prove so essential in restoring the fertility of an exhausted soil, to be discovered at a time when most needed for that purpose.

Because of its bulk, however, marl was costly

to haul for any distance and then to spread on fields via a shovel and wagon. As one New Castle County farmer pointed out, marl was only seven to eight percent lime, "so we had to haul a hundred loads of marl to get seven loads of lime." All of this made the use of marl particularly costly for Kent and Sussex farmers, because they were a considerable distance from the state's most important marl beds, which were in New Castle County. Fortunately, a purer form of lime soon became available and it "became a magic word."

Traveling by train to Middletown a few years after the Civil War, Henry T. Williams found "for a long distance before reaching the station, the track is lined on either side with piles of lime." Indeed, thousands of bushels of pure lime were "used in all parts of Delaware and Maryland...producing splendid results." Make no mistake about it, the use of lime was crucial according to Williams because "the small quantity of barnyard manure produced render the use of artificial fertilizer necessary."[14]

Lime, in both its diluted (marl) and purer forms, was joined by guano from South America, phosphate rock from the Carolinas, crushed horseshoe crabs and oyster shells from Delaware Bay, and pulverized animal bones from everywhere, as just some of the ingredients in the field dressings applied to a number of Delaware farms by the mid-nineteenth century. Also important to the restoration of soil productivity was the growing practice of planting cover crops, which was often called "green manuring," to both anchor and provide nitrogen to the exhausted land. All of this was part of an agricultural revolution that swept through parts of Delaware in the mid-nineteenth century and returned soil fertility to a number of farms, particularly in the northern part of the state. In fact, by adding lime and other minerals to the soil and by cultivating clover as a cover crop, New Castle County's once soil-depleted Red Lion Hundred became "the garden spot of its day in Delaware." To meet the growing demand for specific types of enriching field dressings, at least fifty fertilizer manufacturers were operating in the state by 1880. With such a demand for commercial fertilizer, dishonest dealers set up shop and sold falsely labeled products to unsuspecting farmers. The General Assembly responded, in 1880, by appointing the chemistry professor at Delaware College (University of Delaware) as the state chemist with responsibility for analyzing, for honesty, samples of all fertilizers sold in the state.[15]

Past experience made it increasingly clear to many Delaware farmers that planting legumes, such as clover, also increased soil fertility. But it was not until 1886 that there was a scientific explantion of how the actions of bacteria, located on legume root nodules, added nitrogen to the soil. From 1870 until 1930, red clover became the favored legume of Delaware farmers and, according to agricultural historian Joanne Passmore, it was called the "mortgage lifter," because its presence reduced dependence on particularly expensive fertilizers. The techniques used in the development of nitrogen-based explosives during World War I were subsequently adapted to produce synthetic nitrogen fertilizers. By the late 1920s, nitrogen fertilizers were widely available but not universally used. As late as the 1930s, it was not uncommon to encounter downstate farmers who refused to apply chemical fertilizers of any sort to their exhausted fields. Of course, the deep-rooted reluctance of downstate farmers, particularly those in Sussex, to adopt progressive farming methods is not surprising in view of historic patterns going back to the late eighteenth and early nineteenth centuries. When much of northern Delaware was swept up in the agricultural revolution of the mid-nineteenth century, a Pennsylvania newspaper article pointed out, in 1845, that "as yet the spirit of improvement does not extend to Sussex."[16]

But there was always hope that Sussex and other, less progressive, areas of the state would embrace the new soil-preserving measures. In 1857, Governor Peter F. Causey was euphoric that "neglected fields are now reinvigorated by the use of these new fertilizers and made to yield abundant crops." As the nineteenth century came to an end, however, it was increasingly evident that a large percentage of Delaware farmers, particularly in Sussex, either would not or could not afford to purchase enough fertilizer to make a significant difference in the productivity of their fields. In view of the competition from corn and wheat-producing states further west, the future of Delaware agriculture was again in doubt. In 1895, Governor Robert J. Reynolds admitted that Delaware's farm lands were generally less productive than those of certain other states. This meant that, at best, Delaware soil "yields only a moderate return to the toil of the husbandmen."[17]

Crop Diversification

In the face of increasingly stiff competition in growing corn and wheat, a large number of Delaware farmers turned, for economic salvation, to raising truck crops and producing dairy products to feed the nearby, rapidly-growing urban markets of Wilmington, Philadelphia and New York City. In northern New Castle County after the Civil War, the growing demand for butter caused farmers to increase the size of their dairy herds on the rolling land north of Newark and Wilmington. By the early 1920s, the introduction of pasteurization made possible the production of a considerable volume of milk for nearby urban markets, and milk replaced butter as the major source of income for Delaware's dairy farmers. Further south, commercial peach orchards first appeared in the Delaware City area of New Castle County in the 1830s, and then spread gradually southward through Kent into much of Sussex County by the 1870s. In the last three decades of the nineteenth century, tomatoes, peas, beans, melons, sweet and white potatoes, and, eventually, strawberries were grown from central New Castle to the southern border of Sussex. Theoretically at least, this more diversified crop farming promised to be less soil-depleting than the repeated plantings of corn that marked, in particular, agriculture in ante-bellum Delaware.

Neither dairy nor truck farming was profitable anywhere in the state without an efficient transportation system to speed fruits, vegetables, cheese, and milk to markets in an inexpensive, safe, timely, and dependable manner, and then to deliver fertilizers and other needed agricultural supplies to farmers. The Delaware peach industry, for example, was initially dependent on the steamboat, which meant that raising peaches was profitable along the Chesapeake and Delaware Canal and along other navigable waterways. With the advent of the railroad, landlocked areas of the state also became prime peach-growing centers.

The advent of the railroad also made it possible to grow other fruits and vegetables profitably on soil where corn and wheat once had been the primary crops. Indeed, the presence of the railroad made it possible for Sussex to raise more strawberries than any other county in the nation at the beginning of the twentieth century.[18]

Over the long run, however, heavy dependence on certain truck crops such as peaches and strawberries proved to be problematic. In the 1840s, a virus-produced leaf blight called "the yellows" attacked peach orchards in the Delaware City area of New Castle County and then slowly worked its way south until it reached Sussex in the 1880s. Because of the blight, Delaware's great peach boom came to an end by the 1890s. As for strawberries, by the late 1920s they were suffering from a fungus disease that raised questions about the wisdom of continued heavy dependence on them as an important cash crop. Driven by the pressing need to develop a more dependable alternative cash crop, downstate farmers gradually turned to broilers—chickens specifically bred and raised to produce meat rather than eggs. The remarkable growth of the broiler industry brought new life to Delaware's lagging agricultural economy and, eventually, to the played-out soil that would grow the food necessary to support this rapidly expanding broiler industry.[19]

Nature and the Rise of the Poultry Industry

Two very different acts of nature helped set the stage for the early development of the broiler industry, a new concept in poultry farming that originated in 1923, near the tiny hamlet of Ocean View in southeastern Sussex. From there it expanded nation-wide and changed the eating habits of America. From a diet that once depended overwhelmingly on beef and pork, by 1997, Americans had switched to chicken as their favorite meat.

Prior to the 1920s, chickens were primarily bred and raised to produce eggs, not meat. In fact, most hens and roosters did not carry very much flesh; nor was their meat all that tender to eat. In the mid-1920s, the flocks of Delaware's commercial egg producers were decimated by the outbreak of a disease called "range paralysis." Hardest hit was Baltimore Hundred in southeastern Sussex, forcing the area's experienced chicken farmers to seriously examine other alternatives to commercial egg production. In one of those acts of nature that would help launch the broiler industry, range paralysis did not strike chickens until they were fairly mature. Because chickens raised for meat were usually sold off at a relatively young age, they generally escaped the disease. By contrast, laying hens were far more susceptible to the

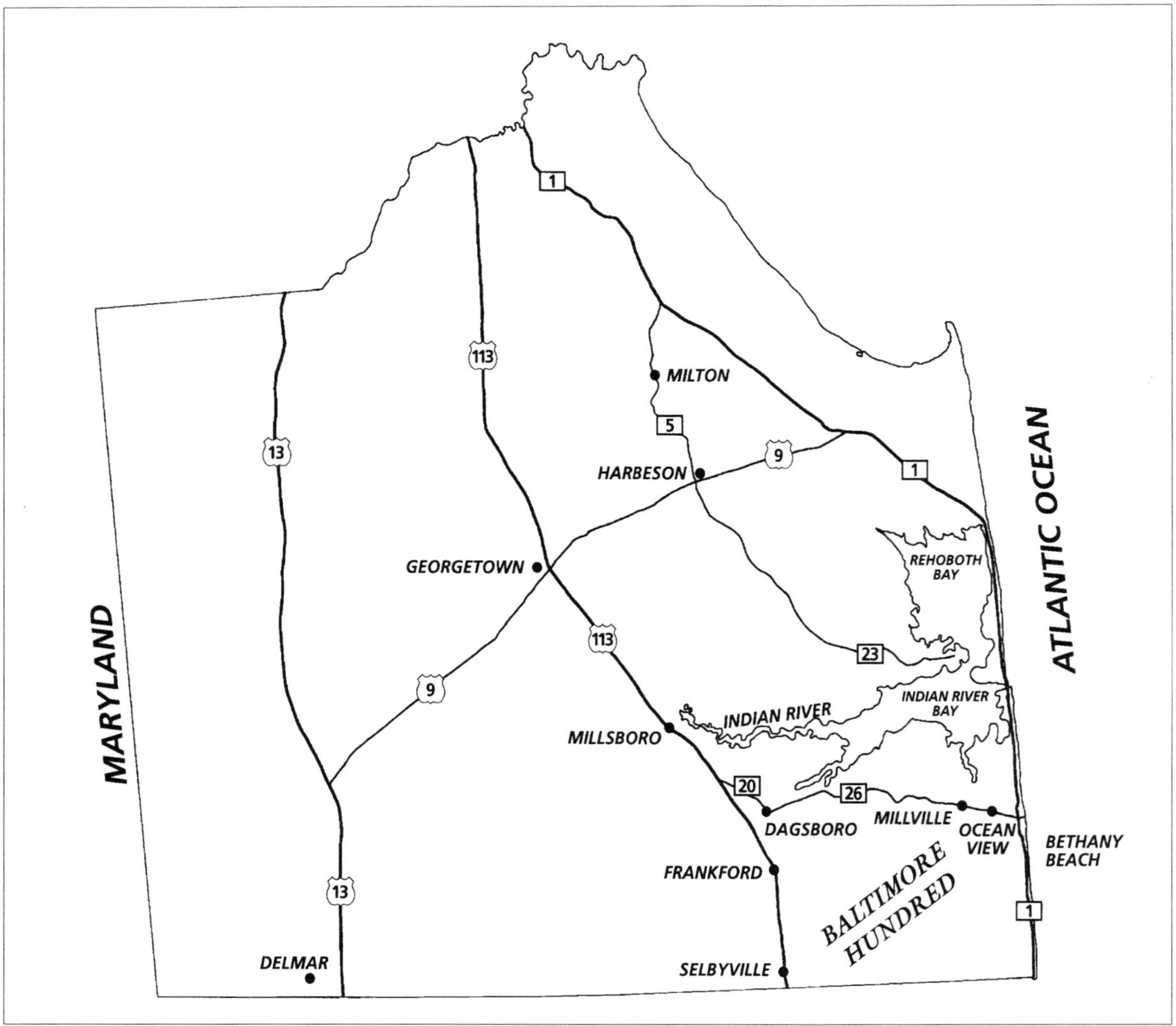

Figure 6: Map of Sussex County showing Ocean View and Indian River Bay, birthplace of America's commercial broiler industry, *map drawn by Jason L. Burleson-Gibson.*

disease because they needed to be kept alive until fully mature in order to produce eggs.[20]

The second act of nature to spur the development of the broiler industry was the dramatic drop in Indian River Bay's salinity level, which decimated the bay's shellfish population. Prior to the twentieth century, freshwater flowing into the bay from Indian River to the west was neutralized by the Atlantic Ocean's saltwater that entered the bay through an inlet to the east. Thus, a delicate balance in water salinity was maintained in Indian River Bay that provided a good natural environment for crabs, clams and oysters. Although shellfish were less numerous in the bay in the early twentieth century than in colonial times, enough remained to provide a number of local farmers, who had also become part-time watermen, with a much-needed second income. In 1908, for example, at least 300 watermen were busy during the warm months, catching crabs from the bay and from the mouth of the Indian River. A large percentage of them were farmers from

Baltimore Hundred. As late as 1924, during high season, as many as 60,000 crabs were taken daily from Indian River Bay, along with a considerable number of clams and oysters. There are records from as far back as 1800 of the inlet from the Atlantic to Indian River Bay shifting its location during major storms. Beginning in 1911, however, the inlet from the Atlantic Ocean became even less stable and closed up entirely for brief periods. This led to a significant drop in the bay's salinity, particularly between 1925 and 1928. Large quantities of crabs, clams and oysters died, forcing the frustrated farmer–watermen of Baltimore Hundred to look for a new source of income.[21]

In 1923, in the midst of an increasingly gloomy economic setting, Cecile Steele, an Ocean View housewife, received 500 chicks from a Dagsboro hatchery. This amounted to five times the number she ordered to replace losses in her laying flock. Rather than send the excess back, she kept the surplus chicks until they were just beginning to mature and then sold them to a local buyer who, in turn, shipped them north to be served in restaurants. This transaction turned a handsome profit for Mrs. Steele, and the next year she ordered 1,000 chicks. Her neighbors took notice. The rest, as they say, is history. By 1928, there were 500 broiler growers in Baltimore Hundred and the adjacent areas of Sussex County and Maryland's Eastern Shore. Within five more years, specifically bred, meat-type chickens were being raised throughout most of Sussex and in much of Kent County.[22]

Just as New Castle County farmers turned to dairy products and truck crops for economic salvation, downstate farmers turned to meat-type chickens. Broiler production offered economic hope and sustenance to thousands of rural people who, otherwise, would have been forced to leave the land in much the same manner as downstate farmers during the economic hard times of the first half of the nineteenth century.

The broiler industry also largely determined the type of field crops grown in Kent and Sussex. With the broiler business expanding dramatically—Delaware's annual production jumped from two million broilers in 1928 to sixty million in 1944—there was money to be made in planting and harvesting the key ingredients in chicken feed. The amount of feed consumed by Delaware's broilers in 1943, for example, would have filled a freight train approximately 110 miles long. Three years earlier a miller pointed out that corn and soybeans made up 50 percent and 25 percent, respectively, of the content of the feed eaten by broilers. But how could downstate Delaware's depleted fields that produced only twenty to thirty bushels of corn per acre in the late 1920s be made more productive? This was a crucial issue, because the only alternative to increased local production was the importation of large amounts of corn and soybeans from a considerable distance. That would have been so expensive that, eventually, it would have rendered Delaware's broiler farms unable to compete with newer broiler farms in other states.[23]

Despite a growing commitment to truck crops during the second half of the nineteenth century, corn continued to be Delaware's most important field crop. Almost four million bushels were produced in 1880 and more than four and one-half million bushels in 1900. Corn was always needed to feed beasts of burden—oxen, mules and horses—as well as other farm animals raised to produce meat, milk, wool, and eggs. Nevertheless, by the 1920s and 1930s the internal combustion engine was rendering beasts of burden obsolete, because they were no longer needed to pull passenger vehicles, freight wagons, plows, harrows, cultivators, and reapers. With the numbers of oxen, mules and horses now dramatically declining, Delaware's corn production should also have declined. Reflecting the surging numbers of Delaware broilers, however, the demand for corn increased dramatically and production kept pace. From approximately four and one-half million bushels in 1900, Delaware's annual corn production reached almost seven and one-half million bushels in 1950 and approximately twenty-five million bushels in 2000.[24]

The real surprise was not the increasing demand for corn, which simply paralleled the stunning increase in the number of broilers, but the fact that, for a while at least, downstate Delaware's exhausted soil could produce most of the corn necessary to feed its chickens and therefore avoid depending too heavily on expensive imported corn. It all was possible, in part, because of the successful use of hybrid corn and expensive chemical fertilizers after 1940. But at least as important was human adaptation and ingenuity in downstate Delaware that, as early as the late 1920s, began turning accumulating mounds of poultry manure into a positive use that insured the future growth of the corn-dependent broiler industry. In the 1920s and 1930s, when rural Delaware faced economic hard times, chemical fertilizers were often too expensive to purchase. But chicken manure, it was observed, had a remarkable impact on soil productivity and cost nothing but sweat to move it from the chicken house to the corn field.[25]

By the mid-1930s, chicken manure was commonly spread on downstate crop land and the results were amazing. In 1935, fields in Sussex which produced only about twenty-five bushels of corn per acre, tripled production after the application of this organic fertilizer. Right up until the end of the twentieth century, the liberal use of chicken manure continued to rejuvenate depleted downstate soil, and helped it to produce annually most of the millions of bushels of corn and soybeans necessary for feeding the increasing flocks of broilers. In turn, the growing number of chickens ate and digested corn and soybeans, and then produced more manure which was spread as field dressing to insure bumper crops of corn and soybeans.

Moreover, using chicken manure also cut back on annual farm expenses. By 1982, some Delaware farmers were saving as much as $5,000 a year by opting for chicken manure over commercial chemical fertilizer. It seemed a wonderfully balanced, circular system of production! Not until the late twentieth century did praise for the agricultural miracle produced by the widespread use of chicken manure begin to be tempered by the discovery of an unintended consequence—a nutrient-rich runoff that became a public health concern as it seeped into downstate Delaware's surface and underground water systems.[26]

More Ditching

With the growing demand for locally grown corn and, later, soybeans to feed the rapidly increasing numbers of broilers, Delaware's farmers continued to support the centuries-old practice of ditching the land to make old fields more accessible and to turn swamps and marshes into productive new fields. As a 1959 study pointed out, "drainage is an acute problem in the state—with the exception of piedmont Delaware—and particularly in Sussex, where most of the land has an elevation above sea level of less than 50 feet." In fact, poor drainage in Sussex had "been a major factor" in the average annual "abandonment" of "slightly less than 2,000 acres" of farm fields in that county from 1890 to 1959. One of the problems for most of that period was that a number of downstate farmers could not or would not pay their individual tax-ditch company's assessments to cover the cost of maintaining old ditches and digging new ones. By the onset of the Great Depression, these private ditch companies were so lacking in cash that, during the 1930s and early 1940s, most ditch maintenance and digging work in Delaware was done by the federal government's Civilian Conservation Corps (CCC), or by Delaware's Highway Department.[27]

Drainage ditches dug on flat land—the Atlantic Coastal Plain accounts for 94 percent of Delaware's land area—produce a very gentle water flow that makes silting and clogging a constant problem and continuous ditch maintenance a necessity. Not only were many Delaware ditch company members unable, or unwilling on principle, to pay their ditch taxes; the collection process was further complicated by the practice of assessing individual land owners rather than their property. When some owners subsequently sold their land, the ditch tax was not automatically transferred to the new owners or to their land. In this very murky legal climate, where ditch taxes were often difficult or impossible to collect, already hard-pressed tax ditch companies simply collapsed, leaving their clogged-up ditches untended. Over the years, much of central and southern Delaware's farmland gradually reverted to marsh land. In 1941, convinced that digging new ditches and maintaining old ones was crucial to the well-being of Delaware's agricultural economy, the General Assembly guaranteed the collection of ditch taxes by simply taking that responsibility away from the tax-ditch companies and their treasurers, and positing it in the hands of the county tax collector.[28]

Into the early 1970s, general opinion continued to support the digging of drainage ditches across Delaware, whether under public or private auspices. Indeed, throughout much of the twentieth century the same people who generally labeled the clearing of forests a public offense, saw nothing wrong with converting swamps and marshes into productive farmland. The reason was that the words "swamp" and "marsh" continued to conjure up some long-held negative images, including breeding grounds for bothersome insects, dangerous diseases and disgusting odors. Moreover, even most conservationists were convinced that the efficiency of nature's economy could be improved by human intervention. This meant that swamps and marshes could reach their natural potential only after they were transformed, through ditching, into productive crop land. Forests, on the other hand, could be made more productive by human intervention without changing their essential physical nature. It would take a combination of scientific research on the significant ecological role of a "swamp" or a "marsh" and the substitution of the more positive label "wetland" before there would be an attitude change concerning the value of draining water from Delaware's rural landscape.[29]

Technology played an increasing role in ditching Delaware's rural landscape as the twentieth century progressed. Although shovel-wielding laborers continued to do much of the work, by 1911, steam-powered dredges were in use, and by 1930, dynamite explosions were blasting through particularly stubborn physical barriers. The cost of this new technology was often beyond the financial resources of tax-ditch companies and individual farmers, and a number of needed ditching projects were not attempted. As a result, in many areas of Delaware, particularly in Sussex, land drainage continued to be a serious problem into the 1940s.

In 1942, a national program supported by the United States Soil Conservation Service came to the aid of Delaware farmers by cooperating with the state government in setting up three soil conservation districts that matched Delaware's three counties. State loans made to each soil conservation district were then used to purchase ditch-digging equipment and to hire equipment operators. The federal Soil Conservation Service provided specialists with ditch-engineering skills and other agricultural knowledge to consult with individual farmers and with their tax-ditch companies. When the tax-ditch companies or individual farmers decided to clear old ditches or dig new ones, they then paid the county-wide soil conservation district for the use of its digging equipment and the services of its equipment operators.[30]

By the early 1940s, the advantages of using tractors and trucks on Delaware farms were clear. But these new motorized farm machines were heavier and therefore more difficult to use in wet and soggy fields than the horses and mules that they replaced. Not surprisingly, the greater tendency of tractors and trucks to get bogged down was an added incentive to dig and maintain drainage ditches. Because of the county's high water table, the Sussex Soil Conservation District was initially busier than the two conservation districts further north.

To Sussex Countians, the other two soil-conservation districts "were more talk than action," and were therefore labeled the "Kent and New Castle 'Conversation' Districts." By 1948, however, the two other districts were more proactive in draining their respective landscapes, and by that year the three soil conservation districts were responsible for most of the ditch work in Delaware. There were, of course, some who protested that government had gone too far in getting involved in the

Preparing the charge.

Dynamiting a ditch—
This series of photographs, taken during the early 1930s near the western Kent County town of Marydel, shows the process of dynamiting a drainage ditch to remove a sandbar. Taken at the junction of the Tappahanna Ditch and Harrington Prong, *photographs courtesy of the Delaware Public Archives.*

Setting off the explosion.

Measuring the water level in the reopened ditch.

physical treatment of privately held property. The way the county soil conservation districts made ditch-digging machinery available may have reminded some critics of the machine-tractor stations established to service collective farms in the Soviet Union. But one county agricultural agent from Sussex recalls that "the protesters certainly didn't take time to talk to local farmers whose economic survival depended on ditching for access to their fields after heavy rains." Moreover, as one state official pointed out in 1989, without ditching to draw off the standing water left by spring rains, the late twentieth-century practice of double-cropping during a calendar year would be impossible.[31]

Dealing with Mosquitoes

Draining water from wetlands was also crucial to the future growth of the summer-resort industry along Delaware's coast, of which Rehoboth Beach was the most important anchor. Nearby Lewes was a potential anchor, but until the late 1960s it was not particularly attractive to summer visitors because of unpleasant odors emitted by local fish-processing plants. Back in 1859, one of the few signs of human habitation in Rehoboth Beach and its surrounding countryside was a rustic hunting shack. All of this changed in the 1870s, when Methodists established a camp meeting site in what is now downtown Rehoboth Beach. By the first decade of the twentieth century, this growing seasonal community by the ocean had become a popular secular summer watering hole. A 1911 publication noted that Rehoboth Beach was becoming one "of the most popular summer resorts in the South." Nevertheless, hordes of mosquitoes from adjacent wetlands limited the appeal and growth potential of summer tourism in Rehoboth Beach and all along coastal Delaware. Well into the 1930s, mosquito bites were so bothersome that Rehoboth Beach housewives were often forced to wrap newspapers around their ankles for protection while working in their gardens. In 1933, at the request of the state, the federal government's Civilian Conservation Corps committed itself to rescuing the state's coastal resorts from the insect scourge. For the next four years, as many as a thousand young men from different parts of the United States annually waged all-out war with their shovels against Delaware's salt-marsh mosquitoes. The strategy was simple: deprive the bothersome insects of the still water needed for reproduction. The tactic was equally clear: dig a grid of drainage ditches across the salt marshes to draw off still water with individual ditches measuring 20 inches wide, twenty inches deep, and spaced about 150 feet apart. A total of 44,000 acres of Delaware's salt marsh was ditched in this manner.[32]

After the CCC workers dug more than 2,000 miles of ditches across Delaware's salt marshes, "the great annual mosquito annoyance" temporarily diminished. Because management of where the ditches would be dug was in the hands of engineers rather than biologists, the primary goal had been to design a perfect grid with no deviations so that the ditch-diggers could be employed most efficiently. Last on the engineers' agenda was a real concern for preserving the natural marsh setting and the biodiversity that it supported. The marsh-scape was dramatically altered, and the sloughs and ponds that were once homes and resting places for wading birds, ducks, and muskrats dried up. Moreover, it soon became evident that the CCC-dug ditches had to be cleaned out every five to seven years to avoid silting and clogging. Up until the mid-1950s, state budget restrictions dictated that only a fraction of the mosquito ditches could be maintained, with those nearest Sussex County's coastal resort towns getting preference. Kent was so neglected that when Delaware's Mosquito Commission finally got around to cleaning out some of that county's saltwater ditches in 1957, they were in such bad shape that the "rehabilitation program was tanta-mount to digging new ditches." Under-standably intent on supporting its summer-resort industry and sensitive to increasing complaints from year-round residents, in 1947 the state turned to spraying its salt marshes with the new miracle insecticide, DDT. The initial results were spectacular.[33]

By the early 1950s, however, saltwater mosquitoes were showing a resistance to DDT. Before the Federal Environmental Protection Agency (EPA) outlawed DDT in 1974, Dela-ware recognized its long-term ineffectiveness and abandoned it for benzene hexachloride, which was used for a couple of years, only to have mosquitoes soon develop a resistance to it as well. In short, by the mid-1950s, the state had a serious mosquito problem, with Kent County facing a particularly high concentration of the insects. On Loockerman Street, in the heart of Dover, the measured mosquito landing rate on humans was at least 100 per minute during some summer days. Obviously, with the only available relief depending on state appro-priations for ditching and spraying, the mosquito would become a serious political issue.[34]

Historically, the state had set aside annually only a minimal amount of money for mosquito control, which, of course, angered many of Kent County's long-suffering residents. Into the fray charged the *Delaware State News*, Dover's daily newspaper. It demanded in editorials that the state make a much larger commitment to diminishing the mosquito population. By contrast, Wilmington's two daily papers, initially at least, argued that more spraying and ditching of the marshlands would harm the muskrat and fish populations. Obviously, the *News Journal* headquarters in the heart of Wilmington had only minor mosquito problems.

Crowds jammed the chambers of both houses of the state legislature in Dover as the mosquito issue was debated in late June of 1955 and 1956. In 1956, when a legislator rose to challenge increased spraying and ditching because of what it would do to the fish and muskrats, an irate spectator yelled out, "what's more important, muskrats or people?" Supporters of more appropriations allegedly removed screens from the windows of Legislative Hall. In this era before air con-ditioning, during hot summer days and evenings the windows were often thrown open to cool off both legislative chambers. As planned, so the story goes, swarms of mosquitoes accosted the legislators while they contemplated anti-mosquito legislation. Begin-ning in 1956, the legislature increased its funding for the war on mosquitoes. By the 1970s, mosquitoes were no longer a major problem in central Kent County. As the state's mosquito control program moved into the year 2000, spraying with less threatening chemicals than DDT continued, but was less central to the overall control plan. In the state's Open Marsh Management Program, which was instituted in the 1970s and continued throughout the rest of the twentieth century, marshland was reconfigured to allow killifish and other small piscine species access to eat mosquito larvae. In short, whenever possible, biological controls were being substituted for the application of chemicals.[35]

Trees

From its seventeenth-century beginnings as a European colony, when the Dutch led the way in draining swamps and marshes, to the mid-twentieth century, when the destruction of swamps and marshes was still legal and even encouraged by state and national governments, Delaware probably lost more than half of its wetlands. During that same period, Delaware lost almost two-thirds of its forested lands.[36] The sylvan landscape that remained was different from the tree-covered Delaware encountered by the first European settlers. While the fragmented wood lots and forests of the mid-twentieth century contained most of the same species of trees that graced Delaware's seventeenth-century landscape, the mix and age of the trees, and their pattern of geographic dispersal had been greatly altered by the impact of more than three centuries of settlement by Europeans and their descendants.

No place in rural Delaware was a better example of the extraordinary impact that humans had on the land, when they drained marshes and cut down trees, than Cypress Swamp. In the seventeenth century, Cypress Swamp covered approximately 50,000 acres with about half in southern Sussex and the rest across the state line in northern Worcester County, Maryland. In addition, most of the land beyond the swamp that lay between the future towns of Laurel to the west and Ocean View to the east, bore a striking natural resemblance to the bogs and forests of Cypress Swamp. Across this broad swath of southern Sussex landscape, a "forest culture" emerged that reflected the needs and concerns of a people economically dependent on the sylvan resources of the swamp and its peripheral forests. By the end of the twentieth century, however, despite the fact that it remained "the largest contiguous forest on the entire Delmarva Peninsula," Cypress Swamp had shrunk to only 12,400 acres, or to approximately one-fourth of its original size. A 1999 study of what remained of Cypress Swamp pointed out that "today we have a much different landscape" than the one that existed only 200 years ago.[37]

Beginning in the decade just before the American Revolution, so many bald cypress trees were cut down that, by 1850, "there was hardly a decent-sized" one still standing in Cypress Swamp. By that date, timbermen had long since turned to harvesting the swamp's Atlantic white cedar, whose shingles and boards shared with the bald cypress a strong resistance to rotting. But just before the Civil War it was discovered that only a few feet under the swamp's bog water and peat moss lay gigantic bald cypress logs, felled by natural causes hundreds and even thousands of years ago. Astonishingly, the logs had been well preserved over that time by the minerals contained in the beer-colored water and the peaty muck below. Releasing the huge fallen sentinels from their gelatinous tombs, however, was no easy task. Winter turned out to be the best time to raise them because the water level was high and this meant that once the soggy peat could be scrapped away and the logs freed from the muck, they could be floated to the surface. From there oxen dragged them off to be hand-rendered into thirty inch-long shingles that might wear out with time but would "never decay." Well into the twentieth century, recovering these ancient cypress logs continued to be an important economic activity in the Cypress Swamp area.[38]

The Great Cypress Swamp— This view of cypress trees in the waters of Trussum Pond is typical of those which once grew in profusion throughout the cypress swamp in the days when the swamp was covered with water during much of the year, *photograph by Richard B. Carter, 1976.*

Prior to the colonial era, forest fires routinely played a vital role in the natural rejuvenation of Cypress Swamp's flora. The relatively high water level, however, moderated the intensity of these conflagrations so that, although they might scorch a number of standing trees and low-growing shrubs, they did not threaten the long-term survival of the area's most dominant species of trees and shrubs. But with the arrival of Europeans and their strong inclination to cut down trees, drain wetlands, cultivate fields, and set fire to the local countryside annually, conditions became increasingly ripe for conflagrations so intense and so destructive that they would transform the very nature of the swamp. The crucial precondition to the many highly destructive fires that followed was the human-induced alteration in Cypress Swamp's hydrology that left both wetlands and forest floors "much drier than previously."[39]

By the 1920s, ditching had so dried out a large part of what was left of the swamp that its peat base was highly combustible. Similar drainage projects also made other forested areas of Delaware into potential tinder boxes. During the summer of 1930, an abnormally long drought left the state with a rain deficit of about sixteen inches. The extreme dryness of the landscape combined with careless burning of debris, numerous lightning strikes, and an explosion in a moonshine still in the heart of Cypress Swamp to cause fires that, "for a time, threatened to wipe out the major portion of Delaware's forests." As it was, almost 30,000 timbered acres across Delaware went up in smoke with about 90 percent of them in Sussex. At the very center of this series of infernos was the fire in Cypress Swamp which, over an eight-month period, consumed what little was left of the pre-colonial forest, as well as most of the swamp's peat base and almost all of its remaining cypress logs. Although the flames of 1930 were the most destructive, over the next seven years dried peat continued to smolder and burn underground in the Cypress Swamp area. Periodically, flames would burst through the surface as destructive sheets of fire, "even in the middle of cornfields that had no trees upon them." [40]

The scorching flames of 1930 and the next few years combined with at least 150 years of systematic dehydration through ditching to create a postfire botanical greening of Cypress Swamp that bore little resemblance to the tree cover worn in the distant past. Lost to the fires were almost all of the remnant populations of bald cypress and white cedar that had somehow survived more than 150 years of heavy timbering. Moreover, the swamp was now so dry that bald cypress could no longer naturally regenerate and white cedar could naturally regenerate in only a few locations. It would take thousands of years of natural action to replace the layers of peat lost in the great fire. Moving almost immediately into this ecological vacuum, the quick-growing loblolly pine brought new life to the charred and desolate landscape. By 1945, stands of young loblolly provided the swamp's primary forest canopy. But pines in monotypic stands, such as those in Cypress Swamp, are more susceptible to insect infestation than when they are individually scattered through mixed stands of other trees. In 1993–1994, a major attack by the southern pine beetle killed thousands of loblolly pines and opened up vast areas of the swamp to the expansion of red maple and sweet gum. By the end of the twentieth century, red maple and sweet gum joined with the loblolly pine as the dominant trees in Cypress Swamp's mixed forest. What a contrast with Hickory Point Swamp, located about forty miles to the southwest in Maryland's lower Eastern Shore, where bald cypress and white cedar continued to grow at the end of the twentieth century! The reason was quite simple: although heavily timbered through the years, Hickory Point Swamp "has not been ditched and drained like" Cypress Swamp.[41]

During the twentieth century, timbering continued throughout Delaware, but was concentrated in the southern part because that was where most of the trees were located. In 1906, for example, only 15 percent of Delaware's piedmont—11,000 acres—was forested. From Wilmington south to Felton in southwestern Kent, 20 percent—95,000 acres—was forested. However, from Felton south to Sussex County's southern border, 35 percent of the land—245,000 acres—was forested. Almost no virgin forests and very few old-growth forests remained, and, as pointed out earlier, almost every tree in the state was less than sixty years of age. Fortunately for the timber business, some trees—especially pines which could be cut for timber as early as twenty-five years of age—grew rapidly and were marketable in a comparatively short time.[42]

In 1906, trees in New Castle County and northern Kent were almost exclusively hardwoods. In southern Kent, hardwoods still predominated, but pines were rapidly increasing in both number and commercial value. Mispillion Hundred, in southwest Kent, was the most heavily forested area in the entire county. In Sussex, where pines had become more numerous than hardwoods, the heaviest sylvan concentrations were in the heart of the county along a broad swath that stretched southward down the center of the county from the county seat at Georgetown to Cypress Swamp. In general, because tree-covered land in the state's extreme north was restricted to steep slopes and other, difficult-to-cultivate acres, the resulting woodlots were twenty acres or less. Woodlots and forests gradually became larger as one moved south until reaching Sussex where many forests covered more than one hundred acres.[43]

Until well after World War II, Delaware's timbermen generally ignored the admonitions of the state forester to plant and nurture their own trees in the same manner as a farmer growing corn stalks. Instead, timbermen chose to continue the traditional practice of culling timber from the naturally generated trees of the forest. By contrast, by 1945, planted rows of loblolly pines occupied many acres of the adjacent Maryland Eastern Shore counties of Somerset and Worcester. Although Delaware was a wood-importing state at the beginning of the twentieth century, it did cut more timber in proportion to its size than Pennsylvania or New Jersey. By 1900, there were seventy-six saw mills in the state and 88 percent of the rough lumber they produced came from pine. Because pine trees matured much faster than hardwoods, they formed a disproportionate number of trees annually cut for timber. Delaware was also home to six paper pulp mills, and they favored wood from the tulip poplar tree. As wood of any nature was becoming increasingly expensive, "there was now little waste material left in the woods after lumbering." Tree tops, once left behind after logs were dragged off to a mill, were now often cut up into cordwood and sold in towns and cities as fuel for stoves and fireplaces.[44]

Forest fires were the greatest potential threat to the Delaware timber industry. Surprisingly, fires had not been a serious concern during the late nineteenth and early twentieth centuries. In 1907, a study of the state's forests reported that there "have been no extensive forest fires in Delaware, as the wooded areas are so cut up by cleared land and roads as to make large fires impossible." In truth, the acres of cleared agricultural land in the state reached their zenith in 1880, but then began a gradual decline that continued through most of the twentieth century. Conversely, because most abandoned farm fields reverted to woods, the state's forested acreage, particularly in Sussex, actually increased over the same period. In short, the physical conditions that inhibited large sylvan conflagrations at the turn of the century, gradually eroded in the decades that followed. By the late 1920s, progressively more costly forest fires broke out in southern Delaware, reaching a climax with the terribly destructive conflagrations of 1930. From that date forward, it was clear that fire was the most serious threat to the state's timber resources. In 1931, the Delaware General Assembly responded to the now very clear danger by funding a series of fire look-out towers in southern Delaware and a cadre of forest fire fighters. Subsequent statistics indicate that the new fire prevention and fire fighting measures immediately paid off, and forest fire destruction in Delaware declined dramatically in the early 1930s.[45]

Success in limiting forest fires and their resulting damage was partly reflected in the production of lumber by the state's saw mills. In 1909, Delaware's lumbering industry had its most productive year in the entire twentieth century when 55 million board feet were milled. Its least productive year was 1933, when only 1.5 million board feet were milled. The unusually low figure for 1933 reflected the collapsing market for lumber during the Great Depression, but was also caused by the dramatically diminished amount of timber available after the catastrophic fires of 1930. Unhindered by subsequent major conflagrations, Delaware's forests began the long road back to recovery in the 1930s. The subsequent demands of a war-time economy, the related expansion of the poultry industry and its need for wooden chicken houses, and the willingness to over-cut the state's supply of timber, boosted Delaware's annual lumber production to 25 million board feet by 1943. William S. Taber, the state forester, claimed that at least part of this increase in production was caused by the survival of more trees due to the expansion in the state government's "efforts in forest fire prevention and control since 1931."[46]

As Delaware moved through the first half of the twentieth century, the gradual reduction in farmland led to a small but gradual increase in forested land. Somewhere between 25 to 30 percent of the state was forested in 1906; by 1945 that figure had increased to about 35 percent. The forests of both dates were composed, overwhelmingly, of trees under sixty years of age. A 1945 study noted that although pines were on the increase, approximately 58 percent of the state's trees remained hardwoods, 38 percent were softwoods and the rest—three to four percent—were mixed.[47]

A Preview of the Future

During the first half of the twentieth century, most of Delaware's residents became increasingly aware that their natural environment had taken a severe buffeting from the actions of farmers, timbermen, industrialists, municipal governments, hunters, and fishermen. To the untrained eye of the average citizen, however, the state's water, land, air, flora and fauna seemed healthy enough to serve the essential needs of the state's residents well into the future. For a few, however, the decline in the quality of the natural environment was now so evident that it was very threatening on a personal level. One example was James T. Edmandson, a farmer who lived in northwestern New Castle County, about two miles east of Newark. In 1924, in the Delaware Court of Chancery, Edmandson sued the National Vulcanized Fibre Company (NVF) of Newark for contaminating the waters of White Clay Creek. The subsequent testimony by Edmandson and by six corroborating witnesses was both chilling and sensational.[48]

White Clay Creek rises in Pennsylvania, flows south to Newark and then bends east to join the Christina River. In the early 1920s, Edmandson and several of his corroborating witnesses operated farms that abutted White Clay Creek. A few miles upstream from Edmandson's farm lay Newark, the home of a number of mills and factories including NVF. Edmandson and other witnesses reported that above Newark and its cluster of mills and factories, the waters of White Clay Creek were clear and full of fish. But by the time the creek flowed downstream, past Edmandson's land, the water was sometimes red, sometimes black, and sometimes grayish in color. In fact, when Edmandson attempted to sell his farm in 1923, the potential buyer saw the color of the water in the creek, and "wouldn't have it."[49]

The color of White Clay Creek was symptomatic of other, much more serious problems. When farm animals waded or drank in the creek, they lost their hair and even their lives. One farmer living downstream from Edmandson testified that

> last year we shipped two car loads of dairy cows from the west...and they were unloaded in the morning and they had been without water for a period, and were turned into one of the meadows that White Clay Creek borders on, and before night we lost five.

Several farmers testified that their "ducks and geese would fall over dead after drinking from the creek." Dogs avoided the creek because, if they wadded into the water, they developed "burns" between their foot pads. Children who swam in the creek complained of burns on their skin. Problems arising from contact by both farm animals and children with White Clay Creek had been going on for eight to nine years, but "had been getting worse and worse ever since." Fish and muskrats probably served as good barometers for the health of the creek's waters. In 1914, catfish, perch, sunfish, suckers, bass, mullets and muskrats were plentiful in White Clay Creek or along its banks, downstream from Newark. By 1924, all the fish and the muskrats were gone.[50]

Both the complainant (Edmandson) and the defendant (NVF) agreed that the waters of White Clay Creek below Newark were highly polluted, but they parted company on who was the culprit and what was the nature of the pollutants. A chemist, testifying at the request of Edmandson, reported that an analysis of the liquids discharged into White Clay Creek by NVF disclosed the presence of zinc chloride, which "has a destructive effect on animal and plant life." Moreover, "in sufficiently strong solution, it would destroy flesh."

NVF's attorney responded by blaming much of the creek's pollution on other mills in the Newark area and on the municipal government of Newark, which was dumping the town's untreated sewage into White Clay Creek. The fibre company's attorney also blamed the City of Wilmington for much of the pollution in the lower reaches of White Clay Creek. His argument was quite simple: Wilmington dumped some of its sewage into the Christina and that sewage was backed up by high tides into that part of White Clay Creek located downstream from Newark.[51]

The case of Edmandson v. National Vulcanized Fibre Company was dismissed on June 4, 1924 with no objection from Edmandson.[52] Why this ending? The court records are not helpful in answering this question, leaving us to surmise that there was an out-of-court settlement or that Edmandson's attorney simply advised his client that he would not win before Delaware's Court of Chancery. Nevertheless, the claim that a company discharged a lethal chemical compound into White Clay Creek set the stage for a deepening awareness, particularly after 1945, of the power of unregulated chemistry to threaten the health of Delaware's citizens by polluting its natural environment.

CHAPTER SEVEN:

Crisis and the Rise of Environmentalism, 1945–2000

"Farmers didn't do anything wrong as far as they knew ... the contamination [of surface and groundwater] is a result of accepted practices."

IN 2002, *News Journal* COLUMNIST AL MASCITTI SUGGESTED THAT BECAUSE few states "can top us in the emission of toxic materials," Delaware should celebrate this distinction by having the state tourism office "support 'Toxic Tours' of chemical factories, oil refineries, power plants and Superfund sites." After all, since "people love tours of mob hit sites and celebrity tragedy spots," they will flock with enthusiasm to the many environmental disaster locations that riddle Delaware's landscape. Mascitti maintained that the list of toxic tourist stops stretches almost the length of the state from "the General Chemical Plant on the Pennsylvania Line in Claymont to the Indian River Power Plant in Sussex County's vacation zone." In fact, there are so many that "it's too much for one day so guests will fill plenty of hotel beds."[1]

Mascitti's sarcasm aside, the years 1945 to 2000 witnessed an unprecedented chemical-led assault on the quality of Delaware's air, land and water. That era also featured an unprecedented construction boom that gobbled up wood lots and prime farmland alike, threatening to suburbanize much of the state. The extraordinary alteration of the state's landscape and the release into its air, land and water of toxic chemicals and organic pollutants by industry, agriculture, municipalities, and even homeowners led to the rise of an in-state environmental movement—and to efforts by both state and federal governments to protect Delaware's natural resources.

Some Delawareans, however, remained skeptical about the seriousness of the state's environmental crisis and strongly resisted the efforts of both state and national governments to restrict the right of industry, farms, municipalities, and individuals to deface and debase Delaware's natural environment. Giving some intellectual credibility to the forces leading this resistance was what J. R. McNeill calls, "The most important idea of the twentieth century." Baldly put, it is the firm belief that almost any economic growth that produces jobs and profits represents progress regardless of how that growth might otherwise impact society and the natural world. Implicit in this perspective is the conviction that government restrictions only impede economic growth and, therefore, must be avoided because they stand in the way of progress. Also supporting resistance to government involvement in the protection of the natural environment was the long-held belief, going all the way back to English precedent in the Old World and strengthened by the colonial Chesapeake Bay experience in the New World, that individual freedom was a paramount right that was threatened by the expansion of government power. With individualism's American roots strongest in the colonial Chesapeake world and its adjoining hinterland, it isn't surprising that in nearby Delaware faith in both individualism and in the operation of unrestricted market forces remained strong.[2]

Unrestrained economic growth was useful in the seventeenth century when the first Europeans settled in a Delaware that featured extraordinary stretches of mature forests, clean water and air, abundant wildlife, and a healthy ozone layer. By 1950, this faith in unrestrained economic growth was reinforced by the grim memories of the Great Depression, which resulted in so much unemployment, and recollections of the material shortages during World War II, when the most coveted goods were rationed. Only unfettered economic growth, it was argued, could meet the need for more Delaware jobs and the production of more goods so that the hard times of the past would not be repeated. Overlooked was the fact that unrestrained economic growth could create extraordinary problems when vigorously pursued, particularly within the environmentally sensitive and heavily populated context of late-twentieth-century Delaware. Nevertheless this idea, which had outlived its practical usefulness to society, continued to provide, along with the exaltation of individualism, much of the ideological basis for resisting government steps to protect the environment.

Threats to the Environment

Of course, Delaware had an advantage over many states in preserving its natural heritage because it had no commercial oil wells and therefore no leaks, spills, and fires usually connected with drilling for petroleum.

Moreover, except for a limited amount of iron extraction in the eighteenth and early nineteenth centuries, and some quarry, sand and gravel pits in the nineteenth and twentieth centuries, no significant commercial mining industry developed in Delaware to deface the surface of the land and to release dangerous chemical substances. The lack of commercial coal deposits, for example, enabled Delaware to avoid the legacy of an industry that elsewhere left behind barren landscapes, fouled streams, huge and ugly slag heaps, and dirty air. But in so many other ways, Delaware was not so fortunate.

First, the extraordinary concentration of industrial plants and oil refineries in northern and central New Castle County or further upstream along the Delaware River in Pennsylvania and New Jersey, seriously degraded Delaware's natural environment. By the late twentieth century, well over one hundred manufacturing plants and eight oil refineries—only one of the refineries was in Delaware—were located in the First State or in nearby areas of Pennsylvania and New Jersey, and were venting, dumping, and flushing toxic chemicals that accumulated in the bodies of fish and threatened human health. To the west, as far as the upper Mississippi Valley, hundreds of other industrial plants poured their impurities into the air. Prevailing westerly winds then pushed the fouled air across the Midwest to Delaware and convincingly demonstrated that being downwind from air polluters was every bit as unpleasant as being downstream from water polluters.[3]

Second, Delaware's dramatic population increase from 1945 to 2000, which was stimulated by jobs provided by the remarkable business and industrial expansion of that era, placed pressure on the carrying capacity of the state's air, land, and water. From 1940 to 2000, the number of Delawareans increased from 266,505 to 783,600, causing the state's population density for the same period to jump from 136 to 401 per square mile. (Compare that with an estimated .15 to 1.4 people per square mile in 1610 and 103 in 1910.) By 2000, the state had a population density that was five times the national average and now ranked seventh among all states in people per square mile. This meant that Delaware, thanks particularly to New Castle County, where the population density was fifteen times the national average, on a square-mile basis used more fresh water and produced far more garbage, human waste, and polluted air from automobiles and home heaters than did forty-three other states. In 1989, state wildlife biologist Tony Florio wrote that quality of life in a region was dependent on the carrying capacity of that region's natural environment. He added that once population grew too large for the region's environmental carrying capacity, then "the quality of life declines." Florio was convinced that Delaware was "about to exceed that level."[4]

Third, when compared to its neighbors, Delaware devoted an unusually large percentage of its land surface to agriculture. In 1969, for example, 43.6 percent of the state was in farmland, which was slightly above the national average and considerably higher than the 34.4 percent for Maryland, 24.1 percent for Pennsylvania and 17.7 percent for New Jersey. That same year, by contrast, Delaware was slightly below the national average in percentage of land that was forested and significantly below the percentage of tree-covered land in Maryland, Pennsylvania, and New Jersey. Because so much of its land was committed to agriculture, extensive loads of poultry manure and chemical fertilizer were dumped on Delaware's soil, along with an array of toxic chemicals from the liberal use of pesticides and fungicides. This caused unsafe levels of nitrogen, phosphorus and other chemicals to wash into surface water and into much of the state's ground water and, eventually, into its inland bays. If Delaware had a higher percentage of woodland and a correspondingly lower percentage of farmland—woodland generally received far fewer pounds of fertilizer and fewer gallons of pesticides and fungicides per acre—the environmental threat to its surface water, groundwater, and inland bays would have been significantly less. In sum, when the high proportion of surface area devoted to farming was combined with Delaware's high density of polluting industries and polluting people, the result was a recipe for environmental trouble.[5]

Trouble

During the late twentieth century it became increasingly evident that rapid economic development and related population growth were causing serious environmental problems all over the state. In northern New Castle County in 1994, the Delaware Nature Society's Lorraine Fleming lamented that now, every place in Delaware's natural world bore "man's imprint." A year later, author Jennifer Ackerman of Lewes focused on some of the consequences of that imprint. Noting the correlation between a healthy environment and a healthy population, she pointed out that "Delawareans are known to be among the least healthy people in the nation" and that "we canaries would seem to be saying that this is a troubled mine indeed." In fact, as it entered the last decade of the twentieth century, Delaware contained some of the worst toxic dumps in the United States, possessed air so dirty that it repeatedly fell below ozone standards set by the Environmental Protection Agency (EPA) of the federal government, and possessed ponds, streams and bays full of harmful chemicals. Delaware was "a troubled mine indeed."[6]

Dirty Air

Despite the fact that it takes considerably less to pollute the air than it does to pollute the oceans, fouled air was practically non-existent or, in the immediate post-Civil War era, restricted to the general vicinity of industrial Wilmington or to the factory towns, villages and hamlets numerous in piedmont Delaware but more thinly scattered across the landscape of central and southern Delaware. While the annual practice of burning sections of the rural countryside did pollute the air, it was for only a few days each year. By the twentieth century, however, air pollution had grown exponentially and spread across broad swaths of Delaware and large sections of nearby states. Meanwhile, business and political leaders continued to believe that spewing smokestacks were indicators of progress and prosperity.

By the late twentieth century, hundreds of thousands of cars and trucks joined factories, power plants, a refinery, and homes in pouring impurities into the state's air. In 1908, there were only 431 registered motor vehicles in Delaware; by 1950, there were approximately 108,000; and by 2000, the figure had sky-rocketed to more than 718,000. The extraordinary number of state-registered cars and trucks venting fumes into Delaware's air was joined by many more motorized vehicles from out-of-state that traveled the I-95 corridor every day in New Castle County or jammed up roads to Sussex County's beaches on weekends. The surge in numbers of cars and trucks on Delaware's highways was particularly harmful to air purity. In fact, the worldwide pattern of automobile exhaust replacing smokestacks and chimneys by 1990 as the largest single source of air pollution was also probably true in Delaware.[7]

In the decades after World War II, oil and gas replaced coal for much of Delaware's power generation and domestic heating needs. The result was a modest but only temporary reduction in air pollution. Soon, filling the pollution gap were lead-containing fumes from the exhaust pipes of the increasing numbers of motor vehicles on state highways. Eventually, lead fell to the ground and was then tracked into homes where crawling infants collected the toxic mineral on their hands and then ingested it. Lead could, among other things, arrest intellectual and neurological development and cause hearing loss in children. Since the early 1920s, lead had been added to automobile gasoline because it prevented engine knock and improved engine performance. Despite growing concern that leaded gasoline was a significant health threat, General Motors and DuPont, who had jointly developed tetraethyl lead, working in concert with other petrochemical and auto industry interests, were able to prevent the federal government from regulating the level of lead additives in motor vehicle fuel until the 1970s. Finally, lead additives were completely ended by the switch to catalytic converters in 1986. Subsequently, lead concentrations in the air were dramatically reduced and that, in turn, significantly lowered harmful lead traces in the blood of children both nationally and state-wide.[8]

The combustion of coal, petroleum, and other fossil fuels has saturated Delaware's air with other toxic substances that, in addition to lead, are generally considered to threaten human health. Ground-level ozone, a primary cause of smog, is a major photochemical air pollutant. It is produced when emissions from automobiles, industrial and electric utilities and other sources provide chemicals that react with sunlight. Ground-level ozone is "bad ozone" and must not be confused with "good ozone" which occurs much higher in the atmosphere—six to thirty miles up—and protects us against skin cancers and temperature increases by blocking and filtering the rays of the sun. Elevated levels of ground (bad) ozone can cause serious breathing problems for humans. Even plants are in jeopardy. In 1993, for example, on a downstate farm north of Milford, "watermelons showed extreme symptoms of ozone pollution." Writing in the *Delaware Medical Journal* in 1995, Dr. Alma T. Jolly noted that Delaware had not been in compliance with the national ozone standard since it was first established in 1970. She pointed out that, "ambient air levels of ozone in Delaware present a health risk to the public," and that the situation calls for "reduced use of automobiles and improved automobile pollution control." Dr. Jolly then insisted that her fellow Delaware physicians "must become advocates for the reduction of ozone levels in the ambient air."[9]

One of the problems with calling for significant reduction in air pollution was that, although health consequences of breathing contaminated air were assumed to be serious, the particulars were difficult to measure in a precise manner. It was generally conceded by health officials across the nation, for example, that certain impurities in the air probably increased the incidence of human cancer. The scarcity of specific numerical statistics to support these assertions, however, encouraged the Delaware State Chamber of Commerce, in 1991, to challenge state officials to directly link industrial pollution of Delaware's air and water with cancer.[10]

Most Delawareans continued to breath impure air into the twenty-first century. A 1996 study by the federal Environmental Protection Agency found the air in northern Delaware—where the vast majority of Delawareans lived—contained some of America's "highest concentrations of toxic pollutants." An alarmed Deborah Heaton, then conservation chairwoman for the Sierra Club's Delaware chapter, exclaimed that we are all inhaling "a soup of chemicals." The EPA study measured the presence of thirty-three airborne substances that were considered toxic. For twenty-one of these, including benzene, mercury, vinyl chloride and arsenic compounds, New Castle County ranked in the top five percent of the nation.[11]

Although automobile exhausts from both in- and out-of-state were now the most important single source of Delaware's air pollution, the state's many industrial sites continued to vent an impressive array of toxic substances into the atmosphere. In 1990, for example, seventy-five Delaware companies released 12,455,947 pounds of toxic chemicals, and more than half of that amount was vented into the air. The greatest concentration, of course, was over northern Delaware, with emissions from such sites as the Chrysler plant in Newark, the power plant in Edgemoor, and the Star Enterprise (later renamed Motiva and then Valero) oil refinery in Delaware City leading the way.

Even in southern Delaware, however, where the air tended to be cleaner, the Indian River power plant east of Millsboro and the DuPont nylon plant in Seaford were state leaders in spewing impurities into the air. In fact, towards the end of the twentieth century, the Indian River plant established itself as the state's "largest overall polluter, releasing more than 2.6 million pounds of air pollution alone in 2000." So pervasive was air pollution throughout the state that a report covering 1999–2001 by the American Lung

Association gave all three Delaware counties a grade of 'F' for air quality.[12]

Toxins on the land

While Delaware's polluted air rained down toxins such as lead, mercury and arsenic compounds on the ground below, more concentrated and more direct deposits of pollutants were unloaded from trucks and cars at specific sites across the state. At first, these sites were primarily open garbage dumps, often located in old sand and gravel pits. By the 1960s, however, landfills were judged to be more sanitary and began to replace the traditional open garbage dumps. Because the old town dumps usually did not charge for disposing garbage and the more up-to-date landfills did, there was a noticeable increase in the amount of garbage strewn along the shoulders of Delaware's less traveled roads by residents unwilling to pay dumping fees. Faced with a rapid increase in the amount of trash in the fifty-five years after World War II, public officials around the state wondered, "Where are we going to put it all?"

The volume of solid waste increased in Delaware for at least two reasons: a rapidly growing population, and increasing material consumption by each Delawarean. Although exact figures are not available for Delaware, trash production for the entire nation increased from 3.2 pounds to 4.4 pounds per person just between the years 1970 and 1993.[13] There is no reason to doubt that Delawareans more than matched those figures. Indeed, by the year 2000, so much garbage and rubbish had been dumped at the Sussex County Landfill that it was now the highest point of land in southern Delaware.

The dramatic per-person increase in the production of garbage from the end of World War II to the end of the century is generally blamed on a significant growth in personal income. In earlier periods, Delawareans "could not afford to discard clothes or household furnishings until they were worn out." But now the habit of saving and trying to repair the old was rendered obsolete because, increasingly, Delawareans could afford to buy new.

The state's per capita income, unadjusted for inflation, rose from $2,571 in 1948 to $31,255 in 2000. Even when these numbers are adjusted for inflation, the rise in per capita income was astounding. From 1980 to 2000, for example, the inflation-adjusted figure for the state's per capita income climbed an astonishing 67 percent. Although there were a few sluggish years, particularly in the 1960s and 1970s, Delaware's economy overall was so robust from 1945 to 2000 that it provided its residents with a level of prosperity only dreamed of by previous generations. Indeed, because Delaware led most of the other states in per capita income—by 2000 it ranked eleventh among the states—the average state consumer probably purchased more goods and then threw away more trash than the average American. And, needless to say, the United States led the world in the amount of trash produced and discarded.

Tracking the volume of garbage disposed of in Delaware from 1981 to 1995, N. C. Vasuki, chief executive officer of the Delaware Solid Waste Authority, observed that "the amount of solid waste kept going up when the economy was booming," and that hard economic times led to a declining amount of garbage. Vasuki noted, for example, that in the very strong economic year of 1989, a new record of 860,000 tons of garbage was deposited in refuse sites around the state, only to be followed by a three-year decline in volume which reflected a softening economy.[14]

Into Delaware's earlier open dumps and, later, into landfills went household, yard, business, and industrial refuse of no perceived value. Included were spoiled or excess food, all forms of paper products, and discarded wood and metal objects. The open dumps swarmed with rats, emitted offensive odors, and often darkened the sky with smoke from fires set to reduce the piles of refuse. The landfills, because they covered refuse with layers of dirt, diminished the number of rodents and the constancy of odors, and generally ended the use of fire to diminish the piles of garbage. But some of the materials thrown into the landfills as well as the older open dumps did not decompose readily and were even toxic.

By 1967, garbage dumps and landfills were beginning to be blamed for imperiling Delaware's drinking water. In addition to dumps and landfills, the state was laced with private junkyards full of paper, rags, and scrap metal objects that included thousands of old and rusting motor vehicles. By 1970, there were more than 300 junkyards in Delaware with 90 in New Castle County, 70 in Kent, and 150 in Sussex. Unlike dumps and landfills, most objects deposited in junkyards had at least some theoretical monetary value. But on the whole, the state's junkyards were seen as "forms of aesthetic and visual blight, offensive to man's sense of beauty and order."[15]

In 1990, Joseph Reardon, a former DuPont chemist and Democratic Party leader, told an audience at the University of Delaware that, because of materials thrown into some of Delaware's garbage dumps and landfills, the First State housed some of the nation's most toxic dumping sites. He said that, worse yet, Delaware and Nebraska were the only states without a plan to clean up toxic waste. By 1991, 240 locations in Delaware were identified "as potential hazardous waste sites." By 1996, twenty-one of them were considered so potentially dangerous that they made the EPA's list of Superfund sites. That, according to political scientist William Boyer, ranked Delaware "third among the states in the number of Superfund sites per square mile." What is surprising is that in-state dumps and landfills—some of which gave Delaware this high but dubious ranking—received less than half of the toxic wastes produced by Delaware's homes, businesses, and industries. The rest was exported for disposal to other states such as Pennsylvania, New Jersey, Louisiana, and Ohio.[16]

A dumping ground that caused particular alarm was the Army Creek Landfill, which occupied an old sand and gravel pit a few miles southwest of the town of New Castle. Established by New Castle County and operated as a county public landfill from 1960 to 1968, it was initially called the Black Cat Landfill after a night club on nearby U.S. Route 13, and then, when the night club went out of business, it became the Llangollen Landfill, after a nearby housing development. When trouble developed concerning the landfill, however, Llangollen Estates demanded a name change for the landfill because it wished to disassociate itself from this now infamous nearby source of water pollution. In 1976, the landfill was named for Army Creek, a stream that marks the landfill's southeast border before eventually emptying into the Delaware River. In addition to garbage and other waste left by the general public, seventeen polluting industries left drums of chemicals on the sixty-acre site before it was closed in 1968. John Bryson, then head of water and air pollution control for the state, remembers being told that what Army Creek really needed was for a bulldozer to push fill dirt over it. Bryson responded that, "using a bulldozer on the Army Creek Landfill, with its buried chemical drums, would most likely put the first man with a bulldozer on the moon."

Just across the creek, southeast of the Army Creek Landfill, was a second source of serious water pollution, the Delaware Sand and Gravel Landfill. Immediately after the closing of the Army Creek site in 1968, drums of industrial chemicals were deposited in the Delaware Sand and Gravel Landfill until about 1970, when a bulldozer struck some of the drums and ignited a serious fire. The fire led to the closing of Delaware Sand and Gravel Landfill to liquid chemicals, but by this time each landfill possessed so much toxic material that, when combined with leaching rainwater, they threatened the purity of the underlying Potomac Aquifer, which supplied most of the drinking water in the area.[17]

Mary Reni lived on the edge of Llangollen Estates, just across the stream bed from the Army Creek Landfill and west of the Delaware Sand and Gravel Landfill. Unlike most residents of Langollen Estates, who depended on the nearby wells sunk by the Artesian Water Company for their drinking water, Mary Reni had her own well. Perhaps because she was a heavy smoker, she was not aware of changes in the taste of her drinking water. In January, 1972, a nephew came to live with her and immediately noticed the peculiar flavor and odor of the water coming out of Mary Reni's faucets. Experts were called in to assess the situation and noted, after sinking exploratory wells, that an expanding plume of contaminated water in the Potomac Aquifer underneath the Army Creek Landfill was the villain, and it was moving southward towards Llangollen Estates. Three homes with their own individual wells were already affected, and the Artesian Water Company's wells, which supplied Llangollen Estates and were located just south and southeast of that suburban development, would soon be in jeopardy.

Subsequent leakage was also discovered from the nearby Delaware Sand and Gravel Landfill, which further threatened the quality of the water taken from the Potomac Aquifer. What was happening was that the Artesian Water Company, in pumping a significant quantity of water from the Potomac Aquifer in the area of Llangollen Estates, was gradually drawing the contaminated water plume, that had originated beneath the two landfills, towards its own wells.[18]

What to do to stop the spread of this contaminated water? Hydrologists and geologists convinced the Artesian Water Company to cut back on the water it pumped. In addition, recovery wells were sunk near the two polluting landfills to draw off contaminated water. This had the effect of decreasing the water level in the aquifer near the two landfills and thus pulled the contaminated water back north towards its source along Army Creek. By November, 1973, ten contaminate recovery wells were in operation. Indeed, Army Creek represented the first water contaminant recovery operation in the United States. It also became the most important of the five cases that spurred the United States Congress to pass the Resource Conservation Recovery Act in 1976, which provided "cradle to grave" responsibility by industry for creating and then dispersing hazardous waste. In 1980, the toxic Army Creek Landfill was declared a Superfund site. In subsequent years a heavy-duty half-inch plastic cover, which was supposed to keep rain water from first mingling with buried waste and then seeping into the Potomac Aquifer below, was laid over the sixty-acre site and was then covered with eighteen inches of fill and six inches of top soil. Grass and wild flowers were planted on the surface, and a few low spots were even turned into wetlands.[19]

By 1994, it all looked "like a giant meadow, featuring black-eyed susans, clover, rye and barley." The only reminders of what had once been the Army Creek Landfill were the eighty-five plastic pipes jutting out of the ground to vent methane gas that was building up in the old trash buried beneath the plastic liner. Governor Tom Carper characterized the transformation as turning an "ugly duckling into a swan." But the Governor's statement was a bit premature. On November 1, 2000, the Artesian Water Company had to close some of its nearby wells because "a toxic element, which could have come from the old Army Creek Landfill, was in the water."

The Army Creek and the Delaware Sand and Gravel cases demonstrate that groundwater contamination was and remains a very serious problem that, once discovered, is extremely difficult and costly to contain. For the one-third of Delawareans who live in the piedmont this may not seem particularly alarming because they draw most of their drinking water from White Clay Creek, Red Clay Creek or the Brandywine. The remaining two-thirds of Delaware's population, however, depend entirely on drinking water pumped to the surface from aquifers via corporate, community or privately-owned wells. To them the possibility of groundwater contamination continues to be a matter of great concern.[20]

Problematic Farming Practices

During World War II, there was enormous pressure in Delaware to produce more food. This, in turn, led formerly marginal farmland to be cleared and cultivated. With the end of the war, however, the international market for agricultural products dramatically declined, which led to the abandonment of much of the state's marginal farmland. At the same time, advances in agricultural science in combination with human ingenuity resulted in significant changes in the way farms operated. Farmers now blanketed their fields with more chemicals because, in the very competitive world of post 1945 agricultural markets, increasing the productivity of each acre was crucial. In the standard wisdom of the day, if a little field dressing was good, more was better. The strong desire for greater per-acre productivity may have also been encouraged by the fact that the amount of land devoted to farming in Delaware was shrinking at an alarming rate. From 1945 to 2000, agricultural acreage in the state decreased by 63 percent.[21]

It was also discovered by farmers that commercial efficiency was best served by abandoning the production of a variety of market crops and annually specializing in one or two crops. As a result, in the fifty-five years after World War II, Delaware farmers turned increasingly to raising corn or soybeans, or a combination thereof, and moved away from the more generalized agricultural patterns that had been the norm since the mid-nineteenth century. Any agricultural patterns even close to monoculture, however, invited serious problems. The first was an old one: repeated plantings of the same crop depleted the soil of specific nutrients. By contrast, the greater variety of crops that characterized Delaware farms prior to World War II was less destructive of these same nutrients. To reduce soil erosion and nutrient depletion caused by plowing, many downstate farmers turned to no-till (no turning of the soil) corn and soybean production. By 1983, Delaware led the nation in the percentage of its cropland devoted to no-till farming.[22] The application of additional chemicals, however, was needed in no-till farming to kill the weeds that were normally contained by plowing and cultivating.

Delaware's farmers also found that productivity in their relatively small fields, bounded by hedgerows, could be improved by the consolidation of smaller fields into fewer, but far larger tracts. The latter were more amenable to plowing, cultivating and harvesting by the larger-sized tractors and other new mechanized equipment that was increasingly in vogue. Not since the introduction of plow-pulling oxen in the mid-eighteenth century was the shape of Delaware's crop-producing fields so dramatically altered. A prime casualty of making a few large fields out of many small ones was the hedgerow, which had historically anchored topsoil during heavy rains and housed a vital bio-community that included small game animals and game birds. Economic progress simply dictated that the flood-ameliorating and wild-life hosting hedgerow be cut down and the flora and fauna it once sustained be swept away in the concerted drive in post World War II Delaware to create a more commercially efficient rural landscape.

Encouraged by fertilizer companies and producers of pesticides and fungicides, it was only natural that Delaware farmers responded to the competitive challenges of post-World War II markets by pouring on even more chemicals. It was also only natural that when the first alarms were raised about the growing environmental damage caused by these practices, both Delaware's farmers and agri-businessmen were reluctant to listen. In the 1970s and 1980s, for example, University of

Delaware Professor William F. Ritter was strongly criticized by some members of the agricultural community for research that indicated that the quality of southern Delaware's surface and ground water was adversely affected by local farming practices. According to Ritter, the state's agricultural leaders "did not believe my results and tried to discredit me whenever possible." Fortunately for Ritter he was tenured. This fact, despite attempts from outside the university community to get him fired, allowed him to continue his research.[23]

The problem with animal and chemical fertilizers is that, when applied to fields in large enough quantities, they both become water pollutants. While estimates vary, there seems to be a general consensus that more than half of the organic and inorganic fertilizers that are applied to crop lands end up in ponds, creeks, rivers, and bays. In addition, polluting elements in fertilizers make their way through the soil into the aquifers that provide drinking water for most of the state. Prior to World War II, nutrient loading of surface and ground water could be blamed primarily on untreated sewage from Delaware's municipalities. After World War II, gradual municipal reforms in sewage treatment reduced nutrient loading from this source, only to have the dramatic increase in the application of poultry manure and chemical fertilizers push nutrient loading beyond all previous levels. By the 1990s, for example, approximately two billion pounds of poultry manure were annually spread on Delaware fields. Pound for pound, this substance contained more nitrogen and phosphorous than any other type of animal waste. Among the first, of course, to sound the alarm had been Professor Ritter, who discovered in the 1970s that unusually high concentrations of nitrates were seeping into groundwater and from there into Delaware's three inland bays. He later wrote that, unequivocally, poultry manure was a prime cause of nitrate overload in the inland bays.

Subsequently, very high concentrations of phosphates were discovered in farmland soil samples by J. Thomas Sims, also of the University of Delaware. During the 1990s, Sims' research alerted Delawareans, not only to the danger posed to bodies of water by phosphates, but also to the fact that even after the cessation of phosphorous application to fields through chicken manure and chemical fertilizers, excess phosphates would remain in the soil for decades. Building on Sims' findings, a 1997 report by the University of Delaware concluded that 92 percent of the fields in Sussex County were so saturated with phosphates that they should not receive any more for years to come.[24]

Spreading the excess nitrates and phosphates from fields to surface waterways were networks of agricultural drainage ditches. Rainwater run-off, after absorbing nitrogen and phosphorus from crop lands, was drawn into this vast system of miniature canals that bypassed filtering wetlands and woods, and emptied into Delaware's ponds, streams, and bays. Although the negative impact on Delaware's marine bodies will be dealt with in greater detail later in this chapter, suffice it to say that by the 1990s the *News Journal*, the state's leading daily paper, routinely ran stories about the dramatic decline in Delaware's marine environment caused by fertilizer runoff, particularly chicken manure. In 1998, for example, the *News Journal* reported that scientists and federal regulators viewed the runoff from chicken manure as the "most serious environmental threat facing the state." Moreover, because Delaware's "poultry industry has assaulted the state's water and waterways so long, experts say it would take decades to reverse the damage even if all chicken manure vanished immediately."[25]

But what else, asked the poultry farmers, can we do with the chicken manure other than spread it on our fields? To haul it away would be so costly that it would render Delaware's poultry farmers unable to compete with their counterparts from other states. Poultry farmer Richard S. Cordrey of Millsboro, former president pro tempore of the Delaware State Senate, pointed out in 1998 that a halt to spreading chicken manure on fields would create a "ski slope here in winter, because there is a lot of manure out there. And it is going to start piling up."

That same year, William Satterfield, executive director of Delmarva Poultry Inc., a trade association representing broiler growers and processors, denied that there was any proven link between chicken manure and chemical fertilizer spread on corn and soybean fields, and the serious pollution of Delaware's waterways and aquifers. Satterfield complained that "some politicians have made the connection, and some things have happened that we do not think were based on facts." But the facts seemed increasingly clear to a growing number of Delawareans. The environmental health of much of their state was at risk because of long-established fertilization practices followed by many, if not most, of Delaware's farmers. The irony of unexpected consequences was not lost on state hydrologist A. Scott Andres. In 1992, he pointed out that "farmers didn't do anything wrong as far as they knew...the contamination [of groundwater and surface water] is a result of accepted practices."[26]

Suburbs and the environment

In Delaware, as well as across the nation, suburbanization was the dominant demographic theme from 1945 to 2000. Behind the move of thousands of Delawareans to the state's suburbs was a pent-up demand for new housing—very few new homes were constructed during the Depression and World War II—and the availability of cheap land, inexpensive cars, low-priced gas and very reasonable interest rates. Moreover, some public officials saw inexpensive tract housing as a vital Cold War weapon that would help inoculate Americans from the seditious ideas of radical leftists. Quite simply, expanded home ownership would contradict the charges that Delaware and the rest of America were far from being egalitarian societies. Moreover, the construction and sale of new homes in new suburbs would stimulate and help sustain an expanding, mass-consumption economy. Finally, because most of Delaware is relatively flat and most of its soil is well drained, the state was ideal for the construction of mass-produced, single-family homes with septic tanks in the backyard. In the excitement of the moment, little thought was given to the environmental impact of the construction of so many new homes.[27]

Delawareans could have turned to an alternative local model of a suburban community that was sensitive to environmental concerns. In 1900, Frank Stephens and Will Ross, with the backing of Philadelphia philanthropist Joseph Fels, founded a unique village on an abandoned farm about four miles northeast of Wilmington, and called it Arden. Partly based on the single-tax principles of American reformer and writer Henry George, Arden allowed its residents to own their own homes, but the sites on which those homes were built continued to be owned by the town of Arden and every homeowner paid a rent (tax) to the community on his or her building lot. This land-rent arrangement, along with careful planning, provided Arden's residents with public greens and gathering places, walking paths, and preserved woodland. The adjacent single-tax communities of Ardentown and Ardencroft were founded in 1922 and 1955 respectively. The residents of all three communities were characterized, from the very beginning, as free-spirited, artistic, intellectual, and close-knit. Moreover, they were unusually committed to preserving the natural landscape in a manner that would strike considerable resonance with late twentieth-century environmentalists. So unique were these three communities that, in 2003, they were listed on the National Register of Historic Places by the United States Department of Interior "as the only non-sectarian utopian enclave surviving for over 100 years in the United States."[28]

In their move to the countryside, however, Delawareans were not drawn to the model presented by the three utopian communities of Arden, Ardentown, and Ardencroft. Rather, they were overwhelmingly drawn to the more traditional property-owning patterns found in environmentally destructive suburban housing tracts, where building lots were privately owned, where public greens, walking paths and preserved woodland were conspicuously absent, and where concern, initially at least, for protecting much of the indigenous natural landscape was minimal.

Although a handful of non-utopian Wilmington suburbs such as Holly Oak, Richardson Park and Holloway Terrace existed prior to 1945, it was not until after World War II that a dramatic, indeed extraordinary, explosion took place in the development of suburbs throughout northern New Castle County. At first, the new housing tracts were located primarily in unincorporated areas around Wilmington's periphery or on unincorporated lands near other New Castle County towns such as Newark. One such example was the unincorporated suburban development of Brookside, near Newark, which had become the state's fifth most populous community by 1970.

By the end of the twentieth century, the planting of suburbs had spread even to the periphery of many downstate towns. So rapid was the transformation of northern New Castle's landscape from bucolic fields, wetlands, and woods to housing tracts that New Castle County assumed, by 1980, a definite suburban character. All of this diminished Wilmington's demographic importance. In 1940, only 38 percent of New Castle County's population lived outside Wilmington; by 1980 that figure was 82 percent. The overwhelming majority of that 82 percent were suburbanites. By the 1980s, New Castle County's suburbs had, in the words of historian John A. Munroe, "the numbers, the wealth and educational advantages to dominate Delaware."[29]

Suburbanization made it possible for more Delawareans than ever before to live out the American dream of buying a home for their families. Thanks to the efficiencies of large-volume, tract-house construction, new homes were now available at affordable prices by a generation of married couples eager to leave behind the memories of a more austere past.

In addition, thousands of soldiers and sailors returning to Delaware from the European and the Pacific theaters, came home eager to take advantage of federal programs that specifically rewarded World War II veterans wishing to buy a home. No wonder builders of suburban housing were seen as heroes, meeting a great social and economic need. The ideal place to throw up tract homes was in unincorporated areas because there it was only necessary for builders to comply with the less stringent regulations of the county as they leveled forests, drained marshes, bulldozed top soil, and paved roads.

Although suburbanization—some critics preferred the more pejorative term "urban sprawl"—significantly expanded the number of Delaware's homeowners, it also had some unsettling environmental consequences. Because most suburban homes were not served by public transportation, suburbanites were more likely than urbanites to own and drive cars. This, in turn, increased air pollution and forced Delaware to build and maintain an extensive network of new roads and highways at the sacrifice of both natural and agricultural landscapes.

Moreover, the stripping away of endemic vegetation, which was the first step in constructing suburban developments, led to soil erosion and the loss of wildlife habitat. Although specific erosion statistics are not available for Delaware, some sense of what was happening can be seen by looking at nearby states. The first year of development and construction at one suburban housing tract in Maryland, for example, led to the loss through erosion of 50,000 tons of soil per square mile in one year, which was about 250 times the annual erosion rate for similar acreage in undeveloped rural areas in two neighboring states.[30]

Even the creation of carefully manicured lawns, the very heart of the suburban landscape, became a significant environmental issue. As early as the 1860s and 1870s, pioneer American landscape architect Frederick Law Olmsted and lawn entrepreneur Frank J. Scott launched their earnest and persuasive efforts to convince Americans that a well-manicured lawn was fundamental to suburban living. In 1870, Scott wrote that "a smooth, closely shaven surface of grass is by far the most essential beauty on the grounds of a suburban house." Up until the 1920s, suburban homeowners in Delaware and across the nation spread horse manure on their

Protest against development — Protester against development along Delaware Route One, just north of Rehoboth Beach, in the mid 1990s. The construction of a large K-Mart store was the specific issue, *courtesy of Mike Cleary and* The Delaware Coast Press.

lawns to make them fit the Scott ideal. The subsequent replacement of horses by motorized vehicles, however, forced the suburbs to turn to chemical fertilizers to produce the same lush, green lawns.

Before World War II, the recommended dosage was only one pound of nitrogen fertilizer per 1,000 square feet. By the 1970s, however, the recommended dosage had jumped to eight pounds per 1,000 square feet. In fact, by the early 1980s, according to environmental historian Ted Steinberg, Americans were spreading "more chemical fertilizer on their lawns than the entire nation of India used to grow food for its people." Unfortunately, runoff of lawn chemicals joined the farm runoff of chemical fertilizers as well as animal-waste fertilizer as an important source of contamination of Delaware's underground aquifers and its ponds, creeks, rivers and bays. Moreover, the chemical residue left behind on lawns was thought by some to pose a health risk to Delawareans of all ages, but particularly to children.[31] Because Delaware had become so suburbanized, on a per capita basis its use of lawn chemical fertilizers was probably higher than in most other states.

The use of backyard septic tanks in heavily populated suburban developments presented yet another environmental problem. Because tract housing was generally constructed on land not served by urban sewers, septic tanks became the method of choice to dispose of human wastes. The problem with septic tanks was that, as Secretary of Delaware's Department of Natural Resources and Environmental Control (DNREC) Edwin H. "Toby" Clark II pointed out in 1991, there were thousands in the state and they worked fine for twenty or thirty years. But older backyard septic systems, particularly those built prior to 1985, "are increasingly failing." Septic tank failures usually led to liquid overflows that made their way into ponds and streams, or seeped into the aquifer tapped for drinking water. Clearly, both of these scenarios threatened the well-being of Delawareans.[32]

By the end of the twentieth century, remarkable suburban growth was also taking place in central and southern New Castle County. In fact, houses were going up so fast that the county's population, south of the Chesapeake and Delaware Canal, increased at an estimated four times the growth rate for the entire county during the 1990s. Further south, it was the explosive growth of condominiums and tract houses inside and outside the corporate limits of the Atlantic Coast communities of Lewes, Rehoboth Beach, Dewey Beach, Bethany Beach and Fenwick Island that transformed the natural landscape. The astonishing growth in the number of eastern Sussex County's year-around inhabitants caused the entire county's population to increase by a stunning 38 percent during the 1990s.[33] Accompanying this population explosion in eastern Sussex were predictable traffic jams and dramatically increasing amounts of human waste that had to be disposed of somewhere. Moreover, runoff from the rapidly multiplying numbers of heavily fertilized lawns joined with malfunctioning septic systems and agricultural runoff and seepage in providing almost all of the non-point source pollution in Delaware's three inland bays.

The decline of the Inland Bays

The three inland bays of southeastern Sussex are, from north to south, Rehoboth, Indian River and Little Assawoman. They are shielded on the east from the turbulent Atlantic Ocean by a long, narrow band of sand that stretches south from Dewey Beach to Fenwick Island and beyond. Although diminished in the quantity of fish and shellfish as early as the mid-eighteenth century, and suffering short unproductive periods such as in the 1920s, Delaware's three inland bays served as sources of seafood for proggers and their customers until well after World War II.

The three inland bays are generally less than six feet deep, except where channels have been dredged for boat traffic. Because they are so shallow and thus contain a limited amount of water, the three inland bays are less able to dilute pollutants than such deeper and wider bodies of water as the Delaware Bay and the Atlantic Ocean. Moreover, direct access to the saltwater tides of the Atlantic and the Delaware Bay is limited. The only outlets for the inland bays are the Indian River Inlet, located near the confluence of Indian River and Rehoboth Bays, and, to a much lesser degree, the inlet at the south end of Ocean City, Maryland, and the Lewes and Rehoboth Canal, which links Rehoboth Bay with the waters of the Delaware Bay through Roosevelt Inlet at Lewes. The natural flushing of the three bays is a very slow process and that means that pollutants congregate rather than disperse.[34]

By the last three decades of the twentieth century there were increasing indications that something was seriously amiss in Delaware's three inland bays. In 1956, approximately 18 million clams were harvested from Rehoboth Bay alone. By 1987, however, the annual harvest in Rehoboth Bay had fallen to only 1.1 million clams. Local residents, by 1990, not only talked of the collapse of finfish and shellfish populations in the inland bays, but also spoke wistfully of "a simpler time when the bays were cleaner...and when seeing one's foot clearly in the water" was taken for granted. Undersea eel grasses, which served as habitat and breeding grounds for a variety of marine life, were once very common in all three bays. By 1991, they were gone and the implications for the future were troubling. As a state environmental scientist pointed out, "clams grew thirty times faster when inside a bed of underwater grass."[35] What was the cause of this ecological catastrophe that was turning the once very productive inland bays into stews of algae that were inhospitable to finfish and shellfish?

The most significant cause seemed to be the excess nitrates and phosphates that were pouring into all three bays from a variety of sources. For the period from 1987 to 1989, for example, the phosphorous levels in both Rehoboth and Indian River bays were five times the limit established by the state, while the nitrogen levels for the two bays were, respectively, 50 percent and 100 percent above the state limit. The excess nitrates and phosphates caused abnormal growth in some marine plants, and this along with the subsequent plant decomposition process seriously depleted the water of its oxygen content. This, in turn, caused fish and shellfish to die from suffocation. Of course, no one group accepted the blame for overloading the inland bays with nitrates and phosphates. Farmers blamed home developers, home developers blamed farmers, and the environmentalists blamed both groups and said that "everyone shares the responsibility of misusing the bays." Also strongly criticized for the rising pollution in Rehoboth and Indian River bays was inadequately treated municipal sewage, which, after being discharged from Rehoboth Beach's sewage treatment plant, eventually made its way into Rehoboth Bay.[36]

In the growing summer resort community of Rehoboth Beach, a network of underground sewers and a sewage treatment plant were in place prior to World War II. The plant was a primary treatment facility where human solid wastes were separated from liquids and then the untreated liquids were discharged into the Lewes and Rehoboth Canal, only a very short distance from its outlet to Rehoboth Bay. By 1986, the Rehoboth Beach sewage treatment plant was cited as the largest point source of nitrates and phosphates for the inner bays. Little Assawoman Bay—the Delaware section of Assawoman Bay—is connected by the Assawoman Canal to Indian River Bay. That allows Little Assawoman Bay to receive some of the pollutants that enter the two inland bays to the north. A new Rehoboth Beach sewage treatment plant was completed in late 1987, which, in addition to primary treatment, provided secondary treatment of liquid sewage before it was discharged into the canal. This meant that, in addition to the separation of solids from liquids, bacteria in the discharged effluent was either killed or reduced in number by chlorination. However, liquid sewage loaded with nitrates and phosphates continued to be flushed into the canal. In 1990, the *News Journal* reported that the Rehoboth Beach treatment plant emptied "up to nine times more nitrogen into the Lewes-Rehoboth Canal over the last two years than allowed in the city's discharge permit." [37]

Although, in the early 1990s, improvements in technology did allow the upgrade of the Rehoboth Beach treatment plant to filter out some of the nitrates and phosphates from the fluids it discharged, it was clear that a far more important source of the two pollutants in the inland bays was not subject to the quick fix of technology. Groundwater—rainwater that seeps into the soil—absorbs nitrates and phosphates from fertilized fields and overly-fertilized suburban lawns and carries them into underground aquifers. In southern Delaware most of the groundwater then moves eastward, below the land's surface, at the sluggish pace of between a quarter foot and two feet per day, slowly but inexorably delivering its toxic cargo to the three inland bays. Even if the application of all fertilizers containing nitrates and phosphates was immediately halted in the year 2000, there were enough pollutants already in the groundwater system of southern Delaware to adversely affect the inland bays for decades to come.[38]

Declension in other bodies of water

By contrast with the inland bays, where serious pollution problems did not surface until the 1970s, the Delaware Bay and, particularly, the Delaware River had exhibited significant pollution problems for many decades prior to World War II. Moreover, the physical degradation of the Bay and River continued through the 1960s and into the early 1970s. In 1971, for example, a Port Mahan crabber observed that instead of producing a healthy white color, the wake from any boat cutting through the upper Delaware Bay "was black as could be." He well remembered that piloting a boat on the upper bay in those days meant running an obstacle course full of floating trash. Close to the line of demarcation between the Delaware Bay to the south and the Delaware River to the north, near Woodland Beach, "the water smelled." As already noted, both finfish and shellfish were increasingly difficult to find, with the oyster harvest for the entire Delaware Bay dropping from 300,000 bushels in 1880 to only 4,700 bushels in 1961.[39]

The Delaware River, much more polluted than the Delaware Bay, continued to be overloaded with new contaminants well into the 1970s. In 1962, for example, numerous dead fish and a high bacteria count caused Battery Park, located on the west bank of the Delaware River in the town of New Castle, to be closed to swimming. A state employee who sampled nutrients, bacteria and dissolved oxygen in the Delaware River in the late 1960s, found oil slicks to be common and raw sewage to be "visible at certain locations, the Wilmington [to] Philadelphia stretch, especially."

The problematic nature of water quality in much of the Delaware Estuary, particularly the Delaware River section, was guaranteed by continued dumping and flushing by industries and municipalities located on shore and by vessels plying the river and bay. A major source of down-river pollution was the effluent discharged into the Delaware by Philadelphia's three major sewage treatment plants. By 1970, all three were primary treatment facilities, but only one offered secondary treatment where sewage was treated with chlorine.

What really upset some Delawareans was that, in the late 1960s, much of the sewage sludge from Philadelphia's treatment plants was being illegally dumped, from mafia-controlled barges, into the Delaware Bay just off Lewes, rather than at its legal dumping grounds many miles out in the Atlantic Ocean. When John Bryson, who then headed Delaware's air and water pollution control program, put a stop to it by calling in the U.S. Coast Guard, he was warned by a Philadelphia public official that the mafia had placed him on its "hit list." Bryson, a hunter who kept a pistol in his desk, responded that "they better not miss the first shot." In subsequent years, municipal and industrial sites up and down the Delaware continued to pump large volumes of pollutants into the river. In 1993 alone, 1,450 industries and municipalities discharged their industrial and human wastes into the Delaware Estuary, including an estimated 34 million pounds of chemicals. By that year, the Delaware Estuary contained the highest nitrogen content of any major estuary in the United States.[40]

Most of Delaware's other streams faced serious pollution problems in the post-World War II era. An example in northern New Castle County was Red Clay Creek which rises in the Kennett Square area of Pennsylvania and then carves its way south, through Delaware's piedmont, until it meets White Clay Creek on the northern periphery of the Atlantic Coastal Plain at Stanton. At an early date, mills were attracted to Red Clay Creek because of its fast-flowing waters. By 1814, in the Delaware portion of the creek, there were seven grist-mills, six sawmills, two cotton mills and one snuff mill. In 1856, at least two paper mills, which tend to heavily pollute streams with industrial waste, were operating on the banks of Red Clay Creek. In more recent times, National Vulcanized Fibre, usually portrayed as a corporate villain in the pollution saga, operated plants on Red Clay Creek at Yorklyn and across the Pennsylvania line at Kennett Square (A third plant, located on White Clay Creek in Newark, is mentioned at the end of chapter six). The manager of the Yorklyn N.V.F. plant admitted that, until the early 1950s, "we used Red Clay Creek as a water supply for both production and waste disposal—just like every other plant was doing." Particularly problematic was the disposal into the creek of zinc chloride, a key ingredient in the manufacture of vulcanized fibre.[41]

Zinc chloride discharges were a major contributor to the fact that, prior to 1971, no fish in Red Clay Creek ventured south below the Yorklyn N.V.F. plant. Although, in that year, a recovery system was installed in the Yorklyn facility that annually eliminated "up to a million pounds of zinc," the creek's bed and waters continued to be laced with toxic materials such as PCBs (polychlorinated biphenals) and zinc from industrial dumping during earlier years. In 1986, for example, a high level of PCBs discovered in fish from Red Clay Creek caused the Pennsylvania Department of Environmental Resources and Delaware's DNREC to warn against eating anything caught in its waters. Historically, point-source pollution has always been easier to control than non-point-source pollution. Industrial sites along Red Clay Creek are examples of the former, while farms in general, and mushroom farms along the creek in the Kennett Square area in particular, are examples of the latter.

Indeed, non-point runoff from mushroom farms as well as septic system overflows were blamed for much of the remaining pollution entering Red Clay Creek in the late twentieth century. In 1978, from Yorklyn to Ashland, the stream bed was "glutted with sewage fungus," and the water "frequently" carried "an unpleasant odor." By 1988, however, with much of the clearly visible pollution gone, Red Clay Creek was looking "pretty good," particularly to old-timers who remembered an earlier era when the creek was visibly degraded. Nevertheless, there continued to be significant environmental quality issues with Red Clay Creek. Among those aware of this was Delaware Nature Society's Lorraine Fleming, who summed up the feelings of many when she said: "you wouldn't catch me putting my face in it."[42]

Rising sea levels

The gradual warming of the earth's atmosphere, albeit with both short and lengthy interruptions, has continued since the beginning of the end of the last Ice Age about 15,000 B.C. Since then, the gradual melting of glaciers and polar ice caps, brought on by higher average temperatures, has produced a steady rise in ocean levels of about 400 feet around the globe, and this has caused Delaware's Atlantic shoreline to retreat westward 80 to 100 miles. Sandy beaches, so precious to the summer tourist trade, continue to fall victim to the slow but inexorable inland march of salt water. Prior to the twentieth century, when rising sea levels submerged one beach in Delaware, the ocean generally created a new strip of sand a few yards inland.[43] By the late twentieth century along much of the state's Atlantic coast, however, boardwalks, hotels, condominiums, stores, and houses stood where a new sandy beach would naturally develop.

Driven by economic imperatives and partially funded by state and federal aid, Delaware's ocean communities have temporarily held off the inevitable by replenishing their eroded beaches with sand hauled in by trucks or sucked in from the ocean bottom through huge, flexible pipes. The old pattern of the natural world shaping human destiny was giving way to the more modern pattern of humans shaping and reshaping the natural world. Occasionally, however, some Delawareans were reminded of an earlier era when powerful natural forces asserted a considerable influence over the lives of people.

An aerial view of Rehoboth Avenue and the oceanfront—
This photo, taken on March 13, 1962, shows what was left after the Great March Storm of 1962 had run its course. Visible to the left of Rehoboth Avenue are the ruins of the old Bellehaven Hotel. To the right, across Rehoboth Avenue, is the spot where the venerable Dolle's Popcorn stand had stood (and where a rebuilt Dolle's stands today). *Photo from the Delaware State Highway Department Collection and reproduced courtesy of the Delaware Public Archives.*

In 1878 and in 1954, particularly strong hurricanes swept through the First State and left behind a wake of destruction. Storms called northeasters, that featured very strong winds blowing consistently from the northeast posed at least as great a threat as hurricanes. On Monday, March 5, 1962, the televised weather report out of Salisbury, Maryland, assured viewers along Delaware's Atlantic Coast that Tuesday would be cloudy and cool, but the overcast skies would eventually clear during the day. By Tuesday, however, two low-pressure systems had come together over the Middle Atlantic region and produced a northeaster that moved eastward to the Delaware coast where it was locked in place for three days by a stalled high-pressure system over eastern Canada. The heights of the resulting ocean waves reached an estimated thirty to forty feet offshore and twenty to thirty feet where surf met sand. Because of constant, gale force winds from the northeast that gusted up to sixty and seventy miles per hour, and because of traditionally high spring tides, the storm tide (storm surge minus mean low tide) measured 9.5 feet at Breakwater Harbor just off Lewes, the highest ever recorded at that location.[44]

From March 6 through March 8, 1962,

Storm's Aftermath—
The ruins of the Atlantic Sands Motel, on the oceanfront at Rehoboth Beach, are shown in the days after the Great March Storm of 1962 had run its course. The pilings visible at right are those for the former boardwalk. *Photo from the Delaware State Highway Department Collection and reproduced courtesy of the Delaware Public Archives.*

wind-driven ocean waves continuously battered the coast, breaking through and flattening sand dunes, and rushing westward across Delaware Route One to meet the rising waters of the inland bays. Ocean waters surged through the summer vacation towns of Dewey Beach, Bethany Beach, South Bethany Beach, and Fenwick Island. Rehoboth Beach lost its boardwalk, and most of its oceanfront buildings were either destroyed or badly damaged. Waves demolished 28 of 29 oceanfront homes in Bethany and every oceanfront home in South Bethany. In all, the coastal communities from Dewey Beach south to Fenwick Island suffered damage to more than 1,900 homes. The three-day pounding was the most catastrophic storm in Delaware history, leaving behind seven dead and enormous property loses. The resilience of Delaware's coastal communities, after being struck such a devastating blow, was noteworthy. By July 4, 1962, the Rehoboth Beach boardwalk was completely rebuilt. Further south, all the way to Fenwick Island, the clean-up and rebuilding proceeded at a remarkable pace.[45]

Although the memory of the northeaster of 1962 gradually faded, the unsettling thought persisted: "Given the much greater development of the coastal towns by 2000, wouldn't another powerful northeaster wreck far more havoc and destruction?" Quite simply, proper allowances were not being made for the potentially devastating combination of a slow but

Oil Refinery— A 1981 aerial view of the Getty Oil Refinery complex near Delaware City, *photo courtesy of the Delaware Economic Development Office.*

constant rise in sea level and the storms that periodically lash the Delaware coast.

Delaware City

Despite occasional examples of nature reasserting its primacy, as it did briefly in the hurricanes of 1878 and 1954 and the northeaster of 1962, the dominant theme in the state's recent environmental history continued to be the degrading of a natural inheritance by Delawareans and their institutions. No place was a better example of such environmental degradation than Delaware City. In 1826, Delaware City was founded on the north bank of the Chesapeake and Delaware Canal where it joined the Delaware River. The canal, however, was not completed and opened for through transportation until 1829. In 1927, Delaware City received a significant economic blow when the Delaware River's entrance to the canal shifted two miles to the south. By the next decade, the fishing industry, which was important to Delaware City in the nineteenth century, had almost disappeared due to the high pollution levels in the Delaware River. Delaware City's location on the river, however, offered a prime location for an oil refinery where tankers, pushing up the Delaware Estuary, could easily unload petroleum for processing. By 1957, the giant Tidewater Oil Refinery was in operation just northwest of Delaware City and occupied a site approximately one-half the size of incorporated Wilmington. The refinery dwarfed all other structures in the area, dominated the skyline for miles, and gave the appearance "of a brightly lighted city at night." In 1958, Tidewater Oil merged with Getty Oil and the refinery took on the Getty name. In subsequent years the Delaware City refinery expanded its facilities, and the petrochemical complex was joined by other companies. From the beginning, the refinery posed a serious threat to New Castle County's natural environment because it was built to process petroleum with a high sulfur content. The only way to prevent the refining of this "dirty oil" from polluting the surrounding area was for Tidewater, Getty and other subsequent owners to make a long-term commitment to costly anti-pollution measures. But that would reduce corporate profits. The subsequent record shows that the Delaware City refinery consistently refused to take the necessary anti-pollution steps to make itself a good neighbor.[46]

During the first year of the refinery's operation, Tidewater Oil predicted that it would have a positive impact on Delaware City by reinvigorating its stagnant economy. Delaware City native and environmental activist John Nickle remembers that "just about everyone in the area thought that the coming of Tidewater would be a good thing." A study of the town sixteen years later, however, found no significant rise in real estate values or in the standard of living of its inhabitants compared to the rest of New Castle County. In fact, despite the construction of the oil refinery, Delaware City remained a relatively depressed area. The environmental impact of the vast petrochemical complex on air, land and water was far from benign. Although, from 1957 to 2000, the refinery went through a series of name changes from Tidewater to Getty, to Star Enterprise, to Motiva and finally to Valero as it passed through the controlling ownership of several different petroleum companies, one thing was constant. The refinery and its neighboring chemical and petrochemical companies in the huge complex northwest of Delaware City, could be counted on to foul the air, poison the land, and pollute the water. By 1996, the most toxic emitting site in the state was the Delaware City petrochemical complex. Moreover, there was potential for even greater environmental disasters at the site by the "very nature of the volatile chemicals and the processes involved."[47]

Health Concerns

The environmental decline that characterized the entire state, but was most heavily concentrated in northern Delaware in general and in the Delaware City area in particular, soon drew attention to the linkage of air, land, and water pollution to certain health problems. By 1978, the hospitalization rate and the mortality rate in Delaware City were, respectively, 53 percent and 36 percent higher than for New Castle County as a whole. In 1990, escaping hydrogen sulfide gas at the Star

Enterprise refinery—renamed Motiva in 1998—hospitalized 23 of 640 workers, including four who had to be revived from cardiac arrest. This caused DNREC Secretary Toby Clark to label the refinery an "imminent hazard" to public health. In 1995, Star Enterprise's emissions were blamed by physicians for patient and staff respiratory problems at Christiana Hospital, approximately seven miles from Delaware City.[48]

Extended from Delaware City to the entire state, statistics concerning the prevalence of cancer and other diseases were particularly unsettling. In 1989, a study by a national insurance company ranked Delaware as America's most unhealthy state. Four years later, Health Care State Rankings found only Florida to be statistically more unhealthy to live in than Delaware, and Florida could argue that its poor health statistics simply reflected the elderly nature of its population.

In 1989, the Center for Disease Control in Atlanta found only four other states with a higher death rate than Delaware from preventable chronic diseases including four forms of cancer. From 1928 to 1978, Delaware's per-resident death rate from cancer doubled to the point that, from the latter date forward, it was often among the state leaders in malignant-related morbidity. So depressing were the statistics that, in 1995, Delawareans took solace in the fact that Louisiana was now tied with the First State for having the highest per capita death rate from cancer.[49] Although, on the state level, it continued to be difficult to make specific linkages between the emission of toxic pollutants and the incidence of cancer, more and more Delawareans were convinced that there was a connection.

Aesthetic Sensibilities

The alarming signs, during the late twentieth century, that Delaware's natural environment was facing an unprecedented crisis eventually led to public indignation and to public demands that something be done. This, in turn, caused the state government to take action on a scale previously unknown in Delaware's history. Although the scale of response was hitherto unmatched, earlier chapters of this text indicate that state intervention to deal with certain aspects of environmental degradation certainly had a long history. Most of the demands that sparked political action in the past, however, were not based on the broad ecological perspective that would undergird much of Delaware's drive for environmental reforms in the late twentieth century. Rather, most earlier calls for environmental reforms were primarily driven by the need to protect specific sources of income.

From the colonial period forward, the transforming impact of European guns, fish-traps, axes, hoes, plows, shovels, dams, free-roaming livestock, and the general practice of firing the countryside made inevitable at least some government intervention to attempt to save trees and certain wild birds, animals, finfish, shellfish, and turtles from completely disappearing from Delaware's landscape and seascape. These legislative efforts primarily reflected the broad-based view that nature played an important role as producer of furs, game meat, seafood, firewood, lumber, and waterpower. But for the most part, nature had little other claim to human appreciation.

There were some early Delawareans, however, who valued nature for another reason. They tended to be members of the landed gentry who were set off from others by the considerable acreage that they owned, by the number of slaves that they possessed, and by the handsome brick and wood-frame houses that they built across rural Delaware to assert their economic, social and political dominance. With perceptions and values partially rooted in the traditions of the European Enlightenment, a number of Delaware gentry viewed nature as a source of aesthetic pleasure, provided that it was first properly reshaped to reflect man's mastery over "wildness." Rather than allow the outdoor panorama to mirror nature "helter-skelter," these Delawareans counted on their comparatively large and elegant homes, usually constructed in the Georgian style, to help bring some order to what might otherwise be perceived as a chaotic rural landscape. On grounds surrounding and leading to their homes, they planted trees and shrubs in patterns to mirror both the geometric certainties of the Enlightenment, and related artistic values that reflected their own ordered, rational sensibilities. A few of Delaware's large landowners, such as Caesar Rodney, who lived a few miles southeast of Dover, even established parks where the presence of semi-tame deer further testified to the gentry's control over nature in a manner that made tempered wilderness aesthetically pleasing.[50]

Combining an appreciation of nature as economic provider with a recognition of nature as the source of considerable aesthetic pleasure was Eleuthère Irénée du Pont de Nemours, the French immigrant to Delaware who founded the DuPont powder works along the lower Brandywine in 1802. To E. I. du Pont, not only did nature supply the raw materials for maintaining life and for providing the basis of significant economic activity, it was also the source of considerable beauty that enriched the human spirit. E. I. du Pont's love of nature—on immigrating to the United States he listed his vocation as "botanist"—for both utilitarian and aesthetic reasons was passed on to his children and grandchildren. In 1842, for example, Quaker miller Edmund Canby noted that the homes and gardens of the du Ponts were "very beautiful, all having the advantage of old forest trees about them." Indeed, through the late twentieth century, successive generations of Delaware's du Ponts continued to value the evocative beauty of nature while also recognizing nature's pragmatic uses.[51]

Despite the aesthetic concerns of certain members of the gentry in general and the du Pont family in particular, during the nineteenth and most of the twentieth century the Delaware voices that addressed environmental decline focused primarily on very specific practical issues such as polluted drinking water in Wilmington, the destruction of young trees and field crops by plundering peripatetic pigs, the loss of soil fertility in the countryside, and the dramatic diminution in the numbers of Delaware deer, muskrats, ducks, shad, clams, oysters and other specific fauna. And yet there were a few mid- and late-nineteenth century civic leaders, particularly in Wilmington, who continued the gentry tradition of insisting that aesthetics were central when addressing the relationship between humans and their natural environment. In rapidly growing Wilmington, open lots and green vistas were increasingly rare amidst the expanding blight of congested housing, filthy streets, and smoke-belching factories. Across the Atlantic, many European cities countered the increasingly claustrophobic nature of urban living with public parks that presented green oases that elevated the human spirit by touching certain aesthetic sensibilities. Subsequently, a number of American civic leaders, in concert with leading American urban landscape planners such as Frederick Law Olmsted, transplanted the European city park movement to the United States.

WILLIAM P. BANCROFT
Founder of Wilmington's Park System,
Photo courtesy of the Historical Society of Delaware.

Particularly influenced by these foreign examples and subsequent domestic models of green oases constructed in the very heart of rapidly growing metropolitan centers was William P. Bancroft, a Wilmington Quaker industrialist, who returned from a European trip in the 1870s convinced that urban parks would vastly improve the lives of Wilmington's residents. Over the next two decades his leadership would earn him the title, "Father of Wilmington's Parks." Joining Bancroft in the drive to develop green havens in Delaware's largest city were a number of Wilmington's wealthiest and most civic-minded residents, making that city's park movement another example of the patrician-led urban reform efforts that were common across America during the late nineteenth and early twentieth centuries. By 1900, most of the tracts of real estate that comprise the modern Wilmington park system were in place, making Delaware's leading city the state's pioneer in the development of parkland as well as in the construction of municipal water and sewer systems. By 1929, the Wilmington Park

PRESTON C. LEA
Governor of Delaware, 1905–1909,
Photo courtesy of the Delaware Public Archives.

Commission controlled approximately 620 acres, with much of this land situated along both banks of the Brandywine.[52]

Aesthetic considerations were at the very heart of the impulse to create Wilmington's parks. In 1869, for example, the committee of four citizens set up to investigate the construction of parks in Wilmington, issued a report stating that the most important mission of future parks would be "that of elevating the taste, promoting in the minds of the masses a love for the beautiful, [and] giving a higher moral tone to all grades of society." Frederick Law Olmsted, a sometime advisor to Wilmington's organizers of the push to establish urban parks, simply recommended the lower Brandywine Valley as a park site because of its great natural beauty. In 1929, an assessment of the role played by Wilmington's parks had "shown that access to the fresh air, the restful scenes, and the natural beauty of woods and open country had a beneficial effect upon those engaged in confining indoor occupations."[53]

During the early twentieth century, Governor Preston Lea (1905–1909) strongly advocated the preservation of Delaware's fauna, soil, and trees. In many ways Governor Lea's approach was traditionally utilitarian in that he called for action to reverse the decline in the numbers of clams and lobsters in Delaware waters so that the state's watermen could make a decent living. Moreover, he saw in marsh and swamp draining the salvation of Delaware's struggling agricultural economy. But Lea was also alarmed because environmental degradation threatened certain aesthetic values. He was, for example, strongly in favor of rebuilding Delaware's depleted forests, not only for the money it would generate for the state's timber industry, but also because trees "beautify the landscape." In 1909, a year after attending a very successful conservation conference in the East Room of the White House called by President Theodore Roosevelt, Lea told the Delaware legislature that "the prodigal wasting of our natural wealth and [the] disregard of the rights of posterity had to stop." Lea then

C. DOUGLASS BUCK
Governor of Delaware, 1929–1937
Photo courtesy of the Delaware Public Archives

unsuccessfully urged the Delaware General Assembly to set up a Delaware commission for conservation.[54]

It wasn't until the second half of the twentieth century that aesthetic concerns became an increasingly important reason for statewide protection of Delaware's land, sea and air from further physical degradation. In 1958, former Governor and U. S. Senator C. Douglass Buck stirred up resistance to further industrial development when he complained that the newly-constructed Tidewater/Getty oil refinery had changed Delaware City and its immediate hinterland "from a clean, attractive and peaceful farming and residential community, to a smelly, unattractive section of the state." The next year a survey of Delaware's water resources justified preserving wetlands because of the "particular kind of beauty associated" with them. In 1970, a state publication that focused on some selected environmental problems, pointed out that we should not only concern ourselves with the commonly recognized need for clean air and clean water, but we also need to protect Delaware's natural environment because it provides us with the "aesthetic qualities of beauty, variety and order," that are so important to our "emotional well being."

Focusing on problems "primarily aesthetic in nature" means doing something about intrusive utility poles, junkyards, abandoned cars, outdoor advertising, and sand and gravel pits. In 1971, the Coastal Zone Act,* which was aimed at protecting Delaware's coastline from further industrial development, was passed by the Delaware General Assembly. Among the justifications for this landmark environmental legislation were aesthetic concerns. Nineteen years later, Michael Riska, executive director of the Delaware Nature Society, reflected the increasing popularity of the aesthetic rationale

* For a detailed discussion of the Coastal Zone Act, its provisions and its national importance, see pp. 213–218.

for saving the natural environment when he wrote that "the plants, animals and ecological communities that share the state with us bring beauty and pleasure to our lives."[55]

Protests and Protestors

Prior to 1945, with a few exceptions, individual Delawareans who raised their voices to strongly protest the degradation of their natural environment were almost as rare as Delaware's old-growth forests or its white-tailed deer. In the subsequent fifty-five years, however, increasing numbers of Delawareans not only voiced considerable alarm about the escalating destruction of their state's natural environment but also organized protest meetings, issued press releases, formed activist environmental organizations, and lobbied governors and members of the General Assembly. The goal was to use the power of state government to protect Delaware's natural environment from the polluting and destructive forces unleashed by industry, agriculture, suburban housing developments, motor vehicles, and substandard community sewage treatment systems.

While the leading advocates and most of their strong supporters in Delaware's post-World War II environmental movement were generally college-educated members of the state's middle and upper classes, in other ways they were a diverse group. Some were Delaware natives; others were relative newcomers to the state. While many had been science majors in college, others had majored in the humanities, the arts, or other non-science disciplines. Although a preponderance of the leadership in the movement lived north of the Chesapeake and Delaware Canal, a number could be found in southern New Castle or even in Kent and Sussex. Reflecting a pattern that was nationwide, the state's environmental movement's leaders were at least as likely to be female as male. What brought them together was their shared perception that Delaware's natural world was being raped and plundered by the out-of-control forces of modernity, and that only a proactive government prodded by a concerned and informed citizenry could halt the carnage. Although key leaders in Delaware's emerging environmental movement are too numerous to name in a study of this nature, a very brief look at a few gives a sense of the variety of different environmental challenges that they faced and the varied but compelling operating styles that they used to confront the forces of apathy and opposition.

Mathilda "Til" Purnell of Herring Creek in eastern Sussex County, for example, combined a sense of personal outrage, a dogged persistence, and a quotable verbal style and flair to combat the twin scourges of over-development—particularly of wetlands—and water pollution. Jennifer Ackerman's *Notes From the Shore* (1995) captures the essence of Purnell's increasing anger as she attends meeting after meeting where public officials decide the fate of one piece of undeveloped wetland after another in southern Delaware. "She knits while she listens, her needles flashing and clicking harder and faster as her rage mounts..." An example of Purnell's use of earthy verbal images to make a point was on display at a hearing in 1990, when Rehoboth Beach asked DNREC to modify the town's wastewater discharge permit so that Rehoboth could dump seven times as much nitrogen as formerly allowed into the Lewes-Rehoboth Canal near its entrance to Rehoboth Bay. Purnell responded that when bays are already contaminated by nitrates, "the last thing you want to do is add more. I'm no scientist, but I'm a right good cook. And if I have a pot of soup that's already too salty, I most certainly do not put more salt in it."[56]

Wilmington's poet-environmentalist, attorney Jacob Kreshtool, first became alarmed about the quality of Delaware's air in the early 1960s, when emissions from an asphalt shingle plant near his home in Edgemoor seemed to cause an inordinate amount of coughing in his young children. By 1963, he was president of Delaware Citizens for Clean Air. Subsequently he spent much of his time skewering the state government for not acting more vigorously to reduce air pollution. In June 1991, for example, Kreshtool accused DNREC's Secretary Clark of being "Mr. Wimpo. He's great at protecting those piping plovers [on Cape Henlopen]" but "on toxic air I tell you he collapsed." Five months later, he attacked Clark and Governor Michael N. Castle, but this time with four lines of doggeral posted along a road near the sprawling, polluting oil refinery and petrochemical complex at Delaware City:

Hark, hark, Castle and Clark,
Their speeches are convoluted.
They tell you how good they are,
But the air is still polluted.

Secretary Clark fired back with doggeral of his own:

Goodness sake, save us from Jake.
We're shaking clear to our bones.
He's tried jokes and shouts,
cries and pouts,
and now he attacks with bad poems.[57]

Less colorful but at least as effective in pushing forward her environmental agenda was Grace "Bubbles" Pierce-Beck of Dover. After being one of the central players in the passage of Delaware's Coastal Zone Act of 1971, Pierce-Beck moved to Washington, D. C. to lobby Congress for key environmental bills, including the regulation of offshore oil drilling. After her first husband's death, Pierce-Beck returned to the Dover area in 1978 to run a small chain of family-owned pharmacies. In the late 1980s, she sold the pharmacies and became Delaware's only full-time environmental lobbyist.

Pierce-Beck was uncommonly successful in defending Delaware's Coastal Zone Act from attempts to weaken it, and in pushing for a broad environmental agenda because she had a certain toughness that belied her diminutive physical presence. As another environmentalist observed, "most of the people that we confront are men." They felt "much less threatened" by a sincere and soft-spoken female "who was only about five feet tall." As a result, Pierce-Beck got "a great deal done in a quiet, unassuming manner...She's like an iron gnat." Governor Russell W. Peterson (1969–73) saw her as "the epitome of eternal vigilance," in the unending battle to save Delaware's coast.[58]

Offering yet another strategy to preserve at least parts of Delaware's natural environment from further destruction was Edmund H. "Ted" Harvey. Born the son of Leroy Harvey and Renée du Pont Harvey in 1911, Ted was often taken by his father, a prosperous Wilmington merchant who had served as city mayor and as Farmer's Bank president, to Ellendale and Cypress Swamp in Sussex County, where Ted developed a life-long interest in plants and wild places. As a young man operating a fishing camp in the Florida Keys in the early 1930s, he witnessed the harmful effects of over-development on Florida's natural environment. Returning from Florida and reflecting the love of many of his maternal du Pont ancestors for plants as well as their desire to "do something of public benefit," Harvey turned his energies to saving Delaware's coastline, which was threatened by heavy industry, and to preserving remaining forest habitats, which were, as previously pointed out, primarily located in southern Delaware.

In 1961, Harvey convinced relatives and close friends to join him in founding Delaware Wild Lands Inc., which used private funds to acquire and then preserve threatened areas of the sylvan landscape. Harvey was convinced that most of the executives and board members of large corporations would support the preservation of the state's wild lands once they were informed of the consequences of continued ecological degradation. "He sought them out, asked for their money, and taught them about the need for clean air and water, healthy forests and wildlife, good soil management, and preservation of open space."

In short, Ted Harvey built Delaware Wild Lands on the financial contributions and good will of Delaware's corporate elite. He did this by gentle persuasion and social networking rather than by shrill accusation and angry confrontation. Beginning by acquiring Sussex County's Trussum's Pond and then most of the remnants of Cypress Swamp, by 1985 Delaware Wild Lands held 30,000 acres of critical habitat for preservation, mostly in Delaware, and had raised more than six million dollars.[59]

It is impossible to estimate the number of Delawareans who, in the latter half of the twentieth century, actively committed themselves to protecting the integrity of their state's natural environment. The considerable depth of their commitment, however, can be measured by their ability to persuade the General Assembly to pass environmental laws despite opposition from most in the business community. Moreover, the forty-five or so organizations that environmentalists founded and supported, and that continued to exist in Delaware in 2000, further suggests that saving the natural environment had considerable support in the First State. The Delaware Nature Society, Delaware Audubon Society, Delaware Chapter of the Sierra Club, Delaware Wild Lands, Delaware Citizens for Clean Air, Wildlife Federation of Delaware, Green Delaware, White Clay Creek Watershed Association and Delawareans for Energy Efficiency are among the best known. They worked hard to influence state legislation by having members who were constituents of specific state representatives and senators call their legislators concerning bills of great interest. As Senator Steven H. Amick, Republican from the west Newark area noted, "constituents would call me and work to educate me on the environmental issues affected by the bill. When I received those calls I always took them seriously."[60]

Although the statewide environmental movement splintered into many organizations, with each one focusing on different environmental priorities such as preserving wetlands, purifying the air, setting aside green spaces, alerting the public to the dangers of toxic waste, or saving certain species of flora and fauna from statewide extinction, the indispensable mortar that provided unity of purpose to these disparate groups and their members was a general commitment to seeing the natural world through the broad, integrated biological perspective provided by ecology.

Ecology

Ecology is viewed as a science in most quarters, but to its critics it is merely a point of view. Ecology maintains that, in the natural world, everything is connected to everything else. These connections are usually marked by balance, harmony, unity and economy. An ecosystem is a self-sustaining unit of nature that depends on the well-being of all, or almost all, of its members to maintain its equilibrium. The implications of the loss of equilibrium are troubling. As John D. Gates, editorial writer for Wilmington's *Morning News,* pointed out in 1971, "man does not exist apart from his environment but within it. He doesn't just use the environment; he lives in it." But when a vital species of flora or fauna becomes endangered or disappears from the ecosystem, equilibrium is lost and all the other residents of the ecosystem, including man, are threatened. Therefore, as a report on protecting Delaware's natural heritage pointed out in 1999, "the maintenance of diverse, sustainable ecosystems is essential for the future stability of human communities." Put another way, humans are climbers on the face of a cliff, so roped together to the other animal and plant species that make up our particular ecosystem that, if they tumble, we, too, are in trouble.[61]

Critics of ecology maintain that, in a sense, ecology developed as a holistic antithesis to the reductionist tendencies that seemed to infect much of twentieth-century science. In contrast to reductionism, ecology focuses on the general rather than on the specific as it brings together seemingly disparate parts of the natural world and integrates them into an orderly and rational system that can be counted on to operate in a predictable fashion. Ecology's critics accuse ecologists of putting too much emphasis on the belief that balance and order are the hallmarks of the complex and interconnected natural community that we label a healthy ecosystem. They ask how, for example, can the concept of biological evolution be reconciled to an ecological view of nature that maintains that stasis is healthy and change is not? How, indeed, can ecologists present a natural world of balance and harmony when constant flux and transformation is a more realistic description of biological history?[62]

According to biologist Larry Curtis, however, the critics of ecology misrepresent ecology's basic ideas. Quite simply, ecologists seek to know the workings of the natural world. "Sometimes, but only sometimes," Curtis notes, "this does involve dealing with systems of equilibrium." Although equilibrium ecology was popular in the 1960s and 1970s, Curtis maintains that ecologists "are evolutionary biologists." Ecologists continue to see the parts of the natural world as interdependent, "but in long-term or short-term flux," because organisms are evolving and the biological and physical world is constantly changing.[63]

The roots of ecology's broad, systematic approach to the workings of the natural world and its concern for every living organism go far back in history. Even a few Christian leaders such as St. Francis of Assisi, in fourteenth-century Italy, or John Wesley, in eighteenth-century England, attempted to temper the anthropomorphic focus of Christianity by urging empathy and respect for all of God's living creatures. But it was not until 1866 that Ernst Haeckel, a German scientist, became the first to use the term "ecology." During the first half of the twentieth century, ecology emerged as a branch of biology. By the 1960s and early 1970s, ecology was being adopted by environmental activists across the nation as the lodestar of their movement. Boosting ecology's appeal was the publication, in 1962, of Rachel Carson's *Silent Spring*, which suggested that the heavy use of pesticides was upsetting nature's delicate equilibrium. Indeed, to future Delaware activists such as Til Purnell and Russell Peterson, reading *Silent Spring* was a significant and defining personal experience. In 1971, John D. Gates declared that, "ecology is a science, but it should also be a religion." In the last two decades of the twentieth century, the ecological perspective had become so pervasive in Delaware that it was routinely trumpeted in publications dealing with the state's environmental issues. The litany used to address specific environmental goals, which featured such working terms as "biological diversity" and "holistic solution," was really the direct application of ecological principles.[64]

CHAPTER EIGHT:

The Government Steps in, 1971–2000

"...so that people can enjoy the hunting, boating, fishing and peace and quiet..."

THE ENVIRONMENTAL DRAMA THAT UNFOLDED IN DELAWARE FROM 1971 to 2000 featured three major players: the polluters, the environmentalists, and the various levels of government from national and state to county and municipal. Adding significant tension to the drama was the fact that, while Delaware was saluting the chemical industry for its positive impact on the state's economy and Wilmington was proudly wearing the mantle "Chemical Capitol of the World," Delawareans were slowly waking up to the chemical industry's historic role in their state's problematic environmental record. After all, although man-made chemicals had been around since the mid-nineteenth century, it was not until the last four decades of the twentieth century that it became evident that many chemicals were environmentally threatening.

A Reluctance to Act

Despite the fact that polluters were polluting and environmentalists were pointing fingers and hurling jeremiads, Delaware's state government was reluctant to act because of a predisposition to favor business interests that went all the way back to the colonial period. By the end of World War II, this traditional pro-business stance had became so pervasive in the state that Democrats as well as Republicans understood that political victory depended on building a friendly relationship with the business community. According to political scientist William Boyer, after Governor Russell W. Peterson's bid for re-election failed in1972,"avoidance of an anti-business image would be requisite for political survival" in the state.[1]

Despite the fact that the Peterson admin-istration was limited to just one term (1969–73), it marked a dramatic watershed in the state government's role in protecting and conserving Delaware's natural environment. A brief look at the state government in general and at the makeup of the General Assembly in particular—as they existed prior to 1969—is essential to understanding the impact of Governor Peterson's administration. According to the present Delaware Constitution, which was adopted in 1897, responsibility for protecting the state's natural resources rests in Dover and not in the hands of county and municipal governments. Even zoning powers, which are usually exercised, albeit sometimes reluctantly, by county, city, and town governments, "are largely subordinate and derived from the state." But it was one thing for the state to have constitutional respon-sibility and quite another to possess the vision and wisdom to act constructively. During the mid-twentieth century, typical members of Delaware's General Assembly were rural in their roots, parochial in their interests, and, because they tended to serve for only one or two terms, inexperienced and unskilled in the legislative process. In short, most members of the state legislature were ill-equipped to recognize the looming environmental crisis and, even if they did, ill-prepared to enact corrective legislation.[2]

Members of the General Assembly continued to be chosen from election districts established back in 1897, but the districts had become increasingly anachronistic because of recent demographic changes that resulted from the extraordinary population growth in New Castle County's suburbs after World War II. By 1960, only 31 percent of the state's population lived in Kent and Sussex, but a clear majority of both houses of the General Assembly was elected from those two counties. By contrast, New Castle County held 69 percent of the state's population but elected only a minority of the state's legislators. All of this made a mockery of the democratic concept of "one man, one vote." In practical terms, this meant that central and southern Delaware, the region of the state least affected by pollution concerns in the immediate post-World War II years and, historically, least sympathetic to political action to protect the environment, dominated the General Assembly. In 1962, in "Baker v. Carr," this perversion of democracy was declared unconstitutional by the U.S. Supreme Court, and Delaware and some other offending states were ordered to redraw election districts. Subsequent redistricting reduced central and southern Delaware's clout in the General Assembly and made it possible for suburban New Castle County to be repre-sented fairly.[3] In short, because of the dramatic change in its geographic roots, the General Assembly of 1969 was far more sympathetic to the revolutionary environmental initiatives about to be introduced by the Peterson administration than it would have been in 1962.

Actions in the Pre-Peterson Era

Prior to the Peterson administration, the state's government took only limited action to protect Delaware's natural environment. Although individual species, such as white-tailed deer, received protection at a relatively early date, it wasn't until the late nineteenth and early twentieth centuries that a more systematic approach to wildlife protection was enacted. In 1879, the General Assembly established the Delaware Game Protective Association, and in 1893, the association helped convince the state legislature to limit the hunting seasons for all game animals. In 1911, a "Board of Game and Fish Commissioners" replaced the Game Protective Association. A year later Delaware had its first chief game warden. By 1930, the Board of Game and Fish Commissioners was acknowledging the importance of maintaining wildlife habitat.

In 1955, the Game and Fish Commissioners began publishing the *Conservationist*, a periodical that celebrated, in both its title and content, the growing public interest in preserving at least some of Delaware's natural heritage. The *Conservationist* elevated the sensitivities of its readers to the growing number of environmental concerns within the state. *Outdoor Delaware*, published by DNREC today, is a direct descendent of the *Conservationist*.[4]

The purity of Delaware's water also became the focus of special attention. As early as 1915, the Delaware Board of Health was deemed responsible for "the sanitary protection" of all water supplied to the public and was, in particular, the guardian of those streams used as a source of drinking water by any municipality in the state. By that same year, the Delaware Game and Fish Commission was responsible for drawing up regulations to halt the dumping of pollutants into creeks, rivers and bays that would adversely impact Delaware's fish. Reflecting the fact that additional measures were needed, in 1949, the General Assembly created the Delaware Water Pollution Commission to better protect the state's streams, rivers, lakes, ponds, and bays. But the effectiveness of all three of these commissions in preserving the state's wildlife and the purity of the state's waters was compromised by the fact that personnel, such as game wardens, were often unqualified political appointees, and by the fact that violators of state regulations faced fines and other punishments that were so mild that they were often ineffective.[5]

During the pre-Peterson years, local governments such as that of the City of Wilmington did take steps to purify their drinking water and then, decades later, to dispose of increasing amounts of sewage in a hygienic manner. After all, impure water and accumulating sewage represented potential health threats to Delaware's largest city and to all other Delaware municipalities. In protecting its potable water, Wilmington again was a pioneer in Delaware but certainly not in the nation. Between 1908 and 1910, Chicago experimented with micro-organism-killing chlorine and with sand filtration to purify its drinking water. By 1914, Wilmington had joined hundreds of other American cities in adopting both procedures to lower its own rates of disease and mortality.[6]

In 1947, Wilmington's sewers were discharging a daily average of twenty-nine million gallons of wastewater, which included raw sewage, into the Brandywine and the Christina. That year, the Delaware State Board of Health ordered the city to reform its wastewater disposal practices because they were in direct conflict with the efforts of the Interstate Commission on the Delaware Basin to control the amount of untreated sewage pouring into the Delaware via its tributary streams.

The Board of Health's directive caused the city to construct interceptor sewers which rerouted the flow of untreated human waste that had formerly headed for the Brandywine and the Christiana, eastward to Cherry Island where, by 1955, a primary treatment plant was in operation. The treated wastewater was then released into the Delaware. But unlike some cities and towns in the Mid-Atlantic region, Wilmington never built a second system of sewers under its streets so that storm water could be separated from raw sewage. Rather, Wilmington's storm water and raw sewage were funneled together through the same sewers.

Most of the time the Wilmington system worked. But when storms dumped really significant amounts of rain on the city, the sudden rush of storm water overwhelmed the city's sewerage system, causing the overflow—storm water mixed with sewage—to pour into the Brandywine and the Christina. This would remain an environmental problem in Wilmington, as well as in Philadelphia and many other older American cities, into the twenty-first century.

Untreated sewage was also a significantly increasing problem for northern New Castle County's smaller but rapidly growing communities. Due to a cooperative arrangement between Wilmington and the New Castle County government, however, a number of those towns were able to solve the problem. By 1955, thanks to the creation of the Wilmington and New Castle County Sewerage System, many of northern New Castle County's municipalities were tied, by cross-county sewers, into Wilmington's wastewater treatment plant at Cherry Island.[7]

Zoning was another way for county and town governments to combat environmental degradation in the pre-Peterson era. Of course, some Delawareans point to certain zoning regulations, like those in Sussex that encouraged ranch houses on rural, half-acre lots, as doing more harm than good. As early as 1927, municipalities were authorized by the General Assembly to pass and enforce zoning ordinances, and by 1931, a few communities had zoning ordinances in place.

It wasn't until much later, however, that Delaware's three counties were empowered by the General Assembly to adopt zoning restrictions. New Castle County was first in 1951, followed by Sussex in 1956 and Kent in 1960. But zoning without planning is like driving the roads of America without a map, or sailing the Atlantic without a compass. Quite simply, a proper sense of direction is missing if planning does not precede zoning. Finally, New Castle County in 1966, and Kent and Sussex a year later, were granted the right to create planning commissions by the General Assembly. In their action or inaction during the rest of the twentieth century, county planning and zoning commissions drew considerable public criticism. One example was the 1999 report, *Protecting Delaware's Natural Heritage*, which found that county comprehensive plans had not yet fulfilled "their promise for planned land use in Delaware."[8]

By the 1960s, there was an awareness that the state's water, air and land were under siege. Wilmington attorney and environmentalist Edward H. "Ned" Cooch, Jr., remembers that, in those days "our rivers were open sewers," and the foul odor from the Delaware City oil refinery "could be smelled ten miles away" in the town of New Castle.

It was no wonder that the plans for constructing Delaware's second oil refinery, this time by Shell Oil, along a bucolic stretch of Delaware's coast produced vigorous opposition. In 1961, Shell asked the New Castle County Levy Court (today's New Castle County Council) to rezone 5,000 acres of agricultural land that Shell had acquired along the upper Delaware Bay, just north of the mouth of the Smyrna River, so that an oil refinery could be constructed on the site. Environmentalists, led by Ted Harvey, quickly organized the Delawareans for Orderly Development (DOD) to better fight against the zoning change. But, as most levels of government did in that era, the New Castle County Levy Court viewed the proposed oil refinery as a source of additional taxes and jobs, and approved the zoning change.

Delawareans for Orderly Development appealed the ruling to the Delaware Supreme Court, only to lose again. Although Shell was now free to build its refinery, for some unclear reason it chose to delay construction for a number of years.[9]

Delaware's governors who immediately preceded Russell Peterson were generally concerned with certain specific challenges to the quality of Delaware's air, land, and water, but were hesitant to embark on major new initiatives that would dramatically change the government's relationship to the environment. Democrat Elbert N. "Bert" Carvel, who served as governor from 1949–53, and again from 1961–65, had serious reservations about the Delaware City refinery complex and its negative impact on the air and water quality of the surrounding area. Carvel, however, took no action to stop expansion of that problematic industrial site. After the end of his second term in office, Governor Carvel did serve as co-chair of Delawareans for Orderly Development in an effort to stop Shell from building Delaware's second oil refinery in southeastern New Castle County.

During Carvel's second term (1961–65), realtors and developers, anxious to make a large profit, pressured the governor to back the sale of Delaware's state-owned beaches along the Atlantic Ocean to commercial corporations. They argued that the sale was in the best interests of the state government because it would transfer public lands to private holdings and thus increase tax revenues for the state's

Governor Carvel—
A portrait of Governor Elbert N. Carvel at age 65, taken in his Laurel office in 1976, some 11 years after he had completed the second of his two terms as governor (1949–1953 and 1961–1965), *photo by Richard B. Carter.*

coffers. Carvel firmly resisted the pressure and Delaware's ocean beaches remained in public hands.

In 1962, also during Carvel's second administration, the Delaware State Planning Office was established. Although it was to help provide "a favorable climate for new industries [and] businesses," in future years the State Planning Office would also concern itself with environmental issues. In 1967, for example, the state's first comprehensive plan was presented by State Planner Rudolph F. Jass. Among other things, it recommended "that all of the marsh areas along Delaware Bay south of the Chesapeake and Delaware Canal be preserved for wildlife conservation purposes."[10]

Carvel's successor, Democrat Charles L. Terry, who served from 1965–69, further protected the state's beaches by transferring their control from the Highway Commission to the Parks Commission where they logically belonged. Governor Terry's administration is also credited with encouraging the State Planning Office to think more deeply about environmental issues, and for tightening certain state environmental controls, specifically concerning the protection of drinking water and the disposal of human waste.[11]

With a few exceptions, such as the hiring of Norman Wilder in game management in 1948, and Wilder's subsequent actions in hiring a better educated and better trained staff for the Game and Fish Commission, there was little concerted effort, prior to the Carvel administration, to fill state environmental positions with trained professionals. Under the second Carvel administration, however, that policy began to change. In 1961, for example, chemist John Bryson was brought to Delaware to head the state's water pollution laboratories. Bryson subsequently headed DNREC from 1972–77, and played a key role in convincing reluctant state legislators to enact more stringent environmental laws.

A second example was N. C. Vasuki, an engineer who was hired in 1962. Through the Carvel, Terry, and Peterson administrations and most of the Sherman Tribbitt (1973–77) administration, Vasuki generally worked as Bryson's lieutenant. The two were particularly skilled in making legislators and governors aware of significant environmental issues. Watching the two administrators work Legislative Hall to persuade senators and representatives to take positive action concerning the environment was, according to a former deputy attorney general, June MacArtor, "like watching a hot knife slice through butter." In 1976, Vasuki was appointed to head the state's newly established Solid Waste Authority and would continue in that post into the early twenty-first century.[12]

The Peterson Administration

The growing sense of urgency concerning environmental issues that characterized the late 1960s was reinforced by events beyond Delaware's borders. Ohio's Cuyahoga River, once a pristine body of water, flows through Cleveland on its way to Lake Erie. By the mid-twentieth century, the river was saturated with oil and sewage. On June 22, 1969, the Cuyahoga burst into flames that leapt five stories into the sky, and sent an environmental message in a far more compelling manner than did scientific papers or investigative newspaper articles. To the citizens of Cleveland and people across the nation, the flame-engulfed Cuyahoga was simply a spectacular example of what happened when human actions pressured nature to the breaking point.

The implications for human life which, ecologists argued, is so dependent on a healthy environment, were very troubling indeed. Less than a year later, an oil spill into a small stream that ran through the suburban Delaware community of Brookside, near Newark, and the leveling of the Great Dune on Cape Henlopen by U.S. Army personnel stirred up considerable controversy. On April 22, 1970, only a few days after the Brookside oil spill and the leveling of the Great Dune, the first Earth Day was held across the nation. Nearly 20 million Americans, or approximately 10 percent of the population of the United States, participated. Historian John Steele Gordon maintains that Earth Day was "one of the most remarkable happenings in the history of democracy," and that "American politics and public policy would never be the same again." The next day, Wilmington's *Morning News* carried the front-page headline, "Earth Day Unites Nation Against Pollution."[13] Simultaneously, it seemed, national and local events occurred that fostered a new openness to environmental initiatives by the state.

Russell W. Peterson was born in 1916 in Portage, Wisconsin, and lived there until he went off to attend undergraduate and graduate school at the University of Wisconsin in Madison. With a PhD. in hand, in 1942 he went to work for the DuPont Company in Wilmington, Delaware, as a research chemist. Both the University of Wisconsin and the state of Wisconsin have long been noted for their interest in and concern for protecting the natural environment. Portage was particularly prescient because this very modest-sized community of only about 5,000 inhabitants was at the center of a very small geographic area that either produced or temporarily hosted some of the giants in the American environmental movement.

RUSSELL W. PETERSON
Governor of Delaware, 1969-1973
Courtesy of Delaware Public Archives

John Muir (1838–1914), who was the nation's most eloquent advocate for preserving wilderness, spent his formative teenage years living and working on two successive farms only about eight miles north of Portage. During those years, Muir became an observer and lover of the natural world. Aldo Leopold (1887–1948), generally credited with founding wildlife ecology, wrote his extraordinarily influential *Sand County Almanac* (1948) at his farm located about fourteen miles southwest of Portage. The *Sand County Almanac* is often compared in importance to Thoreau's *Walden* in raising American sensibilities about the beauty and the importance of the natural world. Russell Peterson called it a pivotal work in alerting Americans to the need for "an environmental revolution." As an adult, Peterson would visit the farms in the Portage area that once served as homes to both Muir and Leopold.[14]

Also born and raised in the town of Portage was Frederick Jackson Turner (1861–1932), perhaps America's most influential historian. Turner, in his "Frontier Thesis," argued that Americans were a unique people and that they owed their uniqueness to having been shaped by their interactions with the American frontier. While not an environmental historian in the strictest sense, Turner dramatically heightened the sensitivities of Americans to the extraordinary exploitation of their nation's natural resources prior to 1890, and to the way that the degradation of those resources affected the human condition. In short, Turner, according to frontier historian

Martin Ridge, made us aware "that there are relationships between society and the physical world that cannot be ignored, except at the peril of the race." It was only fitting that the first class of inductees into the Portage Hall of Fame included Frederick Jackson Turner and Russell W. Peterson. Clearly there was something about Portage and its small geographic hinterland that touched Muir, Leopold, Turner, and Peterson, and made them particularly sensitive to their natural environment. In Peterson's case, he remembers as a boy in Portage that he developed a deep love for sports but also some interest in nature. But it was only after becoming Governor of Delaware that he became a strong environmentalist.[15]

The adult Russell Peterson became physically reconnected to nature when he took up bird-watching while still working for DuPont in Wilmington. Peterson would later maintain that birds serve as barometers, "measuring the health of our ecosystem." While on birding trips up and down the state, Peterson saw the southern part of Delaware as "a relatively unspoiled paradise," particularly when contrasted "to the heavily industrialized Delaware River [Valley]."

In 1968, driven by the desire to apply reform to a number of different aspects of life in Delaware and to the archaic organization of the executive branch of its state government, Peterson accepted the Republican nomination for governor. Thanks to a white-collar constituency in northern New Castle County, Peterson defeated Democrat incumbent Charles L. Terry, but by only two percentage points. This was a narrower-than-usual margin of victory when compared to all but one of the previous gubernatorial races going back to World War II. Peterson then set about producing a series of reforms that would forever change the way that Delaware's executive branch functioned and the role that the state government would play in protecting environmentally sensitive areas.[16]

Russell Peterson brought to Dover the same logical and disciplined approach to problem-solving that served him so well in the corporate world. When faced with long festering problems, he had little use for what was politically expedient. Rather, he favored solutions that were produced by clear, analytical thinking even if the immediate results might be painful. After all, Peterson was a man of science and reason who was far more comfortable with rational discourse than with the back-slapping, relaxed interactions with constituents that were the hallmark of most successful Delaware politicians. His personal traits tended to make him appear to some, as neither "warm and cuddly or even pleasant." A number of key fellow Republicans, as well as most of the Democratic leadership, grew to resent what they perceived to be Peterson's attitude: that he, unlike them, was above politics. Helping fuel this resentment was Peterson's often repeated declaration that "I just didn't get elected to be re-elected," by which he meant that in contrast to politics-as-usual in Dover, the actions of the Peterson administration would not be driven by political expediency. As Delaware political writer Celia Cohen points out, "nothing in politics could have been more heretical."[17]

Strong bipartisan resentment against Peterson did not really surface until late June, 1971. During the honeymoon of his administration's first two years, Peterson moved quickly and successfully to implement a number of important reforms. When he took office, most of the daily operations of the state government were being carried on by an unwieldy number of commissions, boards and agencies numbering more than 140 by 1968. The chairmen and other members were private citizens who were appointed by the governor to staggered terms—or, in the case of the more important boards and commissions, nominated by the governor and confirmed by the state senate. This meant that a newly-elected governor would find many commission chairs and members impossible to remove and, therefore, often unresponsive to the new governor's wishes. Even less reflective of democratic principles was the fact that appointments to these extraordinarily numerous and powerful bodies were allocated on the basis of geography rather than population, causing most to hail from south of the Chesapeake and Delaware Canal. As Russell Peterson pointed out, "70 percent of all commission members lived below the canal, while 70 percent of all Delawareans lived north of the canal."[18]

In his inaugural year, Peterson convinced the newly-elected General Assembly, the first one in Delaware's history to be actually chosen on the principle of "one man, one vote," to replace Delaware's archaic commissions, boards and agencies with a cabinet system of ten departments. By midsummer, 1970, in ten separate legislative acts, all ten departments had been created. Except for public education, which retained its old State Board of Education, the old commission system was dead—education would not be brought into the cabinet system until 1988.

Because the cabinet system would be far more responsive to the governor's will than the commission system, it significantly increased the power of Delaware's governor. But it also made him more directly liable for blame when all was not well in the First State. The ending of the commission system and the court-ordered redrawing of legislative election districts combined to end Kent and Sussex Counties' traditional dominance of the state government. This angered many people in central and southern Delaware. In 1972, they vented their frustrations in the voting booth, and this contributed greatly to Democrat Sherman W. Tribbitt's success in turning back Russell Peterson's bid for a second term.[19]

In 1969, the General Assembly approved the Peterson administration's request to create the Department of Natural Resources and Environmental Control (DNREC), a consolidation of fourteen previous commissions, boards and agencies. Chronologically, DNREC was the second department to be created and would become the state's environmental watchdog and enforcer. Indeed, at its creation, DNREC became the "most encompassing environmental issues department in America." The secretary of DNREC was appointed by the governor and became one of the ten department heads to make up the governor's cabinet. Because of the new, clear lines of authority, it was obvious from the beginning that the attitude of the administration in Dover would dictate how aggressively DNREC would act to protect Delaware from further environmental degradation.

By 1975, DNREC would have 300 full-time and 275 seasonal employees with an annual budget of $30 million. In addition, the department administered an additional $100 million in federal and state government sewage grants. The consolidation of so many former state bodies to create just one department, however, led to certain intradepartmental tensions within DNREC. The tensions were rooted in the independent traditions and the former turf responsibilities of the fourteen former commissions, boards, and agencies that now found themselves under the same roof. In fact, one outside observer notes that, until almost the very end of the twentieth century, "the divisions in DNREC haven't co-operated."[20]

The Coastal Zone Act

With the creation of DNREC, Governor Peterson turned his attention to some specific threats to the natural integrity of Delaware's landscape and seascape. Because Delaware Bay was one of the deepest inland bodies of water along America's Atlantic Coast, industry leaders regarded it and its Delaware coastline as an ideal location for oil refineries, manufacturing plants, steel mills, and for a staging site for the transshipment of such bulk cargoes as oil, coal and iron. In early 1969, while Peterson took office in Dover, Shell Oil made quiet preparations to drill for water as the first significant step in finally constructing its controversial oil refinery in southeastern New Castle County. By then, three other oil companies had discreetly purchased a large parcel of real estate in southeastern Kent, at Big Stone Beach.

Zapata Norness, Inc., a huge bulk shipping company formerly known as Zapata Offshore (and once headed by George H. W. Bush), was secretly laying plans to construct a 300-acre artificial island in the lower Delaware Bay to serve as a transshipping center where coal coming down the bay would be stored prior to being loaded on ocean-going ships. At the same site, other ocean-going vessels would unload and store iron ore to be transferred subsequently to smaller boats and shipped up the Delaware Bay and River to steel plants. Oil, after being brought by ocean-going tankers to the artificial island, would be pumped via a planned network of pipes to the Delaware mainland and then north to the many refineries in the Delaware Valley. The envisioned artificial island was expected to dominate the bay's horizon, its massive piles of stored coal and iron ore making it the highest point of land in southern Delaware.

Also kept quiet was the desire of the Nixon administration (1969-1974) to work with certain businesses to "make Delaware Bay the premier supertanker port and industrial center in the East." Clearly, by the first year of the Peterson administration, plans and designs for unprecedented industrial development of Delaware's coastal area were in the works. But most Delawareans, including their governor, were unaware of the specifics of these plans.[21]

If unaware of the specifics, Peterson

certainly had a clear premonition of some of the alarming possibilities. Only a few weeks after taking office, he shared his growing concerns with his State Council on Planning. He began by asking council members to think hard about what kind of Delaware do "we want to pass on to our children and grandchildren?" He then examined two possibilities that were mutually incompatible. The first was a heavily industrialized Delaware with a coastline featuring a series of oil refineries, port loading and unloading facilities, and all the "benefits and problems" that they would bring. The governor's remarks resonated in the room because of the shared concerns that the environmental problems in Delaware City, and at Marcus Hook in southeastern Pennsylvania would be duplicated along the state's pristine coastal areas. Peterson went on to say that the second possibility was a preserved Delaware coastline, free of industrialization "so people can enjoy the hunting, boating, fishing, and peace and quiet..." In short, Delawareans would continue to bask in "the quality of life the coast now affords us." Peterson was convinced that the second way was "the better way," but insisted on a state-wide discussion of the options. Surprisingly, the lone newspaper account of Peterson's remarks to the State Council on Planning focused on other issues and neglected to carry in print his concerns about the future of the state's landscapes and seascapes. For the rest of 1969, the image of Russell Peterson as revolutionary environmentalist had not yet registered on the business community's radar screen—but that would soon change.[22]

In 1970, Peterson established a moratorium on all new industrial development in Delaware's coastal areas until a task force that he appointed completed its report on how best to protect Delaware's coast. Despite the fact that many of his advisors warned that a moratorium might be unconstitutional, alarm bells still didn't go off in Delaware's business community. The probable reason was that Peterson, a Republican and former DuPont executive, continued to be trusted by industrial leaders because he was "one of their own."

All of this changed dramatically on January 13, 1971, when Peterson gave the governor's annual address to the General Assembly. Although his task force's preliminary report wouldn't be completed until April, 1971, its early findings seemed to "point in one direction." As Peterson stood before the General Assembly, he told the legislators that his administration was "anxious to pump new dollars and new jobs into Delaware's economy, but not if the price is disfigurement and pollution of our environment." He pointed out that one of his most significant responsibilities as governor was the "preservation and enhancement" of Delaware's "beaches, wetlands, fields and forests, parks and recreation areas." As for heavy industries such as oil refineries and steel mills, "they gulp large chunks of land, offer relatively fewer jobs, lower tax revenues and are serious threats to the environment." Peterson reminded the legislators that his task force's study was well underway and that it "should lead to a master plan and controls..."[23]

At last, in corporate headquarters across

Delaware and beyond, it began to register that Governor Peterson might not be "one of us." At that point, Peterson recalls, "all hell broke loose." Opposition to Peterson's proposed restrictions on industrial development was led by major corporations, with the executive committee of the Delaware State Chamber of Commerce acting as their official voice. The State Chamber, subsequently joined by spokespersons for individual companies, charged the governor with giving the state an anti-business image which, they predicted, would seriously hurt Delaware's economy in the years ahead. Subsequently, the business slowdown that hit Delaware and most of the rest of America during much of the 1970s, caused the enemies of the Coastal Zone Act to blame it for Delaware's economic problems. The Building and Construction Trades Council of the AFL-CIO, expecting that the need for new refineries, factories and an artificial island would provide thousands of construction jobs, also opposed the closing of coastal areas to further industrial expansion. With the exception of former Governor Bert Carvel and a handful of others, most leading Democrats, including Sherman Tribbitt, who was then minority leader in Delaware's House of Representatives, either remained neutral or tried to block Peterson's coastal environment initiative.[24]

Crucial support for the governor's position came from Delaware's newly energized environmental activists and their growing organizations, particularly Delaware Wild Lands. Among supportive civic organizations were the American Association of University Women and the Federation of Women's Clubs. In addition, the states' three daily newspapers, Wilmington's *Morning News* and *Evening Journal*, and Dover's *Delaware State News*, supported the governor's position in their editorials. Other support came from the United Auto Workers, a labor organization with a traditional concern for the health of the natural environment, and from the American Federation of State, County, and Municipal Employees. Also standing firm behind Peterson was the leadership of the Republican Party in the General Assembly.[25]

A particularly vocal critic of Governor Peterson's Coastal Zone Bill, however, was Republican Senator J. Donald Isaacs, who chaired the state senate's Natural Resources Committee. On June 15, 1971, fresh from a Shell-sponsored visit with twelve other Delaware legislators to two out-of-state Shell refineries, Senator Isaacs announced that he welcomed the construction of Shell's long-planned refinery in southeastern New Castle County—the proposed location was in Isaacs' senatorial district. After DNREC Secretary Austin Heller, who had also visited the two out-of-state Shell refineries, reported back that those refineries "were not pollution free by any means," and then strongly promoted the Coastal Zone Bill throughout the state, Senator Isaacs attacked Heller for incompetence and called for his dismissal. Next, Isaacs turned on Peterson, charging that, by barring heavy industry from Delaware's coastal areas, the Governor was shirking his gubernatorial "obligation" to provide more jobs for Delawareans.

Peterson, of course, had heard the "jobs

argument" many times before and had always responded that his coastal zone bill would increase rather than decrease employment in Delaware. After all, refineries and steel mills were becoming more automated over time and, except for the initial construction phase, would actually provide fewer jobs than would a growing coastal tourist industry, protected by law from the considerable water and air pollution caused by expanding heavy industry. Moreover, Peterson also spoke of the quality of life issue in Delaware and the vitally related significance of maintaining an aesthetically pleasing landscape. With that in mind, Peterson said that he was determined to stop Delaware's coastline from degenerating into another Marcus Hook, Pennsylvania.[26]

On May 12, 1971, the Coastal Zone Bill, House Bill 300, was presented to Delaware's lower house by first-term Republican Representative Andrew Knox of Greenville, a DuPont employee and a strong Peterson supporter who, like the Governor, held a PhD in chemistry. HB 300, subsequently modified slightly to rally political support, specifically banned all new oil refineries, steel mills, paper mills, petrochemical complexes, and off-shore bulk transfer facilities from the Delaware Bay and from a narrow strip of coastal land, approximately two miles wide, with its western boundaries marked by a series of specifically designated Delaware roads. This narrow strip of coastal Delaware ran from the Pennsylvania line in the north to the Maryland line in the south, and included both banks of the Chesapeake and Delaware Canal as well as the coastline of the three inland bays in southeastern Sussex.

The Nixon administration, viewing HB 300 as a threat to its plans to push for the construction of an off-shore bulk transfer facility in Delaware Bay, sent a letter to the Delaware House of Representatives, urging its members to vote against the Coastal Zone Bill because it threatened to undermine America's leadership position in shipping. On June 4, Peterson was summoned to Washington, D.C., where Secretary of Commerce Maurice Stans accused him of "being disloyal to the country." Governor Peterson replied, "Hell, no. I am being loyal to future generations of Americans." Despite this and subsequent pressure from other officials in the Nixon administration, the process of turning the Coastal Zone Bill into law continued.[27]

Enemies of the Coastal Zone Bill understood that the most effective way to neutralize legislation that has popular support was not to force an up or down vote, which they stood a good chance of losing, but to weaken the bill by adding eviscerating amendments. Opponents decided to launch a major effort to gut the bill by substituting a "case-by-case" approach for the broad ban against specific categories of heavy industry such as oil refineries, paper manufacturing plants and steel mills along Delaware's coast, or the construction of artificial islands for bulk transfer in Delaware's coastal waters.

On June 21, 1971, before a packed visitors' gallery in the Delaware House of Representatives, the Coastal Zone Bill came up for a vote. After a day-long debate that extended into the evening, the case by case amendment, prepared by oil company attorneys, was introduced by the Democratic minority leader, Sherman Tribbitt. At stake was the Coastal Zone Bill's no-exception rule, which was perceived by the Peterson administration and its supporters in the General Assembly as the lynchpin of their legislative effort to protect Delaware's coast. What followed was, in Peterson's words, "an exciting, nerve-wracking moment," as the oil companies' amendment was barely turned back by a 20-to-19 vote. Then, recognizing that the passage of the bill in the House was certain and that going on record against an apparently popular piece of legislation was foolish, Tribbitt and six other Democrats switched sides and sent the Coastal Zone Bill on to the Senate with a favorable vote from 28 of the 39 House members.[28]

The next day, June 22, 1971, the vote in the Senate on the Coastal Zone Bill threatened to be another cliff-hanger. Republican Senator Isaacs introduced four amendments to weaken the bill, but his efforts were beaten back. After much debate during the day and early evening, the Coastal Zone Bill was about to be put to a final vote. But its supporters were uneasy because, although they needed only ten votes for victory, industry lobbyists were very aggressive and had "cornered senators every time they left the sanctuary of the Senate Chamber."

To counteract the influence of lobbyists and other opponents of the bill on some wavering senators, Governor Peterson, Re-publican Lieutenant Governor Eugene Bookhammer, and the Republican leadership in the Senate, led by Reynolds du Pont and Frank Grier, all applied as much pressure as they could. Somehow, the very thin majority of only one senator in support of the Coastal Zone Bill was sustained. With the vote on the Coastal Zone Bill about to commence, it was now apparent that it would win approval, though just barely. This caused a number of Democratic senators, who had previously opposed the bill, to cross over, making the actual vote a 16-to-3 landslide in favor of the Coastal Zone Bill. Rightfully ignoring the final vote tallies in both houses, Peterson accurately summarized the political struggle of the past two days when he pointed out that "our bill had eked by, in both chambers, with only one vote to spare."[29]

At 11:00 a.m., June 28, 1971, the Coastal Zone Bill was signed into law by a triumphant Russell Peterson. In an extraordinary example of how quickly political fortunes can turn, however, that very afternoon Governor Peterson told an emergency joint session of the legislature that his administration's estimate of state income, particularly the state franchise tax, was way off and that Delaware's government was in the red. After Peterson's *mea culpa*, the Democrats gathered in the House minority caucus room where, according to Celia Cohen, Sherman Tribbitt pounded the table and exclaimed, "I just got elected governor."

Although Tribbitt's prediction came true in 1972, the Coastal Zone Act would remain into the twenty-first century as a protective shield against the advent of new heavy industry and the construction of bulk transfer stations in Delaware's coastal areas—but not without numerous challenges! In 1974, 1977, 1985, and 1989, it survived determined but unsuccessful attempts to weaken it. The Coastal Zone Act's place in history is assured because, in terms of its impact on Delaware, it is one of the most remarkable pieces of legislation to be passed by the General Assembly in the past two centuries. The *New York Times* maintained that "Forty-nine other states, including New York, can learn something from this exceptionally perceptive action by Governor Peterson and his responsive legislature." Perhaps Delaware historian Carol Hoffecker best sums up the Coastal Zone Act when she notes that it "remains Delaware's greatest and most comprehensive legislative achievement toward maintaining a livable environment."[30]

To many who like to speculate about the "what-ifs of history," the passage of the Coastal Zone Act stands as the beneficiary of a brief opening of the window of opportunity in the early 1970s. As Dover attorney Mike Parkowski observes, it is important to remember that "then, the environmental movement was really hot." So short in duration was this window of opportunity, however, that David Keifer, Delaware State Planner from 1971–1977, maintains that if the Coastal Zone Act had not gone through in the 1971 legislative session, "it probably wouldn't have gotten through at all." It was just too radical a measure!

Part of the Coastal Zone Act's dramatic departure from past practices lay in the fact that it caused Delaware's state government to practice zoning directly. Although, theoretically, zoning authority generally rested with state governments across the nation, in early 1971, only Hawaii was exercising that power. The other states preferred to leave planning and zoning prerogatives to their county and municipal governments. By enacting the Coastal Zone Act in 1971, Delaware took a truly revolutionary step and became only the second state in which zoning was planned and enforced above the local and county levels. Subsequently, many other states followed the Delaware lead by enacting zoning regulations to protect their coasts. But, as both Governor Peterson and some of his opponents in the business community agreed, "none went so far" as Delaware in banning heavy industry from their coastal areas.[31]

The Federal Government steps in

While the Peterson administration was setting new standards for state involvement in protecting its natural environment, the federal government moved boldly into developing nationwide environmental standards and regulations, and created the Environmental Protection Agency to enforce those regulations. Republican Richard Nixon was then in the White House, and he signed more important environmental acts of Congress into law than any other president before or since. In Peterson's view, however, the environmental movement was then so popular that Nixon, who had no love for environmentalists, simply did what was "politically expedient." Peterson was told by some of Nixon's closest advisors that the president thought that all environmentalists were "kooks," who were primarily interested in destroying "the system" so that everyone could live "like when the Indians were here."[32]

The EPA, created by Congress in 1970, was a patchwork agency put together from parts of existing departments and agencies of the federal government. Like Delaware's DNREC, the EPA's multiple roots initially created an organization divided by many different agendas and concerns. As Daniel J. Fiorino points out, at the end of the twentieth century the EPA continued to function more as a holding company responsible for administering "more than a dozen major laws," than as an "integrated environmental management agency" where decision making was not fragmented. Among the most important early laws that the EPA enforced were the federal Clean Air Act of 1970, which was significantly amended in 1990, and the federal Clean Water Act of 1972. Unlike the unfunded Clean Air Act, the Clean Water Act authorized the federal government to provide 75 percent of the costs of sewers and sewage treatment plants to local communities with the administration of those federal funds the responsibility of the respective state governments.

A number of other environmental protection laws followed, and together they would exert direct or indirect pressure on the individual states to clean up their environments or face disciplinary action from Washington. How vigorously the EPA enforced federal environmental regulations on the states, however, reflected the agenda of the political party that controlled Congress and the White House. On those occasions, when one party exercised considerable power in Congress and the president belonged to the other party, there was a strong tendency by Congress to micromanage by spelling out, in greater detail, the specific responsibilities of the EPA. During most of the 1980s and early 1990s, the environmental acts passed by Congress left little room for EPA discretion because powerful Democrat leaders "no longer trusted the EPA" under the presidency of Ronald Reagan (1981–89) and the final two years (1991–93) of the George H. W. Bush administration.[33]

Beginning with centrist Democrat Jimmy Carter's years in the White House (1977–1981), a softening national economy caused the EPA to be more sensitive to the economic burdens imposed on businesses by environmental regulations. To enhance its image and popularity in an era where at least certain elements of the population had lost some of their earlier enthusiasm for environmental regulations, the EPA recast itself as a government agency more concerned with the nation's health than with broad-based ecological issues. For example, the EPA began to focus particularly on the threat of cancer posed by toxic pollutants in the air, water, soil, and in America's homes.

The irony of this new emphasis on preventing cancer, of course, is that actions by agencies like the EPA and Delaware's DNREC probably had, according to some sources, only a very modest or even a marginal impact on the nation's and the state's cancer rate. While many Americans believed that

SHERMAN W.. TRIBBITT
Governor of Delaware, 1973–1977
Courtesy of Delaware Public Archives

significant environmental cancer risks were caused by polluted water, toxic chemical dumps, and dirty air, a number of medical studies indicate that beyond the issue of genes, most cancers are caused by personal life-style choices. But given the public's beliefs as to the causes of cancer and other diseases, it is not surprising that it was not until the late 1980s that more broad-based ecological issues would again receive some of the same considerations as health issues on the EPA's agenda.[34]

Delaware's Governors After Peterson

From the early 1970s through the rest of the twentieth century, the actions of Delaware's government in protecting the state's environment were heavily influenced by regulations, mandates, guidelines, and funding coming out of Washington, D.C. By the late 1970s and early 1980s, the state's activities in the environmental theater were also influenced by a growing resistance from the business community to the increasing economic and bureaucratic costs of environmental regulations. As David Swayze, a former chief of staff to Governor Pierre S. du Pont, IV (1977–1985), points out, during these years "the cost burden borne by industry because of environmental controls became an increasing issue with the business com-munity."

This growing resistance continued throughout the rest of the twentieth century and impacted the strategies of both Democratic and Republican administrations, because neither party dared to antagonize Delaware's business community. In 1995, DNREC Secretary Christophe A. G. Tulou of Democrat Tom Carper's administration, reflected this political concern when he spoke of abandoning a rate fee increase for industrial polluters because it was "pretty clear [that] the public mood wouldn't support it." Tulou pointed out that resistance to environmental regulations rose out of concern by Delawareans "about the size of government." As for the environment, "while always a priority," it was not at the top of the Carper administration's list. Indeed, except for the Peterson administration, protecting Delaware's natural environment was not a top priority of any governor, Democrat or Republican, for the remainder of the twentieth century.[35]

In 1972, Democrat Sherman Tribbitt was

elected Governor of a state in which environmental policies already were partially dictated by Washington, D.C., where both the state economy and the state government's budget were in considerable disarray, where resistance to further state environmental regulations was stiffening, and yet where state initiatives were needed to meet some serious environmental problems. Because he had attempted to weaken the Coastal Zone Act in 1971, environmentalists looked forward to the Tribbitt administration with considerable trepidation. And certainly Governor Tribbitt (1973–1977) has not been subsequently portrayed as a strong champion of Delaware's environment. Yet, in view of all of the economic woes that threat-ened the state and its government during his one term in office, and the amount of energy and attention that these problems demanded from his administration, Governor Tribbitt established a surprisingly progressive record in protecting Delaware's land and water from the threat of further pollution, despite a certain amount of necessary pro-business rhetoric. In fact, when Republican Lieutenant Governor Eugene Bookhammer, a former supporter of the Coastal Zone Act, led an attempt to change that law to accommodate a huge docking facility for the unloading of coal and crude oil at Big Stone Beach, Tribbitt exercised his political influence to help the Coastal Zone Act survive unscathed.[36]

The Wetlands Act was signed into law by Governor Tribbitt in 1973. Its origins were in the Peterson administration and its justification has a certain familiar ring: "Coastal areas of Delaware are the most critical areas for present and future quality of life in the State and that the protection of the coastal wetlands is crucial to the protection of the natural environment of these coastal areas."

Essentially, the Wetlands Act pretty much ended "dredging, draining, filling, bulkheading, [and] construction of any kind..." unless a permit was first obtained from DNREC. But because the act applied only to tidal wetlands, freshwater wetlands were left unprotected by state law. A second environmental advance during the Tribbitt years concerned strong support from both Dover and Washington for a dramatic expansion of local sewage treatment systems in Kent, Sussex and southern New Castle counties, including Bethany Beach, Fenwick Island, Selbyville, Bridgeville, Greenwood, Laurel–Trap Pond, Bethel, Millsboro, Harrington, and Townsend. A third advance was the establishment of the Delaware Solid Waste Authority in 1975, which soon put an end to private, town-, and county-controlled dumps and landfills.[37]

No matter what advances they made on the environmental front, Tribbitt and subsequent Delaware governors found themselves in the position of a frustrated sailor working hard to bail out a boat springing new leaks. Quite simply, the declared toxic threats to Delaware's natural environment kept getting more numerous with time. By 1976, DNREC Secretary John Bryson was bemoaning the fact that "the multitude and complexity of chemicals considered to harm the environment appear to increase each year."[38]

Republican Pierre S. "Pete" du Pont IV's

PIERRE S. du PONT, IV
Governor of Delaware, 1977–1985
Courtesy of Delaware Public Archives

administration (1977–85) blamed the Tribbitt and Peterson administrations for Delaware's temporarily soft economy and the state govern-ment's deteriorating credit rating. New businesses, so the argument went, were simply not coming to Delaware because of the anti-business image created by the two previous administrations' passage of environmental laws, high taxes, and their practice of fiscal irresponsibility. In the spring of 1977, The *Delaware State News* responded that if Delaware suffers from an anti-industry image, it was because of an "unholy alliance of Big Oil, Big Daddy DuPont and even a few misguided Big Labor leaders" who, "aided by the big-business oriented State Chamber of Commerce," have complained so long "about Delaware's so-called 'anti-industry image...that if there is an image, they've created it."[39]

There was strong suspicion in some quarters that Governor du Pont wished to weaken the Coastal Zone Act to attract more heavy industry and industrial docking facilities. But his former chief of staff, David Swayze, maintains that "there were no conspiratorial meetings that I know of concerning the Coastal Zone Act while I was working for Governor du Pont." As Swayze recalls, "environmental concerns were not an important issue; straightening out the state's economy was." One major part of the du Pont plan to spark economic growth was the passage of the Financial Center Development Act in 1981, which attracted thirteen credit card banks to Delaware by the end of 1984. By 1998, corporate head-quarters of five of the nation's ten largest credit card banks were located in Delaware, and MBNA had become the state's second largest business employer with 11,000 full-time employees.

All of this meant that environmentally "clean" businesses rather than pollution-producing heavy industries were drawn to Delaware through the actions of the du Pont administration. Still, many environmentalists were unhappy with Governor du Pont's decision to abolish the state planning office because they viewed centralized planning as crucial to preserving "a more healthy natural environment." It would be reinstated under the Carper Administration (1993–2001), but only in an attenuated form.[40]

By the time of Republican Governor Michael N. "Mike" Castle's administration (1985–1993), many Republicans in Delaware and in other states were increasingly opposed to strong environmental controls. (Delaware's Republican U.S. Senator, Bill Roth, remained a strong exception to this trend.) But, as Sussex environmentalist Til Purnell pointed out, "Mike Castle was a happy surprise." By his second term in office, his administration was taking a fairly proactive role in guarding the integrity of Delaware's natural environment.

Of course, the realities of Delaware politics demanded, above all else, that business sensibilities be respected. In 1991, for example, the Delaware State Chamber of Commerce presented Castle's appointed Secretary of DNREC, Edwin H. "Toby" Clark II, with its first ever "Environmentalist of the Year" award. This award indicated that Delaware's business community was fairly comfortable with most of the environmental policies of the Castle administration. And yet, one senses in this gesture the concern of the business community to cement ties to a Castle administration that was exhibiting some independence from the State Chamber of Commerce's views on specific environmental laws and regulations.[41]

Michael N. Castle
Governor of Delaware, 1985-1993
Courtesy of Delaware Public Archives

During the Castle years, significant steps were taken to set aside land, either temporarily or permanently, to be preserved in a natural or agricultural condition. In 1988, the Castle administration formed the Greenspace for Delaware's Future Committee to establish a long-term program to protect significant as-yet undeveloped spaces that were "highly important" to the state's "natural and cultural heritage and biological diversity."

In 1990, nine bills protecting the environment were passed by the General Assembly and subsequently signed into law by Governor Castle. Among them was the Delaware Land Protection Act, championed by Democrat Ruth Ann Minner who chaired the Senate's Natural Resources and Environmental Control Committee. The act established the Delaware Open Space Council to review, advise, and make recommendations on land acquisition efforts. Although 250,000 acres in Delaware were identified as worth preserving, by early fall 1999 only 115,000 acres were protected by state purchases or through easements. A second important act of 1990, Delaware's Agricultural

Lands Preservation Act, was signed by Governor Castle in 1991. Its purpose was to preserve "valuable and irreplaceable farmlands and forest lands" from non-agricultural development. As Thomas R. Carper, Governor Castle's successor, would say in 1999, "at the end of the day, one of the secrets, the keys, that makes Delaware so special is the farmland."[42]

In 1986, the state's comparatively poor health record in general and its unusually high cancer rate in particular were made public in an alarming series of reports and studies. Statistics for the period from 1971 to 2000 showed that Delaware led the nation in per-person mortality from cancer, followed by Maryland, Louisiana and New Jersey. Governor Castle and his successor, Democrat Tom Carper, appointed advisory bodies to study the cancer threat and to make recommendations that would improve Delaware's record in this area.

As previously pointed out, most Delawareans were inclined to blame chemicals in the water, on the land, and particularly in the air as the primary controllable causes of cancer. But the reports of the Castle and Carper cancer advisory bodies, which appeared in 1990 and 1995 respectively, pointed the finger of guilt primarily at the unhealthy manner in which many Delawareans lived out their lives. Particularly noted was a much higher rate of smoking among Delawareans than could be found among residents of most other states. Implicit in these two reports was the call to Delawareans to stop blaming their high cancer rate on the acts of others, primarily the so-called polluters, and to take personal responsibility for their health by making good life-style choices.

After the 1990 report, for example, Governor Castle called on Delawareans to "abandon their deadly habits." He added that for every 100 Delawareans, 28 smoke, 10 abuse alcohol, 60 do not wear seat belts, 50 have high cholesterol levels, 26 are obese, and 25 have high blood pressure. After the Carper advisory committee report was released in 1995, incensed *News Journal* columnist, Ralph Moyed, charged that it was a white-wash that simply blamed the victims rather than the perpetrators. Journalist Molly Murray wrote that some Delawareans were unconvinced that personal lifestyle explained their state's high cancer rate. Rather, they continued to see "a connection with the 20 federal superfund sites in the state, the smokestacks that dot the coast from Claymont to Delaware City, the pesticides and herbicides sprinkled on lawns, and the chemicals used on the corn, soybeans and milo fields in the state." By 1999, political scientist William Boyer was arguing that both of the state-sponsored reports on cancer allowed Delaware's "business-friendly state government" to avoid "finger pointing at polluting industries."[43]

And yet those who disagreed with the state's two cancer advisory committee reports had to reconcile their skepticism with the fact that there was no conclusive research to prove that the concentrations of carcinogens in the environment at the end of the twentieth century had more than a very modest or even marginal

THOMAS R. CARPER
Governor of Delaware, 1993–2001
Courtesy of Delaware Public Archives

bearing on the state's high cancer rate. Moreover, a study of Delaware from 1998 to 2002 showed that the state's ranking had fallen to sixth place when compared to cancer mortality rates in the other 49 states. Depending on the locale, the study estimated that only an additional 2.5 to 4.4 people per 100,000 Delawareans contracted cancer due to air pollution. Of the five monitoring stations in the state, it was not surprising that Wilmington and Delaware City registered the highest rates for cancer-causing agents in the air while the lowest rates of carcinogens in the air were in central and southern Delaware.

Given these extremely modest numbers for toxic air-induced cancer and ignoring some other equally significant health problems related to poor air quality as well as to some other degraded sectors of the natural environment, some members of the business community felt that Delaware should be a little more flexible in the application of its environmental regulations, or at least not introduce any new, more stringent pollution controls. Not so, argued Russell Peterson in 1999. The former governor insisted that because so many chemical compounds were so new, they lacked long-term health track records. Therefore, it was foolish to give chemicals the same constitutional rights as humans by declaring them innocent until proven guilty. Rather, even if the scientific risk "may not be scientifically proven," the burden of proof, according to Peterson, should rest on the manufacturers to show that the public benefits of producing new products "outweigh the anticipated risk."[44]

Like that of his Republican predecessor, Mike Castle, Democrat Tom Carper's governorship was characterized by centrist policies. Although the environment may not have been a high Carper priority, by the summer of 1999, *News Journal* columnist Harry Themal characterized Governor Carper as a man "who has evolved into an environmentalist during his two terms."

At first, however, Carper was considered an enemy of the environmental cause. By late June of 1993, there were signs of tension between the Carper administration and many of the state's environmentalists over proposed Carper nominees for the Coastal Zone Industrial Control Board, the enforcing body of the Coastal Zone Act. And during his first term,

Carper seemed to soften the state's stance on environmental matters when he supported a "refinement" of the Coastal Zone Act.

In 1994, Jake Kreshtool complained about Carper's record to date, particularly his inaction in limiting toxic air emissions and his efforts to relax regulations "to guide industrial developent along the environmentally fragile coast." Other environmentalists argued that, as quoted a 1994 article in the *News-Journal*, "only when federal authorities or environmental groups have forced Delaware's hand, have major industrial polluters been prosecuted and forced to stop polluting." A spokesman for the Carper administration answered that it was getting better results with polluters through persuasion than by following the older patterns of prosecution and fines. By 1996, according to Gerard L. Esposito, then director of the Delaware Division of Water Resources, when you consider the opportunities to violate the Federal Clean Water Act, "there's better than 99 percent compliance" in Delaware. "That," said a skeptical Jerry Shields of the Dover-based Watch Our Waterways, "is a real reach."[45]

Governor Carper maintained that he was a long-time believer in voluntary plans and the traditional, free-market methods for pollution control. By 1998, however, because of growing concerns that the excess amounts of chicken manure on Sussex and Kent fields posed a serious threat to the quality of downstate Delaware's underground and surface water systems, Carper admitted that the voluntary and free-market approach had its limits.

By this time, two of Delaware's most important businesses, agriculture and tourism, were locking horns over the disposal of chicken manure. As the *News Journal* reported in 1998:

> Chicken growers, whose political might has long protected them against environmental regulation, face another, increasingly powerful force: the state's booming tourism business, which depends on clean bays and beaches to attract hundreds of thousands of out-of-state visitors.

With Carper's help, a consensus-based nutrient management program was hammered out over the next few years that would become effective in 2004. It did not please a number of environmentalists, however, because of the fifteen voting commissioners who had the power to "develop, review, approve and enforce regulations" for nutrient management, only two would be representatives of environmental advocacy groups. Moreover, the agreement was aimed at maintaining "agricultural profitability" while simultaneously trying to improve water quality. But it was a settlement that most poultry farmers and most members of the tourist industry could live with.[46]

What would have been the outcome of the chicken-manure issue if the powerful tourism industry did not view environmental reform, at least reform to fix this particular problem, as in its own best interests? It was simply a matter of whose ox was being gored that explains most of the responses to environmental issues. Two examples concerned Delaware's ocean beaches, which had never been closed until 1988. That year, despite

Power Plant— An aerial view of the first phase of the Indian River Power Plant as it appeared in 1957, soon after its construction. The plant has since been enlarged several times, *courtesy of the Delaware Public Archives.*

opposition from local merchants and realtors, Fenwick Island's oceanfront was judged polluted by the state and closed in September. In August of 1989, after testing the ocean water off Rehoboth, the Division of Public Health recommended the closure of Rehoboth's ocean beaches. But this time, the outcry from commercial interests in Rehoboth was more effective and caused the beaches to remain open.[47]

Governor Carper was instrumental in another consensus-based solution concerning environmental regulations. Although the Coastal Zone Act became law in 1971, specific regulations for administering it had not been drawn up, thanks to "persistent and successful lobbying" by the Chemical Industry Council of Delaware and the Delmarva Power and Light Company. In 1989, Governor Castle appointed nationally-known environmentalist Edwin H. "Toby" Clark II to head DNREC. After two years of listening to the voices of industrial leaders, environmentalists, and others, Secretary Clark submitted proposed regulations, as required, to the Coastal Zone Industrial Control Board.

After hearings and some alterations, the Control Board declared them in effect in 1993. But the next year the Chemical Industry Council of Delaware and the Delmarva Power and Light

Company challenged the legality of these regulations and were successful in convincing the Delaware Court of Chancery to declare them null and void on a procedural technicality. During the first three years of the Carper administration (1993–95), there seemed to be little interest in reconstituting regulations to put teeth in the Coastal Zone Act. In 1996, however, under prodding by former Governor Peterson, Governor Carper committed himself to seeing regulations enacted, and directed DNREC Secretary Christophe Tulou to begin a consensus process involving a select group of businessmen and environmentalists who would develop guidelines for regulations.

In 1998, after Secretary Tulou submitted a "memorandum of understanding" concerning specific proposed regulations to the Coastal Zone Industrial Control Board, that body held hearings and workshops, and then ruled the regulations legally binding. According to Russell Peterson, the key to success in finally putting specific regulations into the Coastal Zone Act was "Governor Carper's decision to pursue a consensus-building approach."

In addition, by 2000, the Carper administration was receiving praise from the *News Journal* for finally abandoning its soft position on pollution violators. By taking firm enforcement action against Metacham Products, one of many chemical companies clustered near the oil refinery in the Delaware City industrial complex and an egregious polluter over the previous four years, the Carper administration convinced the *News Journal* that it was setting "the right course for the next governor."[48]

As the year 2000 approached, Delaware's environmental health continued to face serious challenges. State and federal efforts to limit pollution and preserve the aesthetic qualities of the state's land and seascapes were countered by continued toxic emissions from many industrial sites, by increasing numbers of exhaust-spewing automobiles, and by the extraordinary transformation of Delaware's Atlantic coast and many inland farms into sprawling housing developments and ugly strip malls. In the face of a generally pro-business tilt by many, if not most, of Delaware's politicians, environmentalists continued their efforts to convince the state to enforce more vigorously old regulations and to place additional new restrictions on those commercial forces that, if left unregulated, promised only further destruction of Delaware's natural environment. Although the Delaware State Chamber of Commerce's chief executive, John Burris, would maintain that by 2000, Delaware's "corporations were working cooperatively with environmental organizations for the common good," there continued to be tension between the two groups. All of this was going on against a backdrop in which new environmental problems were emerging, where older ones rarely disappeared, and where material substances and processes that Delawareans thought that they understood were now taking on "new dimensions."[49]

Delaware's Air in 2000

Thanks to the standards set by the federal government, to the efforts of their state government and state environmental groups, and to the decision by most of their state's industries to abide by environmental reg-ulations, Delawareans were breathing in air that was considerably cleaner in 2000 than the air that they inhaled in 1970. The Clean Air Act of 1970 authorized the Environmental Protection Agency to establish air quality standards to preserve the health and welfare of all Americans, particularly children and asthmatics. The EPA cited first six and then seven principal air pollutants—ozone, carbon monoxide, sulfur dioxide, nitrogen dioxide, lead, and two types of particulate matter based on size—for federal regulation. For each of these pollutants, the EPA then established air quality standards that every state was expected to meet. A state that failed to meet these federal standards in specific areas was judged by the EPA to be in "non-attainment."[50]

Of the seven principal air pollutants, only in its ozone concentrations was the state in non-attainment by the end of the century. Even with ozone levels continuing signif-icantly above EPA limits, Delaware reported that from 1982 until 2000, its ozone levels had declined. Part of Delaware's continuing problem with ozone, as John Burris of the state chamber of commerce pointed out, was that much of the ozone as well as many of Delaware's other air pollutants were generated by emissions beyond Delaware's boundaries and then blown into the state by wind patterns. Because DNREC officials estimated that 60 to 70 percent of the emissions that produced ozone came from car exhaust, it was tempting to point the finger of blame for the state's non-attainment in ozone levels at automobile emissions generated in the Washington-Baltimore and Philadelphia metropolitan areas.

Yet the EPA's seven pollutants were not the only health threats found in Delaware's air. Beginning in 1988, DNREC widened the net of monitored and regulated toxic airborne materials to include asbestos, mercury, beryllium, vinyl chloride, benzene, radio nuclides, and arsenic—a material is categorized as toxic if it can harm living organisms. From 1990 to 2000, the air levels of a number of additional toxic elements, such as benzene, declined significantly. Despite a general decrease in Delaware's toxic air emissions during this time, it was clear that Delaware needed to do more to clean its air. After all, when one focuses on the fact that Delaware ranks 49th in size among the states, it was very troubling that in 2000 it ranked 35th in total pounds of toxic pollution in its air. The latter ranking meant that, on a per square mile basis, Delaware's air was significantly more toxic than average across the nation. As previously pointed out, towards the end of the twentieth century two coal-burning facilities, the Conectiv electric power plant in Edgemoor, near Wilmington, and the Delmarva Power and Light plant in southern Sussex, on the south bank of the Indian River, alternated as the leading single sources of toxic air pollution in Delaware. For the year 2000, the Indian River Power Plant held that dubious distinction.[51]

Delaware's Land in 2000

As with the quality of its air, the use and

abuse of Delaware's land continued to stir controversy. In 1990, about 19 percent of land across the nation was publically owned by national, state, county, and municipal governments. In Delaware, by contrast, only nine percent of the land was publically owned. The lower figure for Delaware is partially explained by the fact that only 2.3 percent of the state—primarily Prime and Bombay Hook National Wildlife Refuges along Delaware Bay—was federally owned, and this placed Delaware, when compared to the other states, near the bottom in the percentage of state acreage owned by the United States government. Thanks primarily to Delaware's Land Protection Act of 1990, however, by 1999 the percentage of publically-held land in Delaware had climbed to 13 percent.[52]

This dramatic change was due in no small part to the work of the Delaware Nature Society. Organized and chartered in 1964 under the leadership of Lynn Williams, who subsequently became one of Delaware's foremost environmentalists, the Delaware Nature Society grew to become the state's premier educational environmental organization. By 2001, it was both a pioneer in land preservation and a powerful advocacy group with about 8,000 members. Pivotal to the group's founding were the dramatic demographic changes in northern New Castle County after World War II, which created a serious disconnect between the natural world and the majority of Delaware's children, who were growing up either in urban Wilmington or in the suburbs.

Lynn Williams, a geology major during her student days at Mount Holyoke College in South Hadley, Massachusetts, was concerned that most of Delaware's young people would henceforth "believe that peas came out of a box and water came out of a spigot." Understandably, the original purpose of the Delaware Nature Society was to reconnect Delaware's youth with nature and then to develop in them an appreciation for the natural world and its important role in their lives. For Williams and other supporters of the mission of the Delaware Nature Society, nature education started "with acquaintance, then understanding and appreciation, and finally a responsibility for nature's world."[53]

In 1971, after retiring from a distinguished career as head of Delaware's Game and Fish Commission, Norman Wilder became the second executive director of the Delaware Nature Society. The next year, he began planning for a study to identify areas in Delaware that "were worthy of preservation because of their exceptional ecological, geological, or archaeological values." In 1974, biologist and future Delaware Nature Society environmental lobbyist, Lorraine Fleming, was hired to coordinate the study. After funding from the Crystal Trust, New Castle County and the State Planning Office, Fleming, Wilder and University of Delaware agronomist Claude Phillips traveled the state in search of places "worthy of preservation."

In 1978, thanks to a second grant from the Crystal Trust and strong support from the president of the Delaware Nature Society, Howard P. Brokaw, and others, Lorraine Fleming brought together the findings of the study and published it in hardback as *Delaware's Outstanding Natural Areas and Their Preservation.* Listing 101 natural areas worth preserving, the book helped convince the General Assembly to pass the Natural Areas Preservation System Act that same year, one of the first state laws of its type in the nation.

In 1988, the natural areas inventory was reduced from 101 areas to 67 but the total acreage targeted for protection was increased. Because, initially, the Preservation System Act did not provide public funding, property was set aside through land gifts and voluntary liens. Additional money for land acquisitions was raised through the 1983 state tax check-off program. Subsequent receipts from a 1988 bond bill allowed the state to be more aggressive about buying lands considered part of Delaware's vital natural areas.

By 1990, ten of the natural areas were removed from the inventory because of site degradation or destruction, while five more sites were added. In 1990, about 2,500 acres of vital natural areas in Delaware were under easement restrictions or state ownership, thanks to the Natural Areas Preservation System Act. That same year, the passage of the previously mentioned Land Protection Act led to a significant increase in state land acquisitions. The passage of the latter act led not only to the state preserving outstanding natural areas but also to setting aside additional acres to protect the sweeping visual perspectives offered by aesthetically pleasing landscapes.[54]

As previously pointed out, beach erosion has been a fact of life in Delaware throughout most of the last 17,000 years. By the late twentieth century, new sand was being pumped routinely onto eroded Atlantic beaches in an effort to put off the inevitable. As John Hughes, then director of DNREC's Soil and Conservation Division, pointed out in 1990, in an era of sea level rise, pumping additional sand "cannot be a successful technique to maintain a beach forever." But it "can buy time, and that's what we are attempting to do."[55] After all, no land in modern Delaware had become more valuable and therefore more worth preserving than the acres of sand and soil that faced the curling, pounding waves of the mighty Atlantic.

Further inland, wind and water erosion continued to diminish the amount and the fertility of topsoil. In northern New Castle County, where the annual erosion rate of a land that was not farmed progressively to prevent soil runoff could reach 100 tons an acre, four tons was a more typical annual rate. South of the piedmont, soil erosion was considerably less, amounting to about two tons per acre. Farmers not only continued to lose topsoil—of course, some of that topsoil washed or blew into nearby farms—they also lost significant amounts of farmland acreage to other economic activities. In 1945, for example, there were 923,000 acres of farmland in Delaware, but by 1999, that had declined to only 580,000 acres.[56]

By 2000, town, county and private dumps no longer littered Delaware's landscape. Instead, trash was now dumped and buried in the three state landfills—one in each county—run by the Delaware Solid Waste Authority. Because state lawmakers were convinced that recycling would reduce dependence on landfills, recycling was encouraged. Through most of the 1970s and 1980s, Delaware became a national leader in recycling, with particular emphasis on collecting paper, plastics, glass, steel, and aluminum. According to the *News Journal,* by 1986, Delaware recycled 40 percent of its trash. In the 1990s, however, enthusiasm for recycling waned and by 1999, only 15 percent of Delaware's trash was being recycled. That placed Delaware last in recycling in comparison to New England and the other Middle Atlantic states (Delaware north to New York.)[57]

Some Delawareans complained that the precipitous decline in recycling was caused by a lack of interest by the Delaware Solid Waste Authority. After all, DSWA was a financially self-supporting institution and was, therefore, dependent on dumpers' tipping fees to meet its expenses. Because no tipping fees were charged for the pre-separated recyclable objects placed in bins at specific sites throughout Delaware, the only significant income to support the program was generated by the sale of the recycled materials. In 1997, for example, DSWA raised only $700,000 through sales of recycled glass, paper, plastics, steel, aluminum, and other items. This was far short of the 1997 recycling program's $2,000,000 cost. For DSWA, collecting pre-separated recyclable objects was a money-losing proposition.[58]

Historically, the Solid Waste Authority's only large-scale recycling attempt—reclamation attempt is a more accurate description—was at Pigeon Point, on the Delaware River only a few miles south of the Port of Wilmington. There, DSWA's Reclamation Plant was built in 1982. It consisted of a group of buildings that were divided into two modules: one to handle solid waste from New Castle County and process it into "beneficial use," and the other to shape into beneficial use sludge from the Cherry Island sewage treatment plant.

When it was built at the cost of approximately $63 million, DSWA's two-module plant was the largest reclamation facility in the world. Financing for the facility came from a variety of sources including the EPA, state government matching funds, and the issuance of bonds by the Delaware Solid Waste Authority. Both modules took a stream of waste and the plant was required by law to turn at least 85 percent of the combined total into reclaimed, but not necessarily recycled, materials. Glass, for example, might be separated out to continue as glass (recycling), but most was transformed into other usable products, such as materials used in insulation or sandblasting (reclamation).[59]

The state contracted with Raytheon Corporation to run the reclamation facility, but in 1993 the plant was closed. Allan J. Muller of the environmental group Green Delaware, interviewed by the *News-Journal*, thought that the huge recycling plant "was very expensive...very complicated...made a lot of noise," and "really didn't work very well."

According to Pat Canzano of DSWA, the plant operated efficiently enough, but complaints concerning the strong odors that accompanied the composting of sewage sludge and solid waste organics caused the facility to shut down. The following year, the United

States Supreme Court decision in C & A Carbone, Inc. v. Town of Clarkston, New York discouraged companies like Raytheon from signing contracts with state governments to treat solid waste and sludge.

Given the realities of DSWA's balance sheet, the complaints about odors, and the Carbone decision, it was understandable that DSWA's head, N. C. Vasuki, would be less than enthusiastic about recycling and even some forms of reclamation. He regarded recyling as a costly "fashion trend," carried on just to make people feel good. Vasuki did favor as a reasonable option the incineration of trash to produce useful gases and electricity to keep trash from piling up in Delaware's landfills. Public concern over increased air pollution and accompanying health problems, however, blocked this solution.[60]

That Delaware faced a serious problem in disposing of mounting amounts of garbage and other waste was acknowledged even in Sussex where, in 1990, County Council President William B. Stevenson blamed some of it on women entering the work force in such large numbers: "If women would only stay home and spend more time canning fruits and vegetables, they would buy fewer packaged goods—and that would lead to fewer containers being thrown away." More than just a prescription for solid waste reduction, Stevenson's words exemplified the emotion-charged language that marked the turbulent "culture wars" of the second half of the twentieth century in general, and the changing role of women in particular. Even the disposal of garbage became part of that ideological battle.[61]

Thus, at the end of the twentieth century, Delaware was choosing to rely less on recycling and more on its three landfills to deal with most of its garbage and other solid waste. That seemed all right as long as the three DSWA landfills were large enough to hold the state's trash into the foreseeable future. Projections in 1999 estimated that the Cherry Island Landfill in Wilmington and the Sandtown Landfill in western Kent would meet Delaware's needs until 2023, while the Jones Crossroads Landfill, in southwest Sussex, would do the job until 2032. But even in 1999, these projections were considered overly optimistic.[62]

While Delawareans debated how best to dispose of garbage and trash across their landscape, they were of two minds concerning political restrictions on new-home locations. In its report, *The Costs of Sprawl in Delaware* that appeared in 2000, the state chapter of the Sierra Club warned that poorly planned development "has eroded the quality of life for residents..." According to an opinion poll that same year, almost 70 percent of the Delawareans sampled agreed that "government should control the location of new home construction to avoid sprawl and the increased cost of public services." And yet, 71 percent of the same poll said that they would "choose to live in a neighborhood where the homes had big yards," rather than in developments "with small yards and shared open space."[63] These contradictory poll results reflected the conflicting views many Delawareans held about their natural environment. While the majority wanted government restrictions that would protect the biological and aesthetic integrity of the state's land, air, and water, they did not want those restrictions to infringe on their own property rights.

Increasing state regulations concerning land use in general and wetlands in particular, sparked another significant controversy over property rights. In 1990, at an informational meeting in Dover sponsored by DNREC, a number of farmers and real estate agents maintained that proposed state efforts to regulate the use of freshwater wetlands "would erode their [farmers'] rights as landowners." Many at the meeting feared that they would be barred from farming "low spots" in their fields. "That's what I call 'em," said John Walton, general manager at Draper-King Cole in Milton, but "some people call 'em wetlands." When faced, in 1992, with a state proposal to protect freshwater wetlands—the Delaware wetlands legislation of 1973 protected only tidal wetlands—Selbyville farmer Clifton R. Parker expressed the views of many when he told members of the General Assembly that this "has the potential to put at risk everything I have worked for." Subsequently, the General Assembly did not act to protect freshwater wetlands—leaving the protection of these particular marshes and swamps in the hands of the United States Corps of Engineers as dictated by the U.S. Clean Water Act of 1972.[64]

At least as concerned about government restrictions compromising their land rights were private property owners Kitty and Johnny Davis who, over fifty years, had accumulated ninety-six acres of farmland in eastern Sussex County's Long Neck as a retirement cushion. At issue in 1997 was a proposed zoning plan by the Sussex County Council, a body that previously had been strongly criticized for its hands-off approach to development. The proposed restrictions on the nature of housing developments in the Long Neck area would limit the price that the couple would get for their land. An irate Kitty Davis demanded to know, "what's wrong with homes being sprawled outside urban areas?" She then compared the proposed restrictions "to communism" and suggested that "it was the work of Satan..."[65]

Although the proposed zoning plan was not passed, those zoning plans and other land regulations that were passed by county and state government created losers as well as winners. The old and challenging conundrum, faced by all levels of government, of how to simultaneously protect the interests of the individual property owner and the well-being of the general public when one is in conflict with the other, was encountered over and over again across the rapidly changing Delaware landscape of the late twentieth century.

Delaware's Water in 2000

Among the traditional sources of pollution of the state's waterways was the old Delaware habit of dumping garbage into the nearest stream. Canneries were particularly prone to disposing of unwanted vegetable and fruit scraps in this manner. The problem was that when organic scraps decomposed, the microbial activity associated with decomposition so depleted the oxygen in the stream that there was not enough left to support fish and other aquatic life.

The number of canneries in Delaware increased dramatically after the Civil War, and by 1909, there were seventy-seven scattered across the state. Although the number would diminish significantly in the decades after World War II—there were only thirty-two in Delaware in 1960—canneries continued to represent a threat to state waterways.

A case in point was Draper-King Cole of Milton in the 1960s. When told by John Bryson, then the state's chief water and air pollution officer, that it must stop dumping into the Broadkill River, the company's chief executive called Bryson a communist and threatened to use his influence to have him fired. In time, however, like a number of other businesses, Draper-King Cole came to realize the environmental problems caused by its disposal procedures and cooperated with the state. "Once you educated the businesses and industries to the harm they were doing," said Bryson, "they generally cooperated by changing most of their practices."

Of course, there were important exceptions such as Delaware's electric power plants and many of the chemical and petro-chemical companies that were located in the Delaware City area and elsewhere. Not only did many of them directly release pollutants into streams and rivers, but they indirectly polluted Delaware's waterways by smoke-stack emissions. According to Russell Peterson, it wasn't until Edgar S. Woolard became chief executive officer in 1987, that the most prominant corporation in Delaware, the DuPont Company, began to show any sensitivity to environmental issues.[66]

Until the late 1980s and early 1990s, according to Kevin C. Donnelly, director of DNREC's Division of Water Resources, most interested parties in Delaware simply did not view good water quality as a priority. Nevertheless, by 2000, many bodies of water bordering on, or wholly within, Delaware were cleaner than in 1945 or even in 1975. And yet, because of the limited quality of environmental testing in some areas and the lack of baseline data to put subsequent environmental testing figures in proper historical perspective, there was a certain frustration in 2000 that reflected fear of the unknown. John Bryson's jeremiad of 1976 that "the multitude and complexity of chemicals considered to harm the environment appear to increase each year," resonated with those concerned about the quality of Delaware's waterways.[67]

In 1996, a study of the Delaware Bay and River Estuary declared that it is "certainly not pristine, but it is much cleaner than at any time in this century." Inland, however, analyses of both surface and ground water were not as sanguine. That year, Delaware's assessment of its bodies of water listed 150 rivers, streams, bays, and ponds in which the water quality was compromised by too much bacteria, too many nutrients and other toxic elements, and too little oxygen. More than half the rivers, lakes and creeks in the state did not support swimming, and one-fourth did not support fish and wildlife.

The polluted waters of Delaware's three inland bays continued to be a particular concern as the century drew to a close. In highlighting

Plowing— Agriculture has long been the backbone of southern Delaware's economy. In recent years, Delaware farmers have had to alter some long-accepted practices, like the widespread use of poultry manure as fertilizer, to reduce their negative impact on the environment, *photo by Richard B. Carter, 1985.*

the top sources of stream, pond, and bay pollution in Delaware, a 1995 DNREC report pointed to agricultural runoff—particularly chicken manure and chemical fertilizer—as the clear leader, followed by urban runoff. Although a livestock nutrient program would be in place in the first decade of the twenty-first century to ostensibly limit the over-loading of fields with chicken manure, a 1999 study maintained that the leaching of farm nutrients from what was already on the fields into both surface and ground water "would continue to occur for twenty to fifty years."

Despite the fact that laws prohibited dumping refuse and other pollutants into Delaware's waters, drainage ditches were exempted. Across the state, the unwanted pollutants thrown into tax and other drainage ditches ranged from general trash to white-tailed deer entrails dumped by hunters.[68]

On the positive side of the water ledger was the fact that sewers were replacing outdated backyard septic systems, and, by 1995, about 80 percent of Delawareans were tied into approximately thirty municipal sewage collection and treatment facilities. But this sense of progress was tempered by the fact that, by 1999, sixty-six industries and municipalities still were dumping domestic and industrial waste-water into the state's water-ways. Although the wastewater was now treated, up until late 1997 it still contained some pollutants. Subsequent enforcement efforts by the state government, however, led one newspaper reporter to write in 1999, that the municipalities and industries that release their treated wastewater into Delaware's streams "aren't perfect." But over the last two years "they have come close... Now the [sewage] treatment plants nearly always comply with their state-issued discharge permits."[69] Environmentally speaking, this was real progress.

Delaware's Flora in 2000

The amount and type of flora that greeted the Dutch when they landed at the future site of Lewes in 1631, was significantly altered by 2000. In the former year, approximately 95 percent of Delaware's landscape was covered by mature forests. By 1910, forests covered only 25 percent of the state's landscape, and almost every tree was less than sixty years of age. At the end of the twentieth century, the area of the state covered by woods and forest had increased to 31 percent with many trees more than sixty years old. Of the 115 tree species found in the state, only 60 were native. Noticeably missing from the 2000 sylvan landscape because disease had almost wiped it out, but very much in evidence in 1631 and 1910, was the American chestnut.[70]

The loss, over the years, of Delaware's native flora was not restricted to the American chestnut. In 1997, biologist Lorraine Fleming declared that Delaware was the leading state in the percentage loss of its native plant species. The next year a study of the state's natural heritage echoed Fleming's statement when it declared that Delaware "has lost a higher percentage of its native plant species than any other state in the United States."

Bill McAvoy, botanist for the Delaware Division of Fish and Wildlife, feels that more research needs to be done before placing Delaware at the very top of the list, but does admit that, by 2005, fully 32 percent of Delaware's native flora was either extirpated, historical (no reported citing for 20 or more years), or rarely seen in the state. Those figures, according to McAvoy, certainly placed Delaware among the leading states in the loss of native flora. Not only did Delaware's native plants disappear at an alarming rate, but new non-native flora moved in to fill the vacuum. The most troublesome of the latter were Norway maple, garlic mustard, Canada thistle, kudzu, hydrilla, varieties of honeysuckle including Japanese honeysuckle, purple loosestrife, Japanese knotweed, Johnson grass, and English ivy.[71]

Particularly bothersome as a colonizer is phragmites, a ten-to-fifteen-foot tall wetland reed which had existed in North America for thousands of years. The type of phragmites that dominated broad stretches of Delaware's tidal and freshwater wetlands in the later half of the twentieth century, however, is probably a hybrid of native European and American species. When it spread into marshland, the nutrient-lacking phragmites displaced cordgrass and other, more nutritious short grasses that provided both habitat and food for a variety of wildlife.

By the end of the twentieth century, an estimated one-third of Delaware's tidal wetlands were a monoculture of this aggressive hybrid reed. Phragmites tends to invade wetlands that have been disturbed by dredging, by the creation of water impoundments, or by artificial alteration of water levels to prevent flooding. And yet, some undisturbed marshes also experienced phragmites incursions.

During the 1950s, the Delaware Game and Fish Commission began to fight back by experimenting with a number of different strategies. By the 1980s, the herbicide "Rodeo" proved to be the most effective weapon in the state's arsenal, but had only partial success in restricting the spread of phragmites.[72]

Delaware's Fauna in 2000

The nature of Delaware's fauna, like its flora, changed dramatically over time. The cougar, wolf, bear, elk, buffalo, white-tailed deer, beaver, turkey, ruffed grouse, passenger pigeon, and rattlesnake were all present in 1630 but not in 1910. By 2000, however, both the previously extirpated white-tailed deer and the beaver were back, and now so numerous that they were a problem. Between 1996 and 2000, there were even a few purported sightings, but no verifications, of cougars in northern Delaware.

Beavers were extinct in Delaware as late as 1935, but by 1991 there were between 750 and 1,500 of them in the state, and they proceeded to cut down trees and dam up streams in a manner that was not appreciated by landowners. But it was the population explosion of white-tailed deer that produced the most havoc in Delaware, because the deer contested roads with motorists, helped themselves to farmers' corn and soybeans, and devoured the vegetable and flower gardens of plant lovers. In 1996, Joseph Calhoun of Sussex complained that "it gets worse every year." That same year, with the estimated deer population in the state reaching 25,000, the Delaware Farm Bureau projected that annual damage from deer "was as high as 25 percent of the total crop on some farms." Not only did deer eat much of the crop, they destroyed grain by lying down in wheat fields.[73] Clearly, Delaware's initial handling of its newly-established deer population is just one more

Year	***Number of Deer Legally Killed in Delaware***
1954	***505***
1965	***749***
1970	***956***
1975	***1,272***
1980	***1,754***
1985	***2,439***
1990	***5,066***
1995	***8,778***
2000	***10,741***

Source: Delaware Division of Fish and Wildlife

White-tailed Deer, *photo by Richard B. Carter, 1985.*

example of unexpected consequences.

The wholesale slaughter of white-tailed deer across the United States from the seventeenth century to the late nineteenth century, caused the national population to hit rock bottom by 1890. One estimate of the combined number of white-tailed deer in Canada and the United States for that year was only 300,000. Subsequently, however, the population statistics began to move upward as recreational hunters put pressure on their state governments to bring back the white-tailed deer. Across the nation, sportsmen's organizations and state agencies imported deer to states and regions where they no longer existed and then, initially at least, banned hunting of the fledgling populations. In 1900, the U.S. Congress passed the Lacey Act which outlawed the interstate shipment of game killed in violation of state laws, and this finally forced commercial deer hunters, who had done so much to decimate deer herds over the centuries, out of business. The restoration of the deer population in neighboring Pennsylvania deserves some mention because most, if not all, of the deer that formed the basis of Delaware's eventual population explosion came either directly or indirectly from the Keystone State. In the 1880s, there were very few, if any, deer in Pennsylvania. By 1907, white-tailed deer had made enough of a comeback for Pennsylvania to introduce a statewide hunting season. Despite legal hunting, the Pennsylvania deer herd increased so rapidly that, by the mid-1920s, it denuded rural areas of certain types of vegetation, and this caused many deer to starve to death.[74]

Further south, however, deer restoration lagged far behind. There were no authenticated deer sightings in Delaware from the 1840s to the 1920s. Former State Representative Charles P. West of Gumboro, in southern Sussex, saw his first deer in 1927, when he almost bumped into a large buck near his family's potato house. However, West did not see another white-tailed deer in Sussex until the late 1930s, when deer were again occasionally sighted throughout Delaware after being conspicuously absent for almost a century. Evidently, the reintroduction of deer to Delaware was not planned and carried out by sportsmen's clubs or state officials as it was elsewhere.

Division of Fish and Wildlife officials maintain that, unlike the beaver, which was artificially reintroduced to the state in the late 1930s, they have no recollection or record of deer being artificially reintroduced. Rather, they think that Delaware was slowly repopulated by deer migrating south from Pennsylvania into New Castle County or indirectly, via Maryland's Eastern Shore, into the western portions of the First State. In any case, just prior to World War II, for the first time in almost a hundred years, an occasional deer was spotted on those vital environmental edges where field met forest.

Subsequently, thanks to the ban against hunting, the deer population grew rapidly. By 1950, Delawareans were being warned that "large numbers of white-tailed deer can be destructive." It was well-known that an exploding deer population could threaten reforestation, agricultural crops, small game habitat and, by denuding much of the forests, its own future food supply. In 1954, Delaware had its first legal deer-hunting season since the shooting of the species was outlawed in 1841. That and the subsequent number of legal annual killings by hunters gives us some sense of the exploding population.[75]

In 1992, with approximately 15,000 deer in the state, Lloyd Alexander, state wildlife administrator, complained that the deer "have adapted to urban-suburban areas better than anybody ever expected." Alexander went on to note that "the state has so many deer now that they are causing crop damage in some areas and road accidents in others." Clearly, "the [deer] harvest has not kept up with production." In 1997, with new, liberalized hunting regulations leading to the killing of a record 10,000 deer, there was hope that, for the first time, the deer population might be stabilized. But others were rightly skeptical. In the meantime, deer continued to wreck environmental havoc by cleaning out young tree growth along forest floors in Delaware and across the nation. Delaware's deer herds were so numerous by the end of the twentieth century that they became "de facto forest managers," their eating habits determining the future appearance of much of the state's woodlands.[76]

Wild turkeys were another once-extinct Delaware fauna. A combination of commercial and subsistence hunters wiped out the state's

Horseshoe Crab Harvest—
Tens of thousands of horseshoe crabs are shown in these 1928 photos, collected on a dock and in an adjacent field near Kent County's Bowers Beach to be sold as fertilizer. Although demand for the ancient, crab-like creatures for that purpose declined in later years, they remained popular for use as bait by local watermen and for use in drug testing by pharmaceutical companies, to the point where their numbers began to decline precipitously, *courtesy of the Delaware Public Archives.*

---

wild turkey population by the early nineteenth century. After a failed attempt in 1980 to reintroduce the bird to Delaware, a 1984 effort was successful, when thirty-four wild turkeys imported from Pennsylvania, Vermont and New Jersey were released. By 1989, there were approximately 350 wild turkeys in the state. Ken Reynolds, who coordinated the wild turkey reintroduction program for the Division of Fish and Wildlife, hoped that the state's wild turkey population would someday reach 3,000 to 5,000 birds. In any event, the gradual growth of the wild turkey population made possible the first hunting season in 1991. Ninety-three birds were harvested. By 1997, there were approximately 2,000 to 2,500 wild turkeys in Delaware, with an all-time high of 152 birds killed that year by hunters. Unlike deer, however, the wild turkey population grew much more gradually up to the year 2000.[77]

Canada Geese— landing in a southern Delaware marsh, *photo by Richard B. Carter, 2006.*

The interrelationship of certain migrating birds and Delaware Bay horseshoe crabs caused DNREC Secretary Clark to exclaim in 1990 that "one of the most significant ecological events in North America, if not the world," was happening along part of Delaware's coastline in the late spring of each year. For approximately two weeks in May, red knots, ruddy turnstones, sanderlings, semi-palmated sandpipers, black-bellied plovers, and at least ten other species of shore birds, perhaps 400,000 individual birds in all, paused on their way north from as far away as Tierra del Fuego, on the southern tip of South America, to feed on the eggs of horseshoe crabs freshly laid on the coastal sands of Delaware Bay. Only the vast Copper River Delta in southern Alaska hosted more shore birds each spring. This carefully-timed pause to coincide with the laying of eggs by horseshoe crabs, allowed the migratory shore birds to fuel up for the remainder of their flights north that usually ended in arctic breeding grounds in Canada. By 2000, however, experts were warning that "this delicate [ecological] balance was in jeopardy" because the horseshoe crabs in the Delaware Bay were declining in number.[78]

Horseshoe crabs are not crabs at all. They are more closely related to spiders than to crustaceans, and have been on earth for more than 190 million years. Their presence in 2000 along the New Jersey and Delaware shores of the Delaware Bay represented the greatest concentration of spawning horseshoe crabs anywhere in the world. As early as the nineteenth century, horseshoe crabs were being

Snow geese— shown congregating on a pond at Prime Hook National Wildlife Refuge near Sussex County's Broadkill Beach, *photo by Richard B. Carter, 2007.*

pulverized and used as fertilizer. By the 1990s watermen increasingly were using them for bait to catch eels, conch and catfish, while pharmaceutical companies used their blue blood, which clots on contact with impurities, to test drugs. All of this meant that there was a profitable commercial market for horseshoe crabs. That, in turn, caused commercial catches to rise and overall numbers of horseshoe crabs to decline.

In 1991, aware of the potentially devastating impact that uncontrolled commercial harvests of horseshoe crabs represented, Delaware began to place restrictions on catching horseshoe crabs in its waters. But evidence indicates that the horseshoe crab population declined during the 1990s, causing the Atlantic States Marine Fisheries Commission to demand that Delaware, as well as all of the other eastern states from Maine to Georgia, cut commercial horseshoe crab harvests by 25 percent by May 1, 2000. At stake for Delaware and the world was the continuity of a timeless and marvelous ecological event.[79]

By 2000, many other birds were facing difficulties. Of the 418 species that were either year-round or part-time Delaware residents, 93 were "in trouble." But there were some counter trends among a few species of the larger birds. Geese provide a good example. The number of Canada geese, which began to rise in

Ospreys— A view of an osprey nest on the upper Indian River near Millsboro, *photograph by Richard B. Carter, 1972.*

Delaware during the 1940s, reached about 50,000 by 1974, topped 180,000 by 1980 and then dropped back down to 60,000 by 2000. (These were census estimates in the fall or winter of each year.) At first, the increase in Canada geese was seen as a positive development by both hunters and environmentalists. It was not long, however, before Canada geese were testing the patience of Delawareans by descending on yards, golf courses, and farm fields, tearing up grass and depositing goose manure everywhere. Moreover, an increasing percentage of these geese were remaining in Delaware year-round. Many of the latter were probably descendants of the live-decoy Canada geese used by hunters before that practice was outlawed in the 1930s. A 1998 headline in the *News Journal* stated the obvious: "Canada geese wearing out their welcome."[80]

Even more upsetting was the dramatic increase in the number of Delaware's snow geese because, despite being smaller than Canada geese, they were even more destructive. By 1989, the snow geese were "a terrific nuisance," because they tended to pull up winter wheat, marsh and other grasses by their roots. That year, for example, snow geese turned a salt marsh at Bombay Hook National Wildlife Refuge "into a half-mile-wide lake." Snow geese were spotted in Delaware by Europeans as early as the seventeenth and eighteenth centuries.

In the 1790s, ornithologist Alexander Wilson noted that snow geese lived on Reedy Island in the Delaware River and in the old Duck Creek area of northeastern Kent during the fall, where they fed on reed roots, "tearing them up from marshes like hogs." During the 1920s, however, the snow geese population was so small in Delaware and throughout the world that the bird was almost an endangered species. Increased numbers did not descend on the state until the 1970s, when thousands of snow geese discovered Delaware's grain fields. Fall census estimates had snow geese increasing from 3,000 in 1974 to 10,000 by 1979. The first snow goose hunting season in Delaware was instituted in 1976, but not enough were harvested in subsequent years to head off an ecological disaster. The snow geese annual fall population figures for Delaware sky-rocketed to 86,000 in 1989 and to 185,000 in 2000. By

Great Blue Heron— A great blue heron fishing in the waters of Herring Creek, a tributary of Rehoboth Bay, *photograph by Richard B. Carter, 1975.*

the end of the twentieth century, each fall and winter their shimmering white feathers blanketed great swaths of unlucky wheat fields and saltmarshes throughout the state.[81]

The American bald eagle suffered enormously from the impact of DDT and other pesticides, habitat loss, and illegal hunting. By 1978, there were only four active bald eagle nests in Delaware and only one produced young eagles. Although, by 1999, bald eagles remained on the state's endangered list, state wildlife biologists reported that the state's fourteen nesting pairs of eagles produced twenty-one chicks that survived "at least to banding age." Ospreys (fish hawks) are another indicator species that reflected in their numbers the decline of Delaware's natural environment through the pesticide poisoning that occurred in the years after World War II.

Prior to the use of DDT in the late 1940s, ospreys were commonly seen in Delaware. Proof of this was the sighting, in 1942, of more than ninety osprey nests in Kent County, most located east of Route 113. By the 1970s, however, osprey nests were a rare sight anywhere in the state. By 2003, in a comeback worthy of the eagle, Delaware's osprey population numbered more than 240 adult birds. This population growth was reflected in the Prime Hook National Wildlife Refuge, where, because of pesticide poisoning, no ospreys bred in the refuge from 1972 to 1994. By 1999, however, there were three osprey nests at Prime Hook.[82]

Unlike some birds of prey, many of Delaware's wading birds did not make a numerical comeback in the late twentieth century. In fact, with the possible exception of the cattle egret, the numbers of beautiful and elegant long-legged wading birds, including herons, ibises, and most egrets, were clearly in decline. The largest of Delaware's three rookeries, which was also the largest nesting spot for wading birds along the entire Atlantic Coast north of Florida, was located on the north end of Pea Patch Island, in the lower Delaware River. The wading bird population there declined from 12,000 nesting pairs in the mid-1980s to only half that number in 1996. Experts could not be sure of the cause but did suggest that some of the newer insecticides were causing nerve damage which allowed the tiny dermestid beetle to bite the birds unnoticed. Eventually, the bites became infected and killed the host birds.[83]

Horseshoe crabs, blue crabs, shad, weakfish (sea trout) and striped bass are marine life of special concern to Delawareans. By the end of the twentieth century, except for the horseshoe crab, all were either relatively healthy or seemed to be recovering in Delaware waters. Of seven other finfish and four other shellfish, only the Atlantic sturgeon seemed to be in serious trouble. But, just as with its flora, twenty-one non-native fish were now found in Delaware waters. They include rainbow, brook and brown trout, bluegill, smallmouth and largemouth bass, and walleye. All of this meant that along many of Delaware's freshwater streams, the majority of fish landed by sportsmen were not the same species that were trapped in weirs by native Americans prior to the contact period.[84]

Some Afterthoughts

In looking back over Delaware's environmental history, one is struck by the fact that the state was a very different place in 1631 than it was in 2000. Most significantly, there were few humans in Delaware in 1631, but approximately 784,000 in Delaware in 2000 A.D., which amounted to 401 people per square mile. That, in a nutshell, is a major explanation for what has happened to the state's landscape, seascape, and air in the intervening years. While the increasing numbers of Delawareans rearranged—some would say shattered—their natural world to suit their own economic and lifestyle needs, in the words of environmental writer Tom Horton, they simultaneously "disconnected" themselves "from knowing or even the need to know the land in supersensory ways."

Long before the twentieth century, Delawareans and most other Americans had lost what their earlier ancestors once had: the ability to perceive nature through the prism of certain acute sensibilities and instincts that were unscientific by modern standards. These same unscientific sensibilities and instincts that are so alien to us today stirred the imagination and produced a certain sense of awe in our distant ancestors. Or, as environmental historian Linda Nash points out, the newer

Sandpipers— A pair of sandpipers on a Sussex County beach, *photo by Richard B. Carter, 1998.*

technologies and modern cultural practices "interposed between the individual and [his or] her surrounding environment" to push aside "earlier ways of experiencing a place."[85]

First the Agricultural Revolution and then the Industrial Revolution, with their attendant scientific perspectives on the natural world, effectively purged from society the ancient ways of viewing nature. In the process, a poetic perspective was lost and a literalist/materialist perspective gained. Although the change usually took centuries to occur, occasionally it was telescoped into one lifetime.

After Mark Twain acquired the bits and pieces of practical scientific information about the Mississippi River crucial to successfully piloting a riverboat, he admitted to a personal loss that could never be restored. The day came when Twain "began to cease noting the glories and the charms which the moon and the sun and the twilight wrought upon the river's face." Indeed, "all of the grace, the beauty, the poetry had gone out of the majestic river." Now, the only value that any feature of the Mississippi had for him "was the amount of usefulness it could furnish toward compassing the safe piloting of a steamboat."[86]

Bereft of the supersensory skills and imaginations of their distant ancestors, and intent on manipulating the natural world to meet their own material needs, Delawareans moved into the twenty-first century with tensions continuing between the business community and environmental groups. These tensions were fueled by the fact that most members of each group were suspicious of the programs and agendas of the other. Prior to 1970, environmentalism appealed to only a very small sector of Delaware society. But after that date the environmental movement became increasingly popular because it positioned itself as the antithesis to the individualism and to the growth fetish that historically fueled the business and industrial expansion that threatened the environment.

By the year 2000, probably a majority in Delaware's powerful business community continued to believe that environmental regulations threatened both profits and job growth, while some environmentalists countered that what was good for business was probably bad for the environment. Both positions, of course, were not necessarily so.

Certainly, the economy of Rehoboth Beach owes a lot to the environmental regulations that keep its ocean waters and beaches relatively clean. How many tourists would come to that resort community if the Atlantic were full of toilet paper, oil slicks, and other refuse that commonly flowed into the ocean from the Delaware River and Bay, from Delaware's inland streams and inland bays, and from its municipal sewers prior to the era of environmental regulations? For that matter, who would move to northern Delaware to fill industrial jobs if the air were considerably more polluted than it already was, or if the drinking water from both surface and aquifer sources was highly contaminated by toxic substances?

Both business and environmental groups need to recognize that what is good for business or for the environment does not necessarily threaten either the environment or the economy. Business success in Delaware can be compatible with a relatively clean, healthy natural environment. But unreasonable business demands or extremist-driven environmental regulations can threaten that compatibility. Cost and risk assessments that evaluate the financial burden placed on the private sector by environmental regulations must be weighed against the public harm caused by the lack of those regulations. This will probably be the primary modus operandi of the future as the state examines the need for additional environmental regulations and their resulting impact on Delaware's economy.

The future health of Delaware's natural environment is partially in the hands of forces beyond the First State's borders and therefore, beyond the state's direct control. The gradual destruction of the earth's (good) ozone layer, which shields us from the sun's harmful rays, and the emissions of greenhouse gases from other states and nations as well as from Dela-ware, which trap heat on the earth's surface, have created some very significant global-warming problems that will extend well into the future. Only efforts of a national and international scope can meet these challenges.

As J. R. McNeill points out, once upon a time there existed ecological buffers such as open land, unpolluted water, and clean air, and they "helped societies weather" environmental difficulties. Unfortunately, by the end of the twentieth century, these buffers were extremely

Inland Bay Sunset— Delaware's future holds both promise and danger. This pristine view of a kingfisher flying in a summer sunset near Burton's Island, between Indian River and Rehoboth Bays, shows what is at stake, *photograph by Richard B. Carter, circa 1985.*

rare in Delaware or, for that matter, in the rest of the world.[87] Instead, polluted air from nearby states and the industrial Midwest continued to roll over Delaware, adding to our own state's emissions of dirty air. Even in rural southern Delaware the air was not clean.

By the year 2000, it was impossible to know with certainty whether or not Delaware was following an environmentally unsustainable path, but there was plenty of evidence to indicate that it was. Continuing to follow an unsustainable path places the future health and quality of life of Delaware's citizens at considerable risk.

William Henry Williams
Georgetown, Delaware — 2007

Reference Notes

Some Abbreviations used: D.P.A. — Delaware Public Archives; H.S.D. — Historical Society of Delaware; DNREC — [Delaware] Department of Natural Resources

Quotation from the *Tao Te Ching*:

1. Stephen Mitchell, *Tao Te Ching* (New York, 1988), 39.

INTRODUCTION:

1. Dates of the *News Journal* quotations are in the text. The Molly Murray quote is from a phone conversation between Ms. Murray and the author on Aug. 30, 2000. For a brief discussion of research on what exactly caused the fish kills of 2000 and of other years see Molly Murray, "Scientists Want to Know Why Sussex Fish are Dying," *News Journal* (July 18, 2005).

2. Donald Worster, "Transformations of the Earth: Towards an Agroecological Perspective in History," *Journal of American History*, vol.76 (March, 1990), 1090.

PROLOGUE:
Before the Arrival of Europeans

1. Horace C. Richards, "The Geological Story of Delaware," H. Clay Reed, ed., *Delaware, A History of the First State*, (New York, 1947), vol. I, 12; John C. Kraft, "Geology," Tracy L. Bryant and Jonathan R. Pennock, eds., *The Delaware Estuary: Rediscovering a Forgotten Resource* (Cinnaminson, NJ, 1988), 32–33; Jay F. Custer, *Delaware Prehistoric Archaeology: An Ecological Approach* (London, 1984), 31.

2. Kraft, "Geology," 32–33; Richard J. Dent, Jr., *Chesapeake Prehistory; Old Traditions*, New Directions (New York, 1995), 77–82; Custer, *Delaware Prehistoric Archaeology*, 31, 44–46; Jay F. Custer, *Prehistoric Cultures of the Delmarva Peninsula; An Archaeological Study* (London, 1989), 90-91; David W. Steadman, "A Long-Term History of Terrestrial Birds and Mammals in the Chesapeake-Susquehanna Watershed," Philip D. Curtin, Grace S. Brush and George W. Fisher, eds., *Discovering the Chesapeake: The History of an Ecosystem* (Baltimore, 2001), 99.

3. Custer, *Delaware Prehistoric Archaeology*, 45–46; Dent, *Chesapeake Prehistory*, 81.

4. Dent, *Chesapeake Prehistory*, 94, 192.

5. For a brief discussion of competing theories on the destruction of global and North American megafauna at the close of the Ice Age see Anthony J. Sutcliffe, *On the Track of Ice Age Mammals* (Cambridge, Mass., 1985), 200–209; Anthony Stuart, "Mammalian Extinctions in the Late Pleistocene of Northern Eurasia and North America," *Biological Reviews*, vol. 66 (1991), 453–462; Shepherd Kirch III, *The Ecological Indian: Myth and History* (New York, 1999), 38–43; and Richard Klein, "The Impact of Early People on the Environment: The Case of Large Mammal Extinctions," J.C. Jacobsen and J. Firor, eds., *Human Impact on the Environment* (Boulder, Colo., 1992), 13–34. A few examples of those strongly supporting the view that Native American hunters were "super predators," at the end of the last Ice Age are Bernard

Powell, "Were These America's First Ecologists?" *Journal of the West*, vol. XXVI, no.3 (1989), 17–25; and Tim Flannery, *The Eternal Frontier* (New York, 2001), 194–205. The pioneer of the view that Native Americans were "superpredators" at the end of the last Ice Age was Paul S. Martin, "Pleisticine Overkill," *Natural History*, vol. LX (Dec., 1967), 32–38.

6. Dent, *Chesapeake Prehistory*, 82, 142.

7. Custer, *Delaware Prehistoric Archaeology,* 41–42; Custer, *Prehistoric Cultures of the Delmarva Peninsula*, 86. For a quick summary of alternative views concerning the geographic origins of North and South America's first human inhabitants see *Smithsonian*, vol. XXXV, no. 8 (Nov., 2004), 90–98.

8. John C. Kraft, "Geology," 32; Richards, "The Geological Story of Delaware," 12.

9. Jay F. Custer, *A Guide to Prehistoric Arrowheads and Spear Points of Delaware* (n.p., 1996), 2, 39; Custer, *Delaware Prehistoric Archaeology*, 148; Dent, *Chesapeake Prehistory*, 245, 247, 268; J.H. Blitz, "Adoption of the Bow in Prehistoric North America," *North American Archaeologist*, vol. 9 (1988), 123–147.

10. Custer, *Prehistoric Cultures of the Delmarva Peninsula*, 54, 126–127, 184; Custer, *Delaware Prehistoric Archaeology*, 50, 63-64;

11. Custer, *Prehistoric Cultures of the Delmarva Peninsula*, 97; Custer, *Delaware Prehistoric Archaeology*, 50, 97; Dent, *Chesapeake Prehistory*, 130, 143.

12. Helen C. Rountree and Thomas E. Davidson, *Eastern Shore Indians of Virginia and Maryland* (Charlottesville, Va., 1997), 1.

13. Custer, *Delaware Prehistoric Archaeology*, 91; Dent, *Chesapeake Prehistory*, 91, 190–192, 208; Rountree and Davidson, *Eastern Shore Indians of Virginia and Maryland*, 12.

14. Custer, *Delaware Prehistoric Archaeology*, 91, 97: Dent, *Chesapeake Prehistory*, 189, 191–192.

15. For his personal explantion of geographic boundaries of the Lenape see Marshall Joseph Becker, "Lenape Population at the time of European Contact...," *Proceedings of the American Philosophical Society*, vol. CXXXIII, no. 2 (1989), 116–117; and Becker, "A Summary of Lenape Socio-Political Organization and Settlement Pattern at the Time of European Contact," *Journal of Middle Atlantic Archaeology*, vol. IV, (1988), 81. Rountree and Davidson, *Eastern Shore Indians of Virginia and Maryland*, 26, make the point that Delmarva Indians of the Late Woodland period (1000 A.D.–1600 A.D.) were the ancestors of the Indian peoples encountered on the Peninsula when the first Europeans arrived. The low population estimate for the Lenape in Delaware in 1600 A.D., (approximately 350) are the author's based on the 360 to 500 for Pennsylvania and Delaware combined for the Late Woodland period in Marshall Joseph Becker, "Lenape Population at the Time of European Contact...," 113, 116; and a more recent e-mail from Becker (June 30, 2006) that maintains that there were, perhaps, only 150 Lenape in Delaware in 1600. The higher population estimate is simply an abitrary figure based on a few conversations with people interested in the population of precontact Delaware. Dent, *Chesapeake Prehistory*, 264, maintains that there were only about 655 Nanticokes living on the entire Delmarva Peninsula by the contact period. Probably only a small percentage of that total lived in Delaware.

For a brief study of the "Siconese," but with no population estimates, see C.A. Weslager, *The Siconese Indians of Lewes, Delaware* (Lewes, DE., 1991). This relatively low population estimate of Native Americans in Delaware just prior to the contact period is supported by inference in Rountree and Davidson's *Eastern Shore Indians of Virginia and Maryland*, 20. They maintain that only a few thousand Indians were living on Virginia and Maryland's Eastern Shore, a land area much larger than Delaware, just prior to the arrival of Europeans. For smallpox among native peoples of the Delaware Valley, see Herbert C. Craft, *The Lenape-Delaware Heritage, 10,000 B.C. to A.D. 2000*, (Summit, NJ, 2000), 389; Samuel Smith, *The History of the Colony of Nova-Caesaria, or New Jersey: Containing an Account of the First Settlement...and Other Events in the Year 1721*, (Burlington, NJ, 1975), 101–102. Thomas J.

Sugrue, "The Peopling and Depeopling of Early Pennsylvania: Indians and Colonists," *Pennsylvania Magazine of History and Biography*, vol. CXVI, no.1 (Jan., 1992), 11–12.

16. The 2000 population per square mile statistics are found in *Statistical Abstract of the United States, 2001*, (Washington, D.C., 2001), 21. For estimates of Native American populations in regions north of Delaware see Gordon J. Whitney, *From Coastal Wilderness to Fruited Plain* (Cambridge, U.K., 1994), 101.

17. Jared Diamond, *Guns, Germs and Steel: The Fates of Human Societies* (New York, 1997), 88–89; Carolyn Merchant, "Gender and Environmental History," *Journal of American History*, vol. LXXVI (March, 1990), 1120; Marshall Joseph Becker, "A Summary of Lenape Socio-Political Organization and Settlement Pattern at the Time of European Contact," (1988), 79–82.

18. Dent, *Chesapeake Prehistory*, 222; Marshall Joseph Becker, "A summary of Lenape Socio-Political Organization and Settlement Pattern at the Time of European Contact," 79–80; Weslager, *The Siconese Indians of Lewes, Delaware*, 17.

19. Becker, "A Summary of Lenape Socio-Political Organization...," 80; Becker, "Lenape Population at the time of European Contact...," 113, 118; Custer, *Delaware Prehistoric Archaeology*, 156; Thomas J. Sugrue, "The Peopling and Depeopling of Early Pennsylvania: Indians and Colonists, 1680–1720," 7, n.7; Herbert C. Kraft, *The Lenape-Delaware Heritage*, 283-284; Marshall Joseph Becker, "Lenape Maize Sales to the Swedish Colonists," Carol E. Hoffecker et al., eds., *New Sweden in America*, (Cranbury, N.J., 1995), 121–133; Adolph B. Benson, ed., *Peter Kalm's Travels in North America*, I, (New York, 1964), 259, 269.

20. Herbert C. Kraft, *The Lenape-Delaware Heritage*, 206; Dent, *Chesapeake Prehistory*, 246–247, 268; Custer, *Delaware Prehistoric Archaeology*, 97, 148.

21. Rountree and Davidson, *Eastern Shore Indians*, 46, 35; Custer, *Delaware Prehistoric Archaeology*, 162.

22. Albert Cook Myers, ed., *Narratives of Early Pennsylvania, West New Jersey and Delaware, 1630-1707* (New York, 1967), 15; C.A. Weslager, *Delaware's Forgotten River* (Wilmington, 1947), 63; Rountree and Davidson, Eastern Shore Indians, 12.

23. Rountree and Davidson, *Eastern Shore Indians*, 10, 12, 17, 18.

24. Rountree and Davidson, *Eastern Shore Indians*, 19, 20; J. Thomas Scharf, *History of Delaware, 1609-1888*, (Philadelphia, 1888), vol. II, 1169n.

25. Custer, *Prehistoric Cultures of the Delmarva Peninsula*, 127, 184, 327–329.

26. Thomas Hobbes, *Leviathan*, Nelle Fuller, ed., *Great Books of the Western World*, vol. XXIII (Chicago, 1952), 85; Diamond, *Guns, Germs and Steel*, 104–105; Mary A. Stewart, *What Nature Suffers to Grow; Life, Labor, and Landscape on the Georgia Coast,1680–1920* (Athens, Georgia, 1996), 23.

27. Diamond, *Guns, Germs, and Steel*, 127–128, 157–164.

28. Rountree and Davidson, *Eastern Shore Indians*, 10, 12, 17, 18, 26; Custer, *Prehistoric Cultures of the Delmarva Peninsula*, 316.

29. Custer, *Prehistoric Cultures of the Delmarva Peninsula*, 345, 329; Custer, *Delaware Prehistoric Archaeology*, 85–87; Dent, *Chesapeake Prehistory*, 267.

30. Becker, "A Summary of Lenape Socio-Political Organization and Settlement Pattern" 82; Becker, "Lenape Population at the Time of European Contact," 113.

31. Rountree and Davidson, *Eastern Shore Indians*, 35–37. Although Rountree and Davidson do not write specifically about the Lenape, their description of seminomadic movement of other Delmarva Indian peoples in the Precontact Period probably fits the Lenape as well.

32. Ibid.

33. Ibid.

34. For general information on the Nanticokes see C.A. Weslager, *The Nanticoke Indians—Past and Present* (Cranbury, NJ, 1983). For general information on the Siconese see Weslager, *The Siconese Indians of Lewes, Delaware*. Weslager maintains that the Siconese were really part of the Lenape people, but some other archaeologists are skeptical. For the view that the Nanticokes and Siconese were more dependent on agriculture than the Lenape of central and northern Delaware, see C.A. Weslager, "The Indians of Delaware," H. Clay Reed, ed., *Delaware: A History of the First State* (New York, 1947),47–48, 49; Weslager, *The Siconese of Lewes, Delaware*, 23, 25; discussion with Marshall Joseph Becker.

35.For the view that the Nanticokes and Siconese were probably less egalitarian during the contact period than the Lenape, see Rountree and Davidson, *Eastern Shore Indians*, 43; Jay F. Custer, "Late Woodland Cultural Diversity in the Middle Atlantic; An Evolutionary Perspective," Custer, ed., *Late Woodland Cultures of the Middle Atlantic Region* (Cranbury, N.J., 1986), 151; Becker, "Cultural Diversity in the Lower Delaware River Valley, 1550–1750, 94, 91–92, 140–142; Becker, "Lenape Maize Sales to the Swedish Colonists," 128; Dent, *Chesapeake Prehistory*, 275. Herbert C. Kraft, *The Lenape-Delaware Heritage*, 389.

CHAPTER ONE:
European Ways, 1631–1683

1. Joanne O. Passmore, *Three Centuries of Delaware Agriculture* (n.p., 1978), 19; Peter Lindestrom, *Geographia Americae* (Philadelphia, 1925), 173.

2. L.J. Cotnoir, *Soils of Delaware* (Newark, DE., 1966), 5.

3. Francis Vincent, *A History of the State of Delaware* (Philadelpia, 1870), 41.

4. Cotnoir, *Soils of Delaware*, 5: Flannery, *The Eternal Frontier*, 320; F. Lucas, Jr., *Delaware* (1823), Map Collection, Special Collections, Morris Library, University of Delaware, Newark, DE.

5. William H. Williams, *Slavery and Freedom in Delaware* (Wilmington, DE., 1996), 11–14, 45; R.O. Bausman, *An Economic Study of Land Utilization in Sussex County, Delaware*, Bulletin no. 233 (Newark, DE., 1941), 190.

6. Passmore, *Three Centuries of Delaware Agriculture*, 49. For climatic differences in Delaware see Jeanette Eckman, Anthony Higgins and William Connor, eds., *Delaware: A Guide to The First State* (New York, 1938), 10.

7. Myers, ed., *Narratives*, 25–26; John A. Munroe, *Colonial Delaware* (Millwood, NY, 1978), 25; Lindestrom, Geographia..., 175; Johnson ed., "The Journal of Andreas Hessalius, 1711-1724," *Del. Hist.*, vol. II, no. 2 (Sept., 1947), 107.

8. Mary A. Stewart, *What Nature Suffers to Grow: Life, Labor and Landscape on the Georgia Coast* (Athens, Georgia, 1996), 34; William H. Williams, *The Garden of American Methodism: The Delmarva Peninsula* (Wilmington, DE., 1984), 132.

9. Lindestrom, *Geographia*, 175–176.

10. Ibid., 183; Bausman, *An Economic Study of Land Utilization in Sussex County, Delaware*, 190; David Grettler, "The 'Other' Chesapeake: Family, Debt, and Tobacco on the Seventeenth-Century Delaware Frontier," *Del. Hist.*, vol. XXX, no. 3, 165.

11. Lindestrom, *Geographia*, 176; J.R. MCNeill, *Something New Under The Sun: An Environmental History of The Twentieth Century World* (New York, 2000), 39.

12. Ibid., 177; John B. Linn and William H. Egle, eds., *Pennsylvania Archives, Second Series*, vol. V, (Harrisburg, PA, 1877), 290; Johnson, ed., "The Journal of Andreas Hesselius," 87.

13. Harold B. Hancock, "Descriptions and Travel Accounts of Delaware, 1700–1740," *Delaware History*, vol. X, (Oct., 1962), 150; Lindestrom, *Geographia ...*, 178; Benson, ed., *Peter Kalm's Travels in North America*, I, 79; Linn and Egle,

eds., *Pennsylvania Archives, Second Series*, vol. V, 11, 770; Vincent, *History of Delaware*, 52.

14. Frank Morton Jones, ed., "Description of Cypress Swamp in Delaware and Maryland States," *Del. Hist.* vol. III, no.3, (March, 1949), 125.

15. Benson, ed., *Peter Kalm's Travels in North America*, I, 76–77.

16. H. Clay Reed, ed., *Delaware, A History of The First State*, I, 3–4, n12; Anonymous, "Mosquitoes at Cape Henlopen," *Columbian Magazine* (Feb., 1788), 108.

17. Johnson, ed., "The Journal of Andreas Hesselius," 2, 6, 86–87, 91, 97, 99, 114.

18. Lindestrom, *Geographia*, 133–134. Although fireflies were not unique to the New World, evidently the soldier had never seen them in Old Sweden.

19. Myers, ed., *Narratives*, 25, 229, 15, 18, 21: Lindestrom, *Geographia*, 188.

20. Myers, ed., *Narratives*, 322; Steadman, "A Long-Term History of Terrestrial Birds and Mammals in the Chesapeake-Susquehanna Watershed," Curtin et al., eds., *Discovering the Chesapeake*, 102; J. Thomas Scharf, *History of Delaware*, vol. II, (Philadelphia, 1888), 1169n; Steven M. Pulinka, "Success and Failure in Cantwell's Bridge, Delaware," *Del. Hist.*, vol. XXI, no.1 (Spring-Summer, 1984), 64; Pemberton Carlisle (1818), Charles Fisher (1843), John Goslin (1845), Jesse Green, Esq. (1834), *Sussex County Inventories*, Georgetown, DE; Benson, ed., *Peter Kalm's Travels*, I, 110. Effectively challenging older versions of bison history that spoke of large herds of buffalo in eighteenth century Pennsylvania is Ted Franklin Belue, *Death of the Buffalo East of the Mississippi* (Mechanicsburg, Pa., 1996), 135–147; and Frank Gilbert, *The North American Bison* (Toronto, Canada, 1970), 204–227.

21. Myers, ed., *Narratives*, 322; Jane Scott, *Between Ocean and Bay: A Natural History of Delmarva* (Centreville, MD., 1991), 69; C.H.B. Turner, ed., *Some Records of Sussex County, Delaware* (Philadelphia, 1909), 70; Reed, ed., *Delaware*, vol. I, 1; Steadman, "A Long-Term History of Terrestrial Birds and Mammals in the Chesapeake-Susquehanna Watershed," 102; John B. Linn and William H. Egle, eds. *Pennsylvania Archives, Second Series*, vol. 5 (Harrisburg, PA, 1877), 283.

22. Reed, ed., *Delaware*, vol. I, 1; Johnson, ed., "The Journal of Andreas Hesselius," 90. For an example of the ease with which hunters brought down wild turkeys in seventeenth century Delaware see Johnson, ed., "The Journal of Andreas Hesselius," 90.

23. Myers, ed., *Narratives*, 398.

24. Lindestrom, *Geographia*, 186. For a discussion of Native Americans firing the landscape see Whitney, *From Coastal Wilderness to Fruited Plain*, 107–120.

25. Howard L. Harrod, "Is There a Native American Environmental Ethic?" *Journal of Intergroup Relations*, XVIII, no. 2 (summer, 1991) 54–64; Powell, "Were These America's First Ecologists?" 17–24; Tim Flannery, *The Eternal Frontier*, 194–205; Lindestrom, *Geographia*, 164; Miller, "Living Along the Great 'Shellfish Bay'," Curtin et. al. eds., *Discovering the Chesapeake*, 121. For the view that Native Americans did not always walk "lightly on the Earth," see Shepherd Krech III, *The Ecological Indian: Myth and History* (New York, 1999), 41, 76, 98–99, 110, 118–122, 158, 163, 212–213, 227.

26. William Beinart and Peter Coates, *Environment and History: The Taming of Nature in the U.S.A. and South Africa* (London, 1995), 5.

27. "Genesis, I:26–29, in *The New English Bible* (U.S.A, 1961); Whitney, *From Coastal Wilderness to Fruited Plain*, 122.

28. For the need of European colonists to domesticate wilderness see Keith Thomas, *Man and Natural World: A History of Natural Sensibility* (New York, 1983), 194–197; Timothy Silver, *A New Face on the Countryside: Indians, Colonists and Slaves in South Atlantic Forests, 1500–1800* (New York, 1990), 104, 138; Roderick Nash, *Wilderness and the American Mind* (New Haven, Conn., 1982), 10–43. For an analysis of cultural replication in colo-

nial Anglo-America see Jack P. Greene and J.R. Pole, eds., *Colonial British America* (Baltimore, 1984), 13–16.

29. Munroe, *Colonial Delaware*, 3–13.

30. Ibid., 13–18.

31. Ibid., 13–27.

32. Ibid., 18, 56; Amandus Johnson, *The Swedish Settlements on the Delaware, 1638–1664* (Baltimore, 1969), 289; Peter O. Wacker and Paul G. E. Clemens, *Land Use in Early New Jersey* (Newark, NJ, 1995), 54.

33. Myers, ed., *Narratives*, 265.

34. Terry Jordan and Matti Kaups, *The American Backwoods Frontier* (Baltimore, MD., 1989), 33–35.

35. For the arrival of Africans and African-Americans in early Delaware see William H. Williams, *Slavery and Freedom in Delaware, 1639–1865* (Wilmington, DE., 1996), 12–17.

36. Jordan and Kaups, *The American Backwoods Frontier,* 96.

37. Ibid., 49–56; Munroe, *Colonial Delaware*, 41. The estimate that perhaps 30 percent of the total Finnish population in the lower Delaware River Valley lived in Delaware in the 1670s is partially based on a reading of two maps in Jordan, *American Backwoods Frontier*, 54, 80.

38. Jordan and Kaups, *American Backwoods Frontier*, 7, 101, 228; Wacker, *Land Use in Early New Jersey*, 72; Johnson, *The Swedish Settlements*, 147–150.

39. See, for example, Wilbur Zelinsky, *The Cultural Geography of the United States* (Englewood Cliffs, N.J., 1973), 13.

40. Jordan and Kaups, *The American Backwoods Frontier*, 233, 59.

41. Ibid., 57, 229; Johnson, *The Swedish Settlements on the Delaware*, 147–150; John B. Linn and William H. Egle, eds., *Pennsylvania Archives, Second Series*, vol V, (Harrisburg, PA, 1877), 470.

42. Johnson, *The Swedish Settlements on the Delaware*, 148–150; Jordan and Kaups, *The American Backwoods Frontier*, 53, 57–58, 105; Munroe, *Colonial Delaware*, 69.

43. Jordan and Kaups, *The American Backwoods Frontier*, 72–73, 95–97.

44. Ibid., 97–100, 136; John R. Stilgoe, *Common Landscape of America, 1580–1845* (New Haven, CT), 175–176.

45. Jordan and Kaups, *The American Backwoods Frontier*, 115–116, 120–121; Johnson, *The Swedish Settlements on the Delaware*, 527–528.

46. Jordan and Kaups, *The American Backwoods Frontier*, 211–232; Michael Williams, *Americans and Their Forests: A Historical Geography,* (Cambridge, U.K., 1989), 121.

47. Jordan and Kaups, *The American Backwoods Frontier*, 102.

48. Lindestrom, *Geographia Americae*, 172–173.

49. Wacker, *Land Use in Early New Jersey*, 258; Jordan and Kaups, *The American Backwoods Frontier*, 96: Passmore, *Three Centuries of Delaware Agriculture*, 22.

50. Jacob Ernest Cooke, ed., *Encyclopedia of the North American Colonies*, vol. III, (New York, 1993), 240.

51. Passmore, *Three Centuries of Delaware Agriculture*, 22; Benson, ed., *Peter Kalm's Travels in North America*, I, 308.

52. Benson, ed., *Peter Kalm's Travels in North America*, I, 55, 307.

CHAPTER TWO: *Chopping Down Trees and Decimating Wildlife, 1631–1867*

1. Hoffecker, ed., "The Diaries of Edmund Canby, A Quaker Miller, 1822–1848," *Del. Hist.*, vol. XVI, no. 3 (Spring-Summer, 1975), 199.

2. Jordan and Kaups, *The American Backwoods Frontier*, 58.

3. Stewart, *What Nature Threatens to Grow*, 70, 74.

4. Johnson, ed., "The Journal of Andreas Hesselius, 1711–1714," 94; *Laws of the State of Delaware*, I, 232.

5. *Laws of the State of Delaware*, I, 217; VII, 273.

6. Jordan and Kaups, *American Backwoods Frontier*, 132.

7. George T. Welch, ed., *Memoirs of Mary Parker Welch*, (Brooklyn, NY, 1947), 102. For the increasing use of oxen in Delaware during the eighteenth century, see R.O. Bausman and John A. Munroe, eds., "James Tilton's Notes on the Agriculture of Delaware, 1788," *Agriculture History,* vol. XX (July, 1946), 184; and estate inventories at the Delaware State Archives. For the extra feed required by horses and the reluctance to eat horses see A. H. Smyth, ed., *Writings of Benjamin Franklin*, (New York, 1905–07), V, 194–195; Virginia DeJohn Anderson, "Animals into the Wilderness: The Development of Livestock Husbandry in Seventeenth Century Chesapeake," *William and Mary Quarterly* (April, 2002), 388.

8. Henry T. Williams, *Delaware: The Garden State of the Union*, (New York, 1868), 8–9, 18, 36.

9. Lois Green Carr, Russell R. Menard, and Lorena S. Walsh, *Robert Cole's World* (Chapel Hill, N.C., 1991), 68; Michael Williams, *Americans and Their Forests*, 17, 75, 77, 78; Timothy Silver, "A Useful Arcadia: European Colonists as Biotic Factors in Chesapeake Forests," 158–159.

10. Michael Williams, *Americans and Their Forests*, 73, 78, 137, 138; J. Ritchie Garrison, "Heating and Cooking Systems in the George Read II House, New Castle Delaware, 1802–1839," unpublished paper, (New Castle, DE., 1990), 5, 7, 10; Estate Inventories, Delaware State Archives, Dover, DE., passim; Peter Dalleo, "'South of the Canal' in the 1840s as Described in the Blue Hen's Chicken," *Del. Hist.*, vol. XXIX, no. 2 (Fall-Winter, 2000–2002), 121; Henry T. Williams, *Delaware: The Garden State of the Union*, 9, 18; Carolyn Merchant, *The Columbia Guide to American Environmental History* (New York, 2002), 61.

11. Michael Williams, *Americans and Their Forests*, 78–79; Whitney, *From Coastal Wilderness to Fruited Plain*, 213; Petition from Appoquinimink Hundred (1763), D.P.A.

12. Joanne Passmore, *History of the Delaware State Grange, 1875–1975* (Denton, MD, 1975), 3–4; Robert L. Melson, "The Early Iron Industry in Delaware," John C. Kraft and Robert L. Salsbury, eds., *Transactions of the Delaware Academy of Science*, vols. I-II (Newark, DE 1973), 47–48, 58; James R. Owen and James B. Owen, *A History of the Iron Hill Area* (Newark, DE, n.d.), 2–5; James M. Tunnell, Jr., "The Manufacture of Iron in Sussex County," *Del. Hist.*, vol. VI, no. 2, (Sept. 1954), 85–87; William E. Ferguson, "Early Iron Works in Southern Delaware," *Delaware Folklore Bulletin*, vol. I, no. 4 (April, 1954), 1, 14.

13. Melson, "The Early Iron Industry in Delaware," 61; Passmore, *Three Centuries of Delaware Agriculture*, 14; Richard B. Carter, *The History of Sussex County* (n.p., 1976), 30; Donna Stachecki, "Cypress: King of the Sussex Swamp," *Delaware Conservationist* (Spring, 1988), 13.

14. *Petition of inhabitants of Sussex concerning Cape Henlopen* (June, 1781), D.P.A.; *Legislative Act to prevent the destruction of timber on Cape Henlopen* (June, 1781), D.P.A.; John C. Kraft and Robert Caulk, "The Evolution of Lewes Harbor," John C. Kraft and Robert L. Salsbury, eds., *Transaction of the Delaware Academy of Science*, vols. I–II (Newark, DE., 1973), 107–108; Scharf, *History of Delaware* vol. II, 1215.

15. Bausman and Munroe, "James Tilton's Notes on the Agriculture of Delaware in 1788," 185.

16. David de Vries, "Short Historical and Journal

Notes," *Collections of the New York Historical Society*, vol. III, part 1 (New York, 1857), 22; William Ecenbarger, '*Walkin' the Line* (New York, 2000), 44; Francis Vincent, *A History of the State of Delaware* (Philadelphia, 1870), 17; Bausman and Munroe, eds., "James Tilton's Notes on the Agriculture of Delaware, 1788," 185; n.a., *Protecting Delaware's Natural Heritage* (n.p., 1999), 15; William S. Taber, *Delaware Trees* (Dover, DE 1939), foreword.

17. Scott, *Between Ocean and Bay*, 31–32; Passmore, *Three Centuries of Delaware Agriculture*, 11; M. Williams, *Americans and Their Forests*, 144; Jack Temple Kirby, "Virginia's Environmental History," *Virginia Magazine of History and Biography*, vol. IC, (Oct., 1991), 472–473; Vincent, *A History of the State of Delaware*, 17–18; Reed, ed., *Delaware: A History of the First State*, 15, 18.

18. Vincent, *A History of Delaware*, 18; H. T. Williams, *Delaware: The Garden State of the Union*, 121; Passmore, *Three Centuries of Delaware Agriculture*, 13.

19. Passmore, *Three Centuries of Delaware Agriculture*, 12; William S. Taber, *Delaware Trees* (Dover, DE, 1939); Frank Morton Jones, ed., "Description of the Cypress Swamps in Delaware and Maryland States," *Del. Hist.*, vol. III (1948–49), 126. Charles Mason speaks of cedar trees as the tallest in Cypress Swamp and doesn't even mention the presence of bald cypress trees. As previously pointed out, the immediate area's residents lumped together both white cedar and bald cypress as simply "cedar," and Mason was simply emulating the local practice. See Vincent, *A History of the State of Delaware*, 17. According to Susan M. Laporte and Walter Gabel, "From Little Acorns," *Delaware Conservationist*, vol. XXXIII, no. 3 (1985), 22, there was a bald cypress at 401 Market Street, in Laurel, Delaware that towered 120' in 1985. According to Molly Murray, "Cypress Takes Crown as Top Delaware Tree," *News Journal*, (May 13, 1994), in 1994 a second 120' tall bald cypress was measured along James Branch, a stream that empties into Broad Creek east of Laurel.

20. Herman, *The Stolen House*, 51–91; Donna Stachecki, "Bald and Beautiful," *Delaware Conservationist*, vol. XXXI (1988), no. 1. 13–14; Anthony Higgins, "The Pleasuring Ground," unpublished manuscript in possession of Judith Showell Loeber, White House Farm, Long Neck, DE.; Jones, ed., "Descriptions of the Cypress Swamps," 128; Taber, *Delaware Trees*, 39.

21. Taber, *Delaware Trees*, 41; C. A. Weslager, "Cedar Swamp and the Lost Cedar Creek," *Del. Hist.* vol. VII, no. 1 (March 1956), 9; Tony Florio, *Progger: A Life on the Marsh* (Dexter, Michigan, 2002), 220–221.

22. M. Williams, *Americans and Their Forests*, 72; *Laws of the State of Delaware*, vol. I, 180; VII, 268; Jordan and Kaups, *American Backwoods Frontier*, 105, 107, 109; *Petition to put a gate on the road between Mispillion and Cedar Creek* (Nov. 10, 1783), D.P.A; Dec. 2, 1803 indenture, John Dickinson Folder, 1800-1810, box 80, Laudon Papers, Historical Society of Pennsylvania.

23. *Laws of the State of Delaware*, vol. I, 226–227, 232; III, 283; Hendrick Hendrickson (1759), *Kent County Will Book K*, 215; Whitney, *From Coastal Wilderness to Fruited Plain*, 236; Stevenson W. Fletcher, *Pennsylvania Agriculture and Country Life*, 1640–1840 (Harrisburg, Pa., 1950), 86–87.

24. Bernard L. Herman, "Fences," J. Ritchie Garrison et al., eds., *After Ratification: Material Life in Delaware, 1789–1820* (Newark, DE, 1988), 9–11, 14, 18–19; David J. Grettler, "The Nature of Capitalism: Environmental Change and Conflict over Commercial Fishing in Nineteenth Century Delaware," *Environmental History*, vol. VI, no. 3 (July, 2001), 458.

25. Jones, "Description of the Cypress Swamps," 126–128; 1796 *Map of Delaware*, Historical Society of Delaware, Wilmington, DE; *1876 Map of Delaware*, Map Collection, Special Collections, Morris Library, University of Delaware, Newark, DE.

26. Jones, "Descriptions of the Cypress Swamps," 129.

27. Bartlett Burleigh James and J. Franklin Jameson, eds., *Journal of Jaspar Danckaerts, 1679–1680* (New York, 1913), 139; Cowdrey, *This Land, This South*, 56–57; *Laws of the State of Dela-*

ware, vol. I, 191–192, 349, 390; IX, 432; X, 564; passim; *Minutes of House of Representatives of the Government of the Counties of New Castle, Kent and Sussex Upon Delaware, 1765–1770* (Wilmington, 1770), 209, 259.

28. *Laws of the State of Delaware*, I, 260, passim; Benson, ed., *Peter Kalm's Travels in North America*, 79; Christopher P. White, *Endangered and Threatened Wildlife of the Chesapeake Bay Region: Delaware, Maryland and Virginia* (Centreville, Maryland, n.d.), 40.

29. Benson, ed., *Peter Kalm's Travels in North America*, 306; 1822 map of "Brandywine Mill Seats," P.S. du Pont Collection, Eleutherian Mills-Hagley Foundation Library, Wilmington, DE; Elizabeth Montgomery, *Reminiscences of Wilmington* (Philadelphia, 1851), 30, 42,; *Pennsylvania Gazette*, 12/20/1752. For a more recent account of copperheads in Delaware see Molly Murray, "Copperheads Are Causing High Anxiety," *News Journal* (July 26, 1989). There is a reference to "moccasins," presumably "water moccasins," in Cypress Swamp according to Howard Pyle, "A Peninsular Canaan," *Harper's New Monthly Magazine*, vol. L, (May 1879), 203.

30. Because the laws of Pennsylvania applied to Delaware from 1682 to 1704, bounties established during that period by Pennsylvania also applied to Delaware. Lemon, *The Best Poor Man's Country*, 40; Benson, ed., *Peter Kalm's Travels in North America*, 153, 164–166, 168; *Laws of the State of Delaware*, I, 349, 390; Johnson, ed., *The Journal of Andreas Hesselius, 1711–1724*, 89.

31. Benson, ed., *Peter Kalm's Travels in North America*, 150; Scharf, *History of Delaware*, vol. II, 611; Horle, ed., *Records of Sussex County*, 104; Weslager, *Delaware's Forgotten River*, 63; *Laws of the State of Delaware*, I, 256–257. Virginia bounty hunters also dug deep pits to trap wolves. See Kirby, "Virginia's Environmental History," 457.

32. For bounty hunters in nearby colonies see Lemon, *The Best Poor Man's Country*, 12, 157; Benson, ed., *Peter Kalm's Travels in North America*, 168; Kirby, "Virginia's Environmental History," 457; Anderson, "Animals into Wilderness," 377, 380. For Native American involvement in hunting wolves see Myers, ed., *Narratives*, 321, and Horle, ed., *Records of Sussex County, Delaware*, 104, 303, 321, 637, 695. For a few examples of Indians selling bounty "heads" to white middlemen see Horle, 303, 695. For the cutting off of ears on wolves' heads see *Laws of the State of Delaware*, vol. I, 257; Timothy Silver, "A Useful Arcadia: European Colonists as Biotic Factors in Chesapeake Forests," 161.

33. Jordan, *American Backwoods Frontier*, 211, 213, 215; Beinart, *Environment and History*, 28; Cowdrey, *This Land, This South*, 46. For slave population of Delaware in 1770 see Williams, *Slavery and Freedom in Delaware*, 99, 142.

34. Beinart, *Environment and History*, 28; Vincent, *A History of the State of Delaware*, 17.

35. Ted Steinberg, "Down to Earth: Nature, Agency and Power in History," *American Historical Review*, vol. 107, no. 3 (June, 2002), 807; *Laws of the State of Delaware*, vol. VII, 552, I, 97, 384. For a similar pattern concerning anadromous fish on the New Jersey side of the Delaware Bay during the eighteenth century see Wacker, *Land Use in Early New Jersey*, 59–61. For another example of protecting communal fishing rights on a specific rivers (St. Jones in 1803 and Indian River in 1817) see *Laws of the State of Delaware*, vol. V, 553.

36. *Laws of the State of Delaware*, vol. VII, 552, I, 101.

37. *Minutes of House of Representatives of the Government of the Counties of New Castle, Kent and Sussex Upon Delaware, 1765–1770* (Wilmington, Del., 1770), 199; *Petition from Sussex County for the Removal of the Obstructions to the Navigation of Indian River* (Jan. 15, 1781), D.P.A.; Benson, ed., *Peter Kalm's Travels in North America*, 154.

38. John A. Munroe, "Pierre Charles Verle and His Map of Delaware," *Del. Hist.*, vol. XXII, no. 1 (Spring–Summer, 1986), see map on pp 27–28; Lindestrom, *Geographia Americae*, 153, 188; C.H.B. Turner, ed., *Some Records of Sussex County Delaware* (Philadelphia, 1909), 198–199; Wacker, *Land Use in Early New Jersey*, 58; *Laws of the*

State of Delaware, IV, 568; John A. Munroe, *Federalist Delaware* (New Brunswick, NJ, 1954), 118. F. Lucas, Jr., (map), "Delaware" (1816), shows oyster beds off Kent County but F. Lucas, Jr., (map), "Delaware" (1823), doesn't show any. Both maps are in Map Collection, Special Collections, Morris Library, University of Delaware, Newark, DE.

39. Benson, ed., *Peter Kalm's Travels in North America*, 152.

40. Ibid., 153; Hess, ed., *Birds of Delaware*, 190–191.

41. Benson, ed., *Peter Kalm's Travels in North America*, 153; Hess, ed., *Birds of Delaware*, 22, 534; Kenneth Reynolds, "Return of the Native," *Delaware Conservationist*, vol. XXXI (1988), no. 3, 4. A few wild turkeys may have found their way back into Delaware by 1911. See *Laws of the State of Delaware*, vol. 26 (1911), 322.

42. Beinart, *Environment and History*, 52.

43. Jones, ed., "Description of the Cypress Swamps in Delaware and Maryland States," 137.

44. William H. Williams, *The Garden of American Methodism: The Delmarva Peninsula, 1769–1820* (Wilmington, Del., 1984), 72–73.

CHAPTER THREE:
The Impact of the World Beyond

1. C. H. B. Turner, ed., *Some Records of Sussex County, Delaware* (Philadelphia, 1909), 224.

2. Bryant, ed., *Delaware Estuary*, 7; Myers, ed., *Narratives*, 138.

3. Linn and Egle, eds., *Pennsylvania Archives, Second Series*, vol. V, 139, 281, 293; Munroe, *Colonial Delaware*, 3; Myers, ed., *Narratives*, 396.

4. Linn and Egle, eds., *Pennsylvania Archives, Second Series*, vol. V, 293, 298; Vincent, *History of the State of Delaware*, 55, 58; Bryant and Pennock, eds., *The Delaware Estuary*, 45; J. Thomas Scharf, *History of Delaware*, vol. II (Philadelphia, 1888), 1226; Iain Murray, ed., *George Whitefield's Journals*, (London, 1960), 338-339; Kraft and Caulk, "The Evolution of Lewes Harbor," 79.

5. Bryant, ed., *The Delaware Estuary*, 18; John A. Munroe, "Pierre Charles Varle and His Map of Delaware," *Del. Hist.*, vol. XXXIII, no.1 (Spring-Summer, 1986), 33. Priscilla M. Thompson, "Navigation on the Delaware Bay, 1790-1830," *Del. Hist.*, vol. XIX, no.2 (Fall-Winter, 1980), 74-75, 81. For details on Joshua Fisher's maps and charts see Lawrence C. Wroth, "Joshua Fisher's Chart of Delaware Bay and River," *Pa. Mag. of Hist. and Bio.*, vol. LXXIV, (1950), 93-104. A copy of the 1776 Fisher map is in the Map Collection, Special Collections, Morris Library, University of Delaware, Newark, DE. For the location of other Delaware Bay and River charts and maps see Thompson, "Navigation on the Delaware Bay, 1790-1830," 84-87.

6. John A. Munroe, *Federalist Delaware* (New Brunswick, N.J., 1954), 34, 110; Howard Pyle, "A Peninsular Canaan," *Harper's Monthly Magazine*, vol. LVIII (May, 1879), 208; Scharf, *History of Delaware*, vol. I, 432, 433; vol II, 1225; Jennifer Ackerman, *Notes From the Shore* (New York, 1995), 32-33; Kraft and Caulk, "The Evolution of Lewes Harbor," 93-94; Hazel D. Brittingham, *Lantern on Lewes: Where the Past is Present (Lewes*, DE, 1998), 35-41.

7. Williams, *Garden of American Methodism*, 169; *Laws of the State of Delaware*, vol. I, 841.

8. Vincent, *A History of the State of Delaware*, 59.

9. Ibid., 60-61; Scharf, *History of Delaware*, vol. I, 433. For a brief description of other storms to strike Delaware's Atlantic Coast and the Delaware Estuary see Bryant and Pennock, eds., *The Delaware Estuary*, 48-49.

10. Myers, ed., *Narratives*., 238; Munroe, *Colonial Delaware*, 200; Lindestrom, *Geographia Americae*, 156.

11. Scott, *Between Ocean and Bay*, 108; Vincent, *A History of the State of Delaware*, 52-55.

12. Scott, *Between Ocean and Bay*, 34; Vincent, *A History of the State of Delaware*, 52-53; Grettler,

"The Nature of Capitalism: Environmental Change and Conflict over Commercial Fishing in Nineteenth-Century Delaware," 458–461.

13. For a brief discussion of the increasing importance of the plow in Delaware agriculture by the mid-eighteenth century see Williams, *Slavery and Freedom in Delaware*, 45–46. For the environmental impact of plows see Lorena S. Walsh, "Land Use, Settlement Patterns and the Impact of European Agriculture, 1620–1820," and Carville Earle and Ronald Hoffman, "The Ecological Consequences of Agrarian Reform in the Chesapeake, 1730–1840," in Curtin et. al. eds., *Discovering the Chesapeake*, 241–242, 287–289, 296–297.

14. *Laws of the State of Delaware*, vol. I, 97.

15. Weslager, "Cedar Swamp and the Lost Cedar Creek," 11; Florio, *Progger*, 219–220; *Laws of the State of Delaware*, vol. I, 850; II, 266; IV, 499, 501; VI, 45, 49–50, 234, 490; VII, 622; "Petitions in Opposition of the Improvement of Duck Creek," (Jan., 1786; Jan. 19, 1786; Jan. 1, 1786; Jan., 1786), D.P.A.

16. Ralph D. Gray, *The National Waterway: A History of the Chesapeake and Delaware Canal, 1769–1965* (Urbana, IL, 1967), 2-3.

17. Ibid., 8, 11–24, 32–42.

18. Ibid., 20–24, 43–66, 67.

19. Ibid., 51–53, 62, 66; Anthony Higgins, ed., "The Memoirs of John Clark Higgins," *Del. Hist.*, vol. XIV, no.1 (April, 1970), 72; James Leston Levingood, *The Philadelphia-Baltimore Trade Rivalry, 1780–1860* (Harrisburg, PA, 1947), 92.

20. Vincent, *A History of the State of Delaware*, 38–39.

21. Scharf, *History of Delaware*, vol. I, 413; Munroe, *Federalist Delaware*, 32. Among the handful of extant Indian trails taken over by Europeans was one that led north along the Delaware River from the town of New Castle to Tinicum Island in Pennsylvania and the one that connected the head of Elk River to the head of the Christina River. See Scharf, *History of Delaware*, vol. II, 904.

22. Munroe, *Federalist Delaware*, 31–32; Scharf, *History of Delaware*, 413, 421; Duc de la Rochefoucault Liancourt, *Travels Through the United States of North America...in the Years 1795, 1796, and 1799* (London, 1800), vol. III, 487; *Court of General Sessions*, Road Papers, (New Castle County), 1760, D.P. A.; *Laws of the State of Delaware*, vol. IV, 372; Hancock, *Descriptions of Travel Accounts of Delaware, 1700–1740*, 129, 133, 140.

23. James and Jameson, eds., *Journal of Jasper Danckaerts, 1679–1680*, 130; Munroe, *Colonial Delaware*, 66, 68; Scharf, *History of Delaware*, 413; John A. Munroe and John C. Dunn, "Benjamin Eastburn, Thomas Noxon and the Earliest Maps of the Lower Counties," *Del. Hist.*, vol. XXI, no. 4, (Fall-Winter 1985), 218–219 .

24. Munroe, *Colonial Delaware*, 200; *Laws of the State of Delaware*, vol. I, 317, 318, 319, 391–394, 402–404, 414–416; *Court of General Sessions*, Road Papers (New Castle County), 1747, 1751, 1758, 1775, 1798, D.P.A.; Scharf, *History of Delaware*, vol. I, 413.

25. *Court of General Sessions*, Road Papers (New Castle County), 1753, 1760, D.P.A.; Petition to put a gate on the road between the Mispillion and Cedar Creek (Nov. 10, 1783), D.P.A.

26. *Court of General Sessions*, Road Papers (New Castle County), 1760, 1798, D.P.A.; Weslager, *Delaware's Forgotten River*, 60. For a specific example of sedimentation causing the expansion of marshland along rivers, see Grettler, "The Nature of Capitalism," 460.

27. Scharf, *History of Delaware*, 419, 420. Johnson, ed., "The Journal of Andreas Hesselius, 1711–1724," 108–109; Quinn, "Traversing the Landscape in Federal Delaware," 46, 49.

28. Scharf, *History of Delaware*, 413; Horle, ed., *Records of Sussex County, Delaware*, 66; *Laws of the State of Delaware*, vol. I, 318, 319, 635, 1005; III, 234; IV, 650–653.

29. *Laws of the State of Delaware*, vol. IV, 178, 183–184; Scharf, *History of Delaware*, vol. I, 417; James C. Booth, *Memoir of the Geological Survey*

of the State of Delaware (Dover, DE, 1841), 173–174; *Journal of the Senate of Delaware* (Dover, DE, 1841), 163.

30. William A. McAvoy, "Disappearing Diversity," *Outdoor Delaware*, vol. II, no. 1 (Spring-Summer, 2002), 15; *Laws of the State of Delaware*, vol. IV, 178, 183, 185, 192, 196, 241, 372, 378, 484, 616, 627; V, 28, 65–66; Scharf, *History of Delaware*, vol. I, 416–418; Henry Seidel Canby, *The Brandywine* (New York, 1941), 159; Ralph Gray, "Transportation and Brandywine Industries," *Del. Hist.* vol. II, no. 4 (Oct., 1961), 310–311.

31. Higgins, ed., "The Memoirs of John Clark Higgins," 181; Welch, ed. Memoirs of Mary Parker Welch (1818–1912), 101; S.H. Black, *Memoirs of the Agricultural Society [of New Castle County*], vol. I (n.p., 1820), 77. For a brief biography of Dr. Samuel Black see Scharf, *History of Delaware*, vol. I, 485. For examples in the neighboring Chesapeake area of the increasing use of the plow a few decades later than in Delaware see Walsh, "Land Use, Settlement Patterns and the Impact of European Agriculture, 1620–1820," 238–243. For examples of the dramatic increase in plow-pulling oxen on Delaware farms by the mid-eighteenth century see estate inventories at D.P.A.; Williams, *Slavery and Freedom in Delaware*, 45–46; and Bausman and Munroe, eds., "James Tilton's Notes," 184. For discussion of significant soil erosion in modern Delaware see *40 years of Soil and Water Conservation...a look to the future New Castle conservation District*, 1983, (n.p. or d.), 19. For a specific area of the Atlantic Coastal Plain in North Carolina see Jonathan D. Phillips, "A Short History of a Flat Place: Three Centuries of Geomorphic Change in the Croatan National Forest," *Annals of the Association of American Geographers*, vol. XXXXVII, no. 2 (June, 1997), 197–216.

32. William Huffington, *The Delaware Register and Farmer's Magazine*, vol. II (Dover, DE., 1839), 82.

33. Henry T. Williams, *Delaware: The Garden State of The Union* (New York, 1868), 38, 39.

34. Alfred Crosby, *Ecological Imperialism* (New York, 1986), 53.

35. C. A. Weslager, "A Digest of Three Presentations," *Lewes Historical Society's Cultural and Historic Seminar of Lewes* (Lewes, DE., 1979), 22; Johnson, *The Swedish Settlements on the Delaware*, vol. I, 126, 319, 320, 338.

36. Myers, ed., *Narratives*, 319, 470; C.A. Weslager, "The City of Amsterdam"s Colony on the Delaware, 1656–1664," *Del. Hist.*, vol. XX, no. 1 (Spring–Summer, 1982), 15; Silver, "A Useful Arcadia: European Colonists as Biotic Factors in Chesapeake Forests," 160; Vincent, *A History of the State of Delaware*, 64. For evidence that foxes continued to be a serious threat to smaller farm animals see "Petition Requesting a bounty on foxes –1841," Legislative Papers, Misc., Folder # 1, 141, D.P.A.

37. Johnson, *Swedish Settlements on the Delaware*, 319–320, 365; *Laws of The State of Delaware*, vol. I, 253; Lemon, *The Best Poor Man's Country*, 163–164; Munroe, *Colonial Delaware*, 198; Bausman and Munroe, ed., "James Tilton's Notes," 185. For specific cropped ear patterns showing ownership see Horle, ed., *Records of Sussex County, Delaware*, 220, 263, 503, 547, 845.

38. Bernard L. Herman, "Fences," Garrison et al., eds. *After Ratification: Material Life in Delaware, 1789-1820*, 7–20; Horle, ed., *Records of Sussex County, Delaware*, 1228.

39. Herman, "Fences," 10–20; David J. Grettler, "Environmental Change and Conflict Over Hogs in Early Nineteenth Century Delaware," *Journal of the Early Republic*, vol. XIX, no. 2 (summer, 1999), 216, 219; Tod Steinberg, "Down to Earth: Nature, Agency and Power in History," *American Historical Review*, vol. CVII, no. 3 (June, 2002), 809–810. For a brief overview of the end of the open range livestock industry on the Great Plains see Ray Allen Billington and Martin Ridge, *Westward Expansion: A History of the American Frontier*, Fifth Edition, (New York, 192), 611–628.

40. Petition to put a gate on the road between the Mispillion and Cedar Creek (Nov. 10, 1783), D.P.A.; "Opposition to Improvement, 1818," Legislative Papers, Petitions, Transportation, Jan., 1818, D.P.A.; *Laws of The State of Delaware*, vol. I, 530–531; vol. III, 67; vol. VI, 343.

41. *Laws of The State of Delaware*, vol. V, 40–41; IX, 167–168; Scharf, *History of Delaware*, vol. II, 1223–1224, 869–870; Garrett Hardin, "The Tragedy of the Commons," *Science*, vol. CLXII (1968), 1243-1258.

42. Grettler, "Environmental Change and Conflict Over Hogs," 209–211; Wacker, *Land Use in Early New Jersey*, 66; Silver, "A Useful Arcadia: European Colonists as Biotic Factors in Chesapeake Forests," 161; Anderson, "Animals into Wilderness," 393–394; Scott, *Between Ocean and Bay*, 36. For a representative sample of complaints about free-ranging pigs and the subsequent responses by the Delaware legislature see *Laws of the State of Delaware*, vols. I–VI, passim.

43. *Laws of The State of Delaware*, vol. III, 283; Legislative Petitions, January, 1810, D.P.A., as quoted in Herman, "Fences," 13. For legislation restricting free-ranging swine see *Laws of The State of Delaware*, vol. I, 109, 275, 324, 441, 508, 561–562, 965–967, 970–971, 1009, 1025–1026, 1218–1219; vol. III, 32, 62, 155, 311, 318, 336, 669; vol. V, 27, 38; vol. VI, 38, 87, 157, 301, 583. Problems with marauding pigs are reported as early as 1656 by the town of New Amstel (New Castle). See Williams, *Slavery and Freedom in Delaware*, 50.

44. Legislative Petitions, January, 1797, D.P.A., as quoted in Herman, "Fences," 12.

45. Grettler, "Environmental Change and Conflict Over Hogs," 210–214, 216; *Laws of the State of Delaware*, vol. XI, 24, 146, 163, 179, 194, 230. Evidently, pigs and other free-ranging livestock continued to be a nuisance in some scattered areas of Delaware after the Civil War and site-specific legislation was passed in 1873 to deal with the problem. *Laws of the State of Delaware*, vol. XIV, 416–447; *Revised Statutes of the State of Delaware* (Wilmington, DE, 1893), 482.

46. Sam Glasscock, "The Closing of the Range in *Colonial Delaware*: An Economic Perspective," unpublished manuscript in author's possession (June 2, 2003), 1–7.

47. Ibid.

48. Ibid.

49. Huffington, *The Delaware Register and Farmer's Magazine*, vol. I, 43.

50. Quinn, "Traversing the Landscape of Federal Delaware," 53; Scott, *Between Ocean and Bay*, 106; Passmore, *Three Centuries of Delaware Agriculture*, 19–20; Wacker, *Land Use in Early New Jersey*, 120, 121, 255; Harold B. Hancock, ed., "Letters to and from Caesar Rodney: The Personal Side," *Del. Hist.* vol. XX, no. 3 (Spring–Summer, 1983), 220.

51. Horle, ed., *Records of Sussex County, Delaware*, 967; *Minutes of House of Representatives of the Government of the Counties of New Castle, Kent and Sussex Upon Delaware, 1765–1770* (Wilmington, DE, 1770), 72–73, passim; Wacker, *Land Use in Early New Jersey*, 125, 127; *Laws of the State of Delaware*, vol. VI, 134–148.

52. Vincent, *History of the State of Delaware*, 44–45; Black, *Memoirs of the Agricultural Society [of New Castle County]*, 77; *Journal of the Senate of the State of Delaware* (Dover, DE, 1841), 154; Scharf, *History of Delaware*, vol. I, 439; Booth, *Memoir of the Geological Survey of the State of Delaware*, 161; James C. Booth, *Diary*, 1838, Special Collections, Morris Library, University of Delaware, 31; Passmore, *Three Centuries of Delaware Agriculture*, 20.

53. Booth, *Memoir of the Geological Survey of Delaware*, 161; Passmore, *Three Centuries of Delaware Agriculture*, 19.

54. Passmore, *Three Centuries of Delaware Agriculture*, 20.

55. Cowdrey, *This Land, This South*, 26–27.

56. Williams, *Slavery and Freedom in Delaware*, 34; Williams, *The Garden of American Methodism*, 131–132; Leonard W. Labaree, ed., *The Papers of Benjamin Franklin*, vol. II (New Haven, Conn., 1960), 222.

57. Anthony Higgins, ed., "The Memoirs of John Clark Higgins," *Del. Hist.*, vol. XIII, no. 3 (April, 1969), 178; Turner, ed., *Some Records of Sussex County*, 240; Williams, *Slavery and Freedom in Delaware*, 34; Ackerman, *Notes From The Shore*,

125. For a discussion of the differences between the *Plasmodium vivax* and the *Plasmodium falciparum* strains of malaria and their respective impacts on the populations of the New World see Douglas H. Ubelaker and Philip D. Curtin, "Human Biology of Populations in the Chesapeake Watershed," Philip D. Curtin et al., eds., *Discovering the Chesapeake: The History of an Ecosystem*, 130–131, 143–144.

58. Hancock, "Descriptions and Travel Accounts of Delaware, 1700–1740," *Del. Hist.*, vol. X, no. 2 (Oct., 1962), 122–123; Williams, *The Garden of American Methodism*, 131–132.

59. Todd L. Savitt, *Medicine and Slavery; the Disease and Health Care of Blacks* (Chicago, 1978) 27–34; Williams, *Slavery and Freedom in Delaware*, 34, xii. Delaware's most famous enslaved African-American, Richard Allen, may have been susceptible to malaria; see *Richard Allen, The Life Experiences and Gospel Labors of the Right Reverend Richard Allen* (New York, 1960), 19.

60. Williams, *Slavery and Freedom in Delaware*, xii, 99, 101. Even Iron Hill, just south of Newark, was considered high enough to be a safe retreat for whites from the fevers of "flat" Delaware. M. M. Merwin, "Thomas Rodney's Diary of a Journey by Carriage from Delaware to New York City," *Del. Hist.*, vol. XVII, no.3 (Spring-Summer, 1977), 204–205. The reference to the role of race in the location of the du Pont family powder works is found in Samuel Bancroft, Jr., to the Hon. John B. Gordon, Wilmington, Feb. 14, 1897, *Bird-Bancroft Collection*, vol. V, box 15, Hist. Soc. of Del. The fact that E.I. du Pont didn't own slaves was provided by Marge McNinch, reference archivist at Hagley Museum and Library, Wilmington, DE.

61. Johnson, ed., "The Journal of Andreas Hesselius, 1711–1724," 90; Jeannette Eckman, "Life Among the Early Dutch at New Castle," *Del. Hist.*, vol. IV, no. 3 (June, 1951), 285, 287.

62. Johnson, ed., "The Journal of Andreas Hesselius, 1711–1724," 90; Benson, ed., *Peter Kalm's Travels*, vol. I, 192 -193.

63. Benson, ed., *Peter Kalm's Travels*, vol. I, 192 –193; Quinn, "Traversing the Landscape in Federal Delaware," 53; Williams, *Delaware: The Garden State of the Union*, 4.

64. Elmer Clark, ed., *Journal and Letters of Francis Asbury*, vol. I, (Nashville, Tenn., 1958), 771, 772; Scharf, *History of Delaware*, vol. II, 639, George A. Phoebus, *Beams of Light on Early Methodism in America* (New York, 1887), 44; Gary Nash, *Forging Freedom: The Formation of Philadelphia's Black Community, 1720-1840* (Cambridge, Mass., 1988), 121–124; John A. Munroe, ed., "James Hemphill's Account of a Visit to Maryland, 1802," *Del. Hist.*, vol.III, (1948–49), 66; Canby, *The Brandywine*, 6, 109; Wade P. Catts, "Slaves, Free Blacks and French Negroes: An Archeological and Historical Perspective on Wilmington's Forgotten Folk," (Master's thesis, University of Delaware, Newark, DE, 1988), passim.

65. Henry Stommel and Elizabeth Stommel, *Volcano Weather: The Story of 1816, The Year Without Summer* (Newport, R.I., 1983), xi, 3, 10, 11, 12, 19–43, 44–51;

66. Grettler, "Nature of Capitalism," 453–454; Grettler, "Environmental Change," 212; Ted Steinberg, *Down to Earth: Nature's Role in American History* (New York, 2002), 47; Brian Fagan, *The Little Ice Age: How Climate Made History, 1300–1850* (New York, 2000), 167–180; Black, *Memoirs of the Agricultural Society*, 77; news clipping, Box 11, folder 29, Bill Frank Collection, Hist. Soc. of Del.; Terry Sharrer, "Farming, Disease and Change in the Chesapeake Ecosystem," Curtin, et al., eds., *Discovering the Chesapeake*, 309; "Memorandum or Common Place Book for Thomas Lea's Mill at Brandywine, Delaware State, Commencing 3 Mo., 1st, 1817," Business, Hist. Soc. of DE., March 1, 1817, March 5, 1817; June 1, 1817; Aug. 1, 1818; Aug. 13, 1818; Norman B. Wilkinson, *E.I. du Pont, Botaniste* (Charlottesville, VA, 1972), 74; Gov. Daniel Rodney, *Journal of the Senate of the State of Delaware* (Dover, DE, 1817), 7.

67. U. S. Census Manuscripts 1810, Delaware (microfilm); U.S. Census Manuscripts, 1840, Delaware (microfilm).

CHAPTER FOUR: *Water Power, Steam Power and the Urban Environment, 1730–1920*

1. Quotes are taken from Scharf, *History of Delaware*, vol. II, 631–632.

2. Ibid.; Carol Hoffecker, *Brandywine Village: The Story of a Milling Community*, 16–17. See also Benjamin Ferris, "My Kith and Kin," manuscript, Historical Society of Delaware, Wilmington, DE.

3. Scharf, *History of Delaware*, 632; Hoffecker, *Brandywine Village*, 16–17.

4. Mileage computed from map in Lester J. Cappon, ed., *Atlas of Early American History* (Princeton, N.J., 1976), 5. Description of grain-laden wagons is found in Scharf, *History of Delaware*, vol. II, 787–788.

5. Scharf, *History of Delaware*, vol. II, 629, 751; Weslager, *Delaware's Forgotten River*, 84; Hoffecker, *Brandywine Village*, 17; Carol E. Hoffecker, *Wilmington, Delaware: Portrait of an Industrial City, 1830–1970* (Charlottesville, VA, 1974), 4–6; Munroe, *Colonial Delaware*, 155; F. Lucas, Jr., "Delaware Map, 1823," Map Collection, Special Collections, Morris Library, University of Delaware, Newark, DE.

6. C.A. Weslager, "Watermills, Windmills, Horsemills, and Tidemills: Early Colonial Grain Mills in Delaware," *Del. Hist.*, vol. XIV (April, 1970) 52–59.

7. Ibid., 53–55, 58. I was unable to find the exact location of Cobb's Creek. For the location of mill sites in northern New Castle County about 1800, see map in Peter C. Welsh, "Merchants, Millers and Ocean Ships," *Del. Hist.*, vol. VII, no. 4 (Sept., 1957), facing 337.

8. J.C. Kraft, "Geologic and Human Factors in the Decline of the Tidal Salt Marsh...," *Sedimentary Geology*, vol. LXXX (1992), 233; Williams, *Delaware: The Garden State of the Union*, 59.

9. William Guthrie, *A New System of Modern Geography*, vol. II (Philadelphia, 1795), 459. Guthrie's quote is also found in Peter Welsh, "The Brandywine Mills: A Chronicle of an Industry, 1762–1816," *Del. Hist.*, vol. VII, no. 1 (March, 1956), 21.

10. Canby, *The Brandywine*, 5–6; Lucas, "Delaware Map, 1823," Map Collection, Special Collections, Morris Library, University of Delaware.

11. Welsh, "Merchants, Millers and Ocean Ships: The Components of an Early American Industrial Town," 335.

12. Scharf, *History of Delaware*, vol. II, 904; Hoffecker, *Brandywine Village*, 21; Canby, *The Brandywine*, 63, 81–82; Munroe, *Colonial Delaware*, 202.

13. Hoffecker, *Brandywine Village*, 19; Scharf, *History of Delaware*, vol. II, 786; Welsh, "Merchants, Millers and Ocean Ships," 331.

14. Hoffecker, *Brandywine Village*, 22–24; Welsh, "Brandywine Mills," 20.

15. Hoffecker, *Brandywine Village*, 24–26; Munroe, *Federalist Delaware*, 119–120; Welsh, "Brandywine Mills," 21.

16. Welsh, "The Brandywine Mills: A Chronicle of an Industry, 1762–1816," 27; Gray, "Transportation and Brandywine Industries, 1800–1840," 305.

17. Welsh, "The Brandywine Mills," 35; Hoffecker, ed., "The Diaries of Edmund Canby, A Quaker Miller, 1822–1848," *Del. Hist.*, vol. XVI, no. 2 (Oct., 1974), 120; Hoffecker, "Nineteenth Century Wilmington: Satellite or Independent City?" *Del. Hist.* Vol. XV, no. 1 (April 1972), 5; Gray, "Transportation and Brandywine Industries, 1800–1840," 310.

18. Hoffecker, ed., "The Diaries of Edmund Canby," *Del. Hist.*, vol. XVI, no. 3 (Spring-Summer, 1975), 186–187, 50, 53; Hoffecker, ed., "The Diaries of Edmund Canby," *Del. Hist.*, vol. XVI, no. 2 (Oct., 1974), 129; Delaware Gazette (Feb. 6, 1790).

19. Tod Steinberg, "Down to Earth: Nature, Agency and Power in History," *Am. Hist. Review*, vol.107, no. 3 (June, 2002), 807.

20. Canby, *The Brandywine*, 62–64, 82; *Laws of The State of Delaware*, vol. I, 97, 384. Because The *Laws of The State of Delaware*, vol. I, does not give the exact year for some of the acts of the early eighteenth century, the date of 1727 for the act that responded to the Lenape complaint was arrived at after examining a number of sources.

21. Canby, *The Brandywine*, 65–66; *Laws of The State of Delaware*, vol. I, 384.

22. Canby, *The Brandywine*, 65–66.

23. Montgomery, *Wilmington*, 29.

24. Hoffecker, ed., "The Diaries of Edmund Canby," 85, 203–206; Scharf, *History of Delaware*, vol. II, 671; Montgomery, Wilmington, 34.

25. Glenn Porter, "Mid-Atlantic Manufacturing in the Jacksonian Era: The McClane Report and Delaware," *Del. Hist.*, vol. XXIV, no. 1 (1990), 66.

26. Welsh, "Merchants, Millers and Ocean Ships," 326; Thompson, "Navigation on the Delaware Bay, 1790–1830," 71; Carol E. Hoffecker "Nineteenth Century Wilmington: Satellite or Independent City," *Del. Hist.*, vol. XV, no. 1 (April, 1972), 6n.

27. Berger, *History and the Human Ecology of the Delaware River*, 51; Hoffecker, "Nineteenth Century Wilmington: Satellite or Independent City," 17–18. In 1837, Wilmington banned the dumping of ballast in either the Christina or the Brandywine. See *Ordinances of the City of Wilmington* (Wilmington, DE, 1841), 96.

28. *Laws of The State of Delaware*, vol. VIII, 366–367.

29. Ibid., 367; Monte A. Calvert, "The Wilmington Board of Trade, 1867–1875," *Del. Hist.*, vol. XVII, no. 3 (April, 1967), 176–177; Hoffecker, "Nineteenth Century Wilmington," 8.

30. Hoffecker, "Nineteenth Century Wilmington," 5–6.

31. Ibid., 7–8; Hoffecker, Wilmington, Delaware, 17.

32. William Cronon, "Modes of Prophecy and Production: Placing Nature in History," *Journal of American History*, vol. LXXVI (March, 1990), 1131; *Second Annual Report of the Board of Managers to the Wilmington Board of Trade, 1869–1870* (Wilmington, 1869), 3; Richard White, "Afterward, Environmental History: Watching a Historical Field Mature," *Pacific Historical Review*, vol. LXX, no. 1 (2001), 103. For the presence of rural bias in the application of environmental history see Donald Worster, "Transformations of the Earth: Toward an Agroecological Perspective in History," *Journal of American History*, vol. LXXVI (March, 1990), 1087–1147; Richard White, "American Environmental History: The Development of a New Historical Field," *Pacific Historical Review*, vol. LIV (Aug., 1985), 300, 330; Elizabeth Blackmar, "Contemplating the Force of Nature," *Radical Historians Newsletter*, no. LXX (May, 1994), 4.

33. Canby, *The Brandywine*, 99; Berger, *History of Human Ecology of the Delaware Estuary*, 50; Henry Seidel Canby, *The Age of Confidence: Life in the Nineties* (New York, 1934), 7–10.

34. Dalleo, "South of the Canal," 121; M. Williams, *Americans and Their Forests*, 138; Estate Inventories, D.P.A., Dover, DE., passim.

35. Hoffecker, ed., "The Diaries of Edmund Canby," 81; Bushman, ed., "Life Along the Brandywine Between 1880–1895 by Samuel Canby Rumford," 126, 138, 186.

36. Martin V. Melosi, *The Sanitary City: Urban Infrastructure in America from Colonial Times to the Present* (Baltimore, 2000), 18–19, 42, 273.

37. Steinberg, *Down to Earth: Nature's Role in American History*, 160; *Ordinances of the City of Wilmington* (Wilmington, DE, 1841), 5, 76; Steinberg, "Down to Earth, Nature, Urgency and Power in American History," 811; *Charter and Ordinances of the City of Wilmington, Delaware to January 1, 1863*, (Wilmington, DE, 1863), 212; *Ordinances of the City of Wilmington* (Wilmington, DE, 1872), 224, 336. Resistance by the poor to attempts to close down urban street "commons" was widespread in American cities. It was not until 1860, for example, that most of New York City was pig free. Steinberg, "Down to Earth, Nature,

Urgency and Power in American History," 811.

38. Steinberg, *Down to Earth: Nature's Role in American History*, 162. "Probable" statistics for Wilmington are based on statistics for Rochester, NY in 1900. That year Wilmington had 41% of Rochester's population. One might therefore assume that it had approximately 41% of the horses, manure, and flies of the upstate New York city. Joel Tarr, "Urban Pollution—Many Long Years Ago," *American Heritage*, vol. XII, no. 6 (Oct., 1971) 65–66.

39. Tarr, "Urban Pollution-Many Long Years Ago," 65, 69; Bushman, ed., "Life Along the Brandywine Between 1880–1895 by Samuel Canby Rumford," 123.

40. Bushman, ed., "Life Along the Brandywine Between 1880–1895 by Samuel Canby Rumford," 196.

41. Tarr, "Urban Pollution—Many Long Years Ago," 68–69, 106; *Ordinances of the City of Wilmington* (Wilmington, DE, 1841), 76; *Ordinances of the City of Wilmington* (Wilmington, DE, 1849), 105–106; Hoffecker, ed., "The Diaries of Edmund Canby," 96; Peter C. Welsh, "A Cardon and Company, Brandywine Tanners, 1815–1826," *Del. Hist.*, vol. VIII, no.2 (Sept., 1958), 124.

42. *Journal of the Senate of the State of Delaware* (Wilmington, DE, 1897), 25; *Journal of the Senate*...(Dover, DE, 1909), 51: Passmore, *Three Centuries of Delaware Agriculture*, 38–39; John A. Munroe, *The University of Delaware: A History* (Newark, DE, 1986), 162, 170.

43 Melosi, *The Sanitary City*, 29, 84; Carol E. Hoffecker, "Water and Sewage Works in Wilmington, Delaware, 1810–1910," *Essays in Public History* (Chicago, July, 1981), 3.

44. Hoffecker, "Water and Sewage Works in Wilmington...," 3–6, 8; Scharf, *History of Delaware*, vol. II, 665–666; *Wilmington Water Works, Report of the Chief Engineer, 1872* (Wilmington,DE, 1872) 8–10.

45. Scharf, *History of Delaware*, vol. II, 665.

46. Melosi, *The Sanitary City*, 30–31; Scharf, *History of Delaware*, vol. II, 665; L.P. Bush, *Some Vital Statistics of the City of Wilmington* (Wilmington, DE., 1877), 5–6, 8; Hoffecker, "Water and Sewage Works in Wilmington," 8, 9. For a discussion of epidemic diseases in Wilmington prior to 1877 see Bush, *Some Vital Statistics*. For Wilmington's annual mortality rate from typhoid fever for years 1881–1906, see *Thirty-Seventh Annual Report, Chief Engineer to the Board of Water Commissioners* (Wilmington, DE, 1906), 63.

47. Bushman, ed., "Life Along the Brandywine Between 1880 and 1895 by Samuel Canby Rumfeld," 185, 193; *Wilmington Daily Commercial*, Feb 14, 1867; Bush, *Some Vital Statistics of the City of Wilmington*, 8; Hoffecker, "Water and Sewage Works in Wilmington," 8–9.

48. Board of Health to E.I. du Pont de Nemours and Co., Nov. 15, 1881, Eleutherian Mills Historical Library, Greenville, DE; Hoffecker, "Water and Sewage Works in Wilmington," 8–9.

49. Bushman, ed., "Life Along the Brandywine Between 1880 and 1895 by Samuel Canby Rumfeld," 186, 184–185.

50. Bushman, ed., "Life Along the Brandywine Between 1880 and 1895 by Samuel Canby Rumfeld," 185; Hoffecker, "Water and Sewage Works in Wilmington," 9; Melosi, *The Sanitary City*, 85. For typhoid deaths in years prior to March, 1877, see Bush, *Some Vital Statistics of the City of Wilmington*, 2–3.

51. Melosi, *The Sanitary City*, 91.

52. Henry Seidel Canby, *The Age of Confidence: Life in the Nineties* (New York, 1934), 10; Hoffecker, "Water and Sewage Works in Wilmington," 9; Edmund Frazer, *Health Department, City of Wilmington, 1882* (Wilmington, DE, 1883), 33.

53. Melosi, *The Sanitary City*, 60, 68–70.

54. Edmund Frazer, *Health Department, City of Wilmington, 1881* (Wilmington, DE, 1882), 15–16.

55. Frazer, *Health Department, City of Wilmington, 1882*, 8, 19; Edmund Frazer, *Health Department,*

City of Wilmington, 1883 (Wilmington, DE, 1884), 9–11; Hoffecker, "Water and Sewage Works," 9–10.

56. T. Chalkley Hatton, *Annual Report of the Board of Directors of the Street and Sewer Department* (Wilmington, DE, 1892), 120–123; Merchant, *The Columbia Guide to American Environmental History*, 111; Hoffecker, "Water and Sewage Works," 9–10. To compare Wilmington's record with other nineteenth century American cities see Melosi, *The Sanitary City.*

CHAPTER FIVE:
Protecting Some Biotic Life, 1825–1945

1. W.D. Sterrett, *Report on Forest Conditions in Delaware, Bulletin no. 82*, Delaware College Agricultural Experiment Station (Newark, DE, 1908), 10, 12, 16, 18, 19.

2. Population statistics for 1910 are in Munroe, *History of Delaware*, 269. Population statistics for Delaware's Native Americans are approximations based on sources used in chapter I, n17.

3. Pyle, "A Peninsular Canaan," 803; Canby, *The Age of Confidence*, 15; Dalleo, "South of the Canal," 135; Williams, *The Garden State of the Union*, 39.

4. Florio, *Progger*, 138, 141, 148.

5. *Laws of the State of Delaware*, vol. IX, 263; Cowdrey, *This Land, This South*, 117; Whitney, *From Coastal Wilderness to Fruited Plain*, 301, 326; John Vaillant, "Overkill,"*The Atlantic Monthly*,(April, 2003), 124; Florio, *Progger*, 23; Passmore, *Three Centuries of Delaware Agriculture*, 67.

6. Jennifer Price, *Flight Maps: Adventures with Nature in Modern America* (New York, 1999), 22; Price, "Hats Off to Audubon," *Audubon*, (November–December, 2004), 46, 48, 50; *Revised Statutes of Delaware* (Wilmington, 1893), 458; Gene K. Hess, Richard L. West, Maurice Barnhill III, and Lorraine M. Fleming, *Birds of Delaware* (Pittsburgh, PA, 2000), 24, 25,n 6; *Laws of the State of Delaware* vol. XXVI (1911), 318, 322.

7. Vincent, *A History of The State of Delaware*, 63, 63n, 64; Kit Hecksher, interview, (July 25, 2005); *Revised Statutes of Delaware* (1893), 457; Johnson, ed., "The Journal of Andreas Hesselius, 1711–1724," 89, n70; Price, *Flight Maps*, 41; James F. Lynch, "Bird Populations of the Chesapeake Bay Region: 350 Years of Change," Curtin, ed., *Discovering the Chesapeake*, 335–336; Steinberg, *Down to Earth*, 67–69; Scott, *Between Ocean and Bay*, 70; Reed, ed., *Delaware: A History of The First State*, 15; Karen A. Bennett et al., *An Ecological Characterization of Delmarva's Great Cypress Swamp* (n.p., 1999), 66.

8. Vincent, *A History of The State of Delaware*, 27; Higgins, ed. "The Memoirs of John Clark Higgins," 218; n.a., *40 Years of Soil and Water Conservation...a look to the future New Castle Conservation District, 1983*, (n.p., n.d.), 14; Florio, *Progger*, 9; *Revised Statutes of Delaware (1893)*, 457; *Laws of The State of Delaware*, vol. XXIV, 288.

9. Passmore, *Three Centuries of Delaware Agriculture*, 67; *Laws of The State of Delaware*, vol. IV, 568; X, 562; XXIV, 285; XXV, 237; XXX, 482; XXXII, 564; *Revised Statutes of Delaware (1893)*, 477; Eckman, *Delaware: A Guide to The First State*, 13–14.

10. Scott, *Between Ocean and Bay*, 132; Passmore, *Three Centuries of Delaware Agriculture*, 67; Higgins, "The Pleasuring Ground," 132; Florio, *Progger*, 22–23. For support of the view that draining marshes posed a serious threat to muskrat populations, see George L. Schuster, "Annual Report of the Director for the Fiscal Year Ending June 30, 1940," *University of Delaware Agricultural Experiment Station, bulletin no. 227* (Newark, 1940), 27.

11. *Laws of The State of Delaware*, vol. XXI, 420; XXII, 728; XXV, 248; XXVI, 322–323; 367.

12. Reed, *Delaware*, 397–398; Florio, *Progger*, 123–124; L.W. Heck et al., *Delaware Place Names* (Washington, D.C., 1960), 64; Scharf, *History of Delaware*, vol. II, 1091; Passmore, *Three Centuries of Delaware Agriculture*, 67; J.L. Miller, "Its

tough keeping a tradition alive," *News Journal* (Feb. 25, 2001).

13. Higgins, ed., "The Memoirs of John Clark Higgins," *Del. Hist.*, vol. XIII, no. 4 (Oct, 1969), 315–316; Whitney, *From Coastal Wilderness to Fruited Plain*, 308; Scott, *Between Ocean and Bay*, 73.

14. Harold Hancock, ed., "Alexander B. Cooper's Civil War Memories of Camden," *Del. Hist.*, vol. XX, no.1 (Spring–Summer, 1982), 58; *Laws of the State of Delaware*, vol XXII, 233–234; XXVI, 327; Passmore, *Three Centuries of Delaware Agriculture*, 47; James L. Latchum, "Recollections of My Mother and Father Growing up in Milford, Delaware," *Del. Hist.*, vol. XXIV, no.3 (Spring–Summer, 1991), 186.

15. Higgins, "The Pleasuring Ground," 3. For location of "Hoss Goin' Over" swamp see Eckman, ed., *Delaware: A Guide to the First State*, 503.

16. *Revised Statutes of Delaware (1893)*, 457; *Laws of the State of Delaware*, vol. XXV, 246; XXVI, 322–323, 324; XXX, 481; XXXII, 564; XXXI, 215; Norman G. Wilder, "The Delaware Board of Game and Fish Commissioners; Its Objectives, History and Problems," (unpublished paper, 1968), 4, 5.

17. Mary Emily Miller, "The Delaware Oyster Industry," *Del. Hist.*, vol. XIV, no. 4 (Oct., 1971), 241; *Laws of the State of Delaware*, vol. IV, 568; VIII, 345. Delaware, in a reciprocal arrangement with Maryland, did allow Maryland oyster boats to work Delaware's waters. Miller, "The Delaware Oyster Industry," 241.

18. *Laws of the State of Delaware*, vol. X, 419, 562–563; Gove Saulsbury, "State of State Messages," 1871, D.P.A.

19. Miller, "The Delaware Oyster Industry," 245–246; Passmore, *Three Centuries of Delaware Agriculture*, 64; *Revised Statutes of Delaware (1893)*, 435–437.

20. Miller, "The Delaware Oyster Industry," 248–249; Passmore, *Three Centuries of Delaware Agriculture*, 65–66.

21. James E. Valle, "Harvesting Oysters," Tracy L. Bryant and Jonathan R. Pennock, eds., *The Delaware Estuary* (1988, Newark, DE), 26; Kent Price, Robert A. Beck, Stewart Tweed, and Charles E. Epifanio, "Fisheries," Bryant and Pennock, eds., *The Delaware Estuary*, 81–86; Miller, "The Delaware Oyster Industry," 253; Passmore, *Three Centuries of Delaware Agriculture*, 66.

22. J.W. Collins, "The Shore Fisheries of Southern Delaware," George Brown Goode, ed., *The Fisheries and Fishery Industries of the United States* (Washington, D.C., 1887), 527–528, 532, 538.

23. Vincent, *History of Delaware*, 63; Chester Hardy III, "Fish or Foul: A History of the Delaware River Basin through the perspective of the American Shad, 1682–Present," *Pa. Hist.*, vol. LXVI (autumn, 1999), 508–509, 514; Jesse M. Malcolm and Jay L. Harmic, "A Springtime Delicacy: American Shad," *Delaware Conservationist* (Spring, 1967), 3; Dennis Forney, "The Inland Bays: Delaware's Special Resource," *Delaware Conservationist*, vol. XXVI (1983), 10; Price et al., "Fisheries," 71; Bryant and Pennock, eds., *The Delaware Estuary*, 90. For an examination of shad fisheries in New England see Daniel Vickers, "Those Damned Shad: Would the River Fisheries of New England Have Survived in the Absence of Industrialization," *William and Mary Quarterly*, Third Series, Vol. LVI, no. 4, 685–712; Gary Kulik, "Dams, Fish, and Farmers: Defense of Public Rights in Eighteenth Century Rhode Island," in Stephen Hahn and Jonathan Prude, eds., *The Countryside in the Age of Capitalist Transformation* (Chapel Hill, 1985).

24. *Journal of the Senate of the State of Delaware*, (Dover, DE, 1901), 34.

25. *Laws of the State of Delaware*, 1835-1843, vol. IX, 263-264; *Biennial Message of His Excellency Gove Saulsbury, Jan. 3, 1871* (Dover, DE, 1871), 12, D.P.A.; *First Biennial Message of His Excellency, John P. Cochran, Governor of Delaware, Jan. 19, 1875* (Wilmington, DE, 1876), 20, D.P.A.; *Revised Statutes of the State of Delaware (1893)*, 472, 475–476.

26. Price et al., "Fisheries," 75; *State of State Bien-*

nial Message of His Excellency, John W. Hall, Governor, Jan. 2, 1883 (Dover, DE, 1883), D.P.A.

27. Hardy, "Fish or Foul," 527.

28. Ibid., 524; Berger et al., *History of Human Ecology of the Delaware River*, 106; Kathleen Jamison, "New Hope for Troubled Waters," *Delaware Conservationist*, vol. XXXII, no.2 (1989), 10; Molly Murray, "Cleaner Delaware: Proof is in the Fishing," *News Journal* (Feb. 12, 1989); n.a., Man and the River: The Story of the Delaware (n.p., 1959), unpaginated.

29. Marshall McDonald, "Fisheries of the Delaware River," George Brown Goode, ed., *The Fisheries and Fishery Industries of the United States* (Washington, D.C., 1887), 654; Hardy, "Fish or Foul," 526; Price et al., "Fisheries," 77.

30. Hardy, "Fish or Foul," 526; Price et al., "Fisheries," 77; Malcolm and Harmic, "A Springtime Delicacy: American Shad," 4; George Alexander, "Dirty Flows the Delaware," *Newsweek* (Nov. 2, 1970), 68–73; A. Joel Kaplovsky, *A Comprehensive Study of Pollution and Its Effect upon the Waters within the Brandywine Creek Watershed* (n.p., 1954), 37–39.

31. Hardy, "Fish or Foul," 520–521.

32. Ibid., 528–529; Molly Murray, "Cleaner Delaware: Proof is in the Fishing," *News Journal* (Feb. 12, 1989).

CHAPTER SIX:
Saving Field and Forest, 1818-2000

1. James C. Booth, *Memoir of the Geological Survey of the State of Delaware* (Dover, DE, 1841), v.

2. Whitney, *From Coastal Wilderness to Fruited Plain*, 227; Williams, *Delaware: The Garden State of the Union*, 38.

3. *Journal of the Senate of the State of Delaware* (Dover, DE, 1807), 8; *Journal of the Senate* (Dover, DE, 1821), 8; *Journal of the Senate* (Dover, DE, 1837), 7; Eckman, ed., *Delaware: A Guide to the First State*, 84; Marie Windell, ed., "James Van Dyke Moore's Trip to the West, 1826–1828," *Del. Hist.*, vol. IV, no. 2 (Sept., 1950), 70.

4. Lu Ann De Cunzo and Ann Maria Garcia, *Neither Desert Nor Paradise* (n.p., 1993), 31; George B. Tindall, *America: A Narrative History* (New York, 1984), 714; Eckman, ed., *Delaware: A Guide to the First State*, 89; William S. Taber, *Annual Report of the State Forester* (Dover, DE, 1941), 15; Passmore, *Three Centuries of Delaware Agriculture*, 6.

5. Black, *Memoirs of the Agricultural Society*, 77; Huffington, *The Delaware Register and Farmer's Magazine*, vol. I (Dover, DE, 1838), 44; Harold B. Hancock, "Agriculture in Delaware, 1789–1900," Reed, ed., *Delaware: A History of the First State*, vol. I, 383; Taber, *Annual Report of the State Forester* (Dover, DE, 1941), 15. For the view that tenant farmers in northern Delaware were younger than farm owners and, therefore, more open to agricultural reform see Lu Ann De Cunzo and Maria Garcia, *The Archaeology of Agriculture and Rural Life, New Castle and Kent Counties, Delaware*, 1830–1940 (n.p., Oct., 1992), 72–73.

6. Huffington, *The Delaware Register and Farmer's Magazine*, vol. I (Dover, DE, 1838), 44.

7. Higgins, ed., "The Memoirs of John Clark Higgins," 181; *Journal of the House of Representatives* (1847), 196–197; *Blue Hen's Chicken*, June 1, 1849; Williams, *Delaware: The Garden State of the Union*, 40.

8. Williams, *Delaware: The Garden State of the Union*, 29, 8–9; Dalleo, "South of the Canal," 122, 126–127; Lu Ann De Cunzo and Maria Garcia, *Historic Context: The Archaeology of Agriculture and Rural Life, New Castle and Kent Counties, Delaware, 1830–1940* (n. p., 1992), 13.

9. Stewart Rafert, "James Curtis Booth: Delaware's First Geologist," *Del. Hist.*, vol XVII, no. 2 (Fall–Winter, 1976), 139–144; *Laws of the State of Delaware*, vol. IX, 163–164; Huffington, *The Delaware Register and Farmer's Magazine*, vol. II, 127–128; Vincent, *A History of the State of Delaware*, 17–18.

10. Scharf, *History of Delaware*, vol. I, 434; Rafert, "James Curtis Booth," 144; Booth, *Memoir of the Geological Survey of the State of Delaware*, 156, viii,

154; Huffington, *The Delaware Register and Farmer's Magazine*, vol. I, 44, 82; Henry T. Williams, *Delaware: The Garden State of the Union*, 39.

11. Huffington, *The Delaware Register and Farmer's Magazine*, vol. I, 45, 47–48.

12. Ibid., 47–48; Passmore, *Three Centuries of Delaware Agriculture*, 17; Eric Paul Wittkopf, "James Curtis Booth: Chemist in Antebellum Philadelphia" (Ph.D. dissertation, University of Delaware, 1994), 196.

13. Scharf, *History of Delaware*, vol. I, 435–436; Passmore, *Three Centuries of Delaware Agriculture*, 17; Robert V. Rimini, *Andrew Jackson and the Course of American Democracy* (New York, 1984), vol. II, 68–69. Other sources credit William Hurlock or James Wilson with being the first to apply marl from the canal to their fields. See Passmore, *Three Centuries of Delaware Agriculture*, 17; and Booth, *Memoir of the Geological Survey of the State of Delaware*, 154.

14. Higgins, ed., "The Memoirs of John Clark Higgins," 182; Henry T. Williams, *Delaware: The Garden State of the Union*, 8.

15. Higgins, ed., "The Memoirs of John Clark Higgins," 182; Passmore, *Three Centuries of Delaware Agriculture*, 18.

16. Passmore, *Three Centuries of Delaware Agriculture*, 18; William H. Williams, *Delmarva's Chicken Industry: 75 Years of Progress* (Georgetown, DE, 1998), 58; "An Historical Appraisal of Poultry Extension," unpublished report found in Delmarva Poultry Industry Archives, Georgetown, DE, 1; Scharf, *History of Delaware*, vol. I, 321.

17. *Journal of the Senate*, 1857, 6; *Journal of the Senate*...1895, 13.

18. Hancock, "Agriculture in Delaware," 388; Thomas A. Baker, "Twentieth Century Agriculture," Reed, ed., *Delaware: A History of the First State*, 391; n.a., *Farming Possibilities of the Delaware, Maryland, Virginia Peninsula* (n.p., 1911), 43.

19. Passmore, *Three Centuries of Delaware Agriculture*, 71-72; Williams, *Delmarva's Chicken Industry*, 7.

20. J. Frank Gordy, "Broilers," Oscar Hanke ed., *American Poultry History* (Madison, Wisconsin, 1974), 378-379; J. Frank Gordy, "The Delaware Broiler Industry," unpublished paper, Delmarva Poultry Industry, Georgetown, DE., 1; Hugh Johnson, *The Broiler Industry in Delaware*, bulletin no. 250 (Newark, DE., 1941), 10.

21. M. M. Dougherty, *A Short History of the Broiler Industry in Delaware*, pamphlet no. 15, July, 1944, (Newark, DE., 1944), 1; Glenn A. Lanon and Robert A. Dalrymple, *A Coastal Engineering Study of Indian River Inlet, Delaware, Ocean Engineering Report no. 14* (Newark, DE, n.d.), 25–27; Williams, *Delmarva's Chicken Industry*, 9–11.

22. W.D. Termohlen, J.W. Kinghorne, E.L. Warren and J.H. Radabaugh, *An Economic Survey of the Poultry Industry* (Washington, D.C., 1936), 19; Williams, *Delmarva's Chicken Industry*, 13.

23. Baker, "Twentieth Century [Delaware] Agriculture," 402, 404; Douglas Hanks, Jr., *Talbot Memories of J. Mckinney Willis, Jr.* (Easton, MD, 1992), 44; Williams, *Delmarva's Chicken Industry*, 25–26.

24. Hancock, "Agriculture in Delaware," 387; "Corn Data for Delaware," *Internet, Economics and Statistics Systems*, Albert Mann Library, Cornell University, Ithaca, NY; *Agricultural Statistics Summary, 2002–2003* (n.p., n.d.), 5.

25. Williams, *Delmarva's Chicken Industry*, 26; Passmore, *Three Centuries of Delaware Agriculture*, 20.

26. Williams, *Delmarva's Chicken Industry*, 26, 91, 109; Sylvester Harmon, interview; William H. Williams, *The First State: An Illustrated History of Delaware*, second edition (Sun Valley, CA, 1999), 164.

27. William Joseph Coleman, "State-County-Local Cooperation Under Delaware Drainage Laws," United States Agricultural Research Service (Washington, D.C., 1959), 7, 18.

28. Ibid., 5, 30. For just a few examples of tax-ditch companies unable to clean out their ditches

because they were collapsing financially see *Laws of the State of Delaware* vol. X, 583; vol. XI, 692–693, 755.

29. William B. Meyer, "When Dismal Swamps Became Priceless Wetlands," *American Heritage* (May/June, 1944), 109–110.

30. *Laws of the State of Delaware*, vol. XXVI, 4; Meeting of Kent County Ditch Companies with the State Highway Department, Department of Natural Resources and Civilian Conservation Corps, and Mosquito Control Photographs, D.P.A.; R.O. Bausman, *An Economic Study of Land Utilization in Sussex County, Delaware*, bulletin no. 233 (Newark, DE, 1941), 4; Coleman, "State-County-Local Cooperation Under Delaware Drainage Laws," 6.

31. Coleman, "State-County-Local Cooperation Under Delaware Drainage Laws," 30; Bill Henderson, interview (Oct. 31, 2003). Henderson began his employment as agricultural agent in Sussex in 1942; Molly Murray, "Sussex Ditches Earn Praise and Blame," *News Journal* (Aug. 25, 1989).

32. Donna Stacheki Sharp, "A Bond with the Bay," *Delaware Conservationist*, vol XXXII, no. 2 (1989), 22; Williams, *The First State: An Illustrated History of Delaware*, 122; Francis Jordan, Jr., *The Remains of an Aboriginal Encampment at Rehoboth, Delaware* (Philadelphia, PA, 1880), 1; n.a., *Farming Possibilities of the Delaware, Maryland, Virginia Peninsula*, 43; Taber, *Annual Report of the State Forester* (Dover, DE, 1935), 12–13; A. Warren Wheatley, "History of Delaware's Mosquito Control Problem," *Delaware Conservationist* (Spring–Summer, 1971), 13. During the period 1934 to 1939, a separate CCC contingent of approximately 200 men was assigned forestry tasks at Redden State Forest in central Sussex, where they also dug and repaired drainage ditches. See Taber, *Annual Report of the State Forester* (Dover, DE, 1935), 7–8; Taber, *Annual Report of the State Forester* (Dover, DE, 1937), 13–14, 21, 23; Taber, *Annual Report of the State Forester* (Dover, DE, 1938),7; Taber, *Annual Report of the State Forester* (Dover, DE, 1939), 7–8.

33. Wheatley, "History of Delaware's Mosquito Control Program," 13–14; Chet Stachecki, interview (July 20, 2005); Schuster, "Annual Report of the Director for the Fiscal Year Ending June 30, 1940," 27; Florio, *Progger*, 27–29.

34. Chet Stachecki, interview.

35. Ibid.; Jim Miller, *Start the Press* (n.p. or d.), 125–135; n.a., *Protecting Delaware's Natural Heritage*, 112; Molly Murray, "Kill-a-bug method: Killifish," *News Journal* (Sept. 7, 1992); Murray, "Del. Mosquito Control Starts Early at Beach," *News Journal* (Jan. 8, 1995); Wheatley, "History of Delaware's Mosquito Control Program," 13–15; William H. Meredith, "Close Encounters of a Good Kind: Mosquitoes and Fish," *Delaware Conservationist,* vol. XXIV, no. 5 (1981), 19.

36. Ackerman, *Notes From the Shore*, 124; William S. Taber, *Annual Report of the State Forester* (Dover, DE, 1945), 7.

37. Bennett, *An Ecological Characterization of Delaware's Great Cypress Swamp*, 4–5. For a perceptive analysis of this "forest culture," see Herman, *The Stolen House*.

38. Anthony Higgins, "The Swamp where They Mined Cypress," *Baltimore Sun* (April 3, 1932), 16; A. O. Tucker and N.H. Dill, "Rafinesques Florula Delawarica," *Bartonia* (1989), no. LV, 5.

39. Bennett, *An Ecological Characterization of Delaware's Great Cypress Swamp*, 5, 10, 21.

40. Ibid., 13; Daniel J. Leathers, state climatologist, "Average Rainfall in Delaware, 1895–2004," copy in author's possession; William S. Taber, *Second Annual Report of the State Forester* (Dover, DE, 1932), 31, 36; Taber, *Annual Report of the State Forester* (Dover, DE, 1944), 23; Donna Stachecki, "Bald and Beautiful," *Delaware Conservationist* (Spring, 1988), no. I, 14; Robert Phillips, *Land of Lost Content* (New York, 1967), 15–16.

41. Bennett, *An Ecological Characterization of Delaware's Great Cypress Swamp*, 14, 16–17, 21–23.

42. Sterrett, *Report on Forest Conditions in Delaware*, 16.

43. Ibid., 16–21.

44. Taber, *Report of the State Forester* (Dover, DE, 1932), 14; Taber, *Report of the State Forester* (Dover, DE, 1934), 14; Taber, *Report of the State Forester* (Dover, DE, 1936), 17; Taber, *Report of the State Forester* (Dover, DE, 1951), 5; John R. Wennerstein, *The Chesapeake: An Environmental Biography* (Baltimore, 2001), 167–168; Sterrett, *Report on Forest Conditions in Delaware*, 2, 12–14.

45. Bausman, *An Economic Study of Land Utilization in Sussex County, Delaware*, 7; R. O. Bausman, *An Economic Study of Land Utilization in Kent County, Delaware*, bulletin no. 224 (Newark, DE, 1940), 7; Taber, *Annual Report of the State Forester* (Dover, DE, 1958), 6; Taber, *Second Report of the State Forester* (Dover, DE, 1932),7; Taber, *Annual Report of the State Forester* (Dover, DE, 1934), 20, 29.

46. Taber, *Annual Report of the State Forester* (Dover, DE, 1944), 7; Taber, *Annual Report of the State Forester* (Dover, DE, 1943), 8; Taber, *Annual Report of the State Forester* (Dover, DE, 1946), 7.

47. Taber, *Annual Report of the State Forester* (Dover, DE, 1945), 7; Taber, *Annual Report of the State Forester* (Dover, DE, 1944), 7.

48. James T. Edmandson v. National Vulcanized Fiber Company, April 26, 1924, Court of Chancery in and for New Castle County, Transcripts, D.P.A.

49. Ibid., 1–5, 12, 26, 34, 43, 66–77, 75, 84.

50. Ibid., 1–8, 35–39, 47, 61–66.

51. Ibid., 17–18, 23–24, 69–71, 142.

52. James T. Edmandson v. National Vulcanized Fibre Co. (1924), Book G, vol. III, *Registry in Chancery Records*, New Castle County, Wilmington, DE. The author thanks Richard Kiger for finding this citation.

CHAPTER SEVEN: *Crisis and the Rise of Environmentalism, 1945–2000*

1. Al Mascitti, "If You've Got It, Flaunt It—Exploiting our Toxic Legacy," *News Journal*, (June 5, 2002).

2. J. R. McNeill, *Something New Under the Sun: An Environmental History of the Twentieth Century World* (New York, 2000), 332, 335–336; David Hackett Fischer, *Albion's Seed: Four British Folkways in America* (New York, 1989), 410–418; William H. Williams, "Delaware and Ratification: A Paradox Examined," *Delaware Lawyer* (Fall, 1987), 9–12.

3. Molly Murray, "Panel sets Delaware River Pollution Hearings," *News Journal* (Oct. 4, 1995); Murray, "Acid Raindrops Keep Falling on Our Heads," *News Journal* (Jan. 29, 1989); Murray, "Delaware Getting Wind of Ozone," *News Journal* (Oct. 23, 1991); William Boyer, *Governing Delaware: Policy Problems in the First State* (Newark, DE, 2000), 162.

4. Kathleen O'Leary Morgan and Scott Morgan, eds., *State Rankings 2000: Statistical View of the 50 United States* (Lawrence, KS, 2001), 430; Eckman, ed., *Delaware: A Guide to the First State*, 8; Munroe, *History of Delaware*, 269; Tony Florio, "Lifeblood of the Estuary," *Delaware Conservationist*, vol. XXXII, no. 2 (1989), 29; *Delaware, 2000 Summary Population Characteristics* (Washington, D.C., 2002), 38–39, provided raw statistics upon which the above calculations and generalizations were made. In 2000, New Castle County was in the top 4% of the most densely populated counties in America, according to Jeff Montgomery, "Delaware among the Worst in Air Quality," *News Journal*, (Aug. 27, 2000).

5. Bruce L. Hudson and Barbara L. Peterson, *The Quality of Life in Delaware* (Dover, DE, 1975), 141,143; Boyer, *Governing Delaware: Policy Problems in the First State*, 22.

6. Molly Murray, "Del. Rare Plants Uncovered," *News Journal*, (June 24, 1994); Jennifer Ackerman, *Notes From the Shore* (New York, 1995), 160.

7. McNeill, *Something New Under the Sun*, 51, 58; Jeff Montgomery, "Del. Among Worst in Air Quality," *News Journal* (Aug 27, 2000). Statistics on motor vehicle registration were in *Statistical Abstract of the United States, 1970* (Washington D.C. 1970), 545, or provided by the Delaware Division of Motor Vehicles, Dover, DE.

8. McNeill, *Something New Under the Sun*, 61–62, 103–104; Steinberg, *Down to Earth*, 209–210.

9. Molly Murray, "A Check on the Air," *News Journal*, (Nov. 30, 1995); Murray, "A Breath of Cleaner Air," *News Journal*, (May 23, 1993); Boyer, *Governing Delaware*, 161; Alma T. Jolly, "Ozone and Human Health: Is the Air in Delaware Safe to Breath," *Delaware Medical Journal*, No. 3 (March, 1995), 175, 177.

10. Molly Murray, "Pollution Fell 13.9% in Delaware During 1990," *News Journal*, (Sept. 3, 1991).

11. Jeff Montgomery, "Del. Among Worst in Air Quality," *News Journal*, (Aug. 27, 2000).

12. Boyer, *Governing Delaware*, 168, 172; Jeff Montgomery, "Connective Issues Data on Pollution," *News Journal* (June 4, 1999); Montgomery, "Over the Years Plant Gave a Lot and Polluted a Lot," *News Journal* (March 22, 2001); Montgomery, "EPA Report Details Delaware's Toxic Record," *News Journal* (June 5, 2002); Molly Murray, "Sussex Air Quality Ok'd," *News Journal* (March 10, 1995); Murray, " Top of the Heap," *News Journal* (May 25, 1994); Victor Gretto, "State Gets another 'F' for Air Quality," *News Journal* (May 2, 2003).

13. Sussan Strasser, *Waste and Want: A Social History of Trash* (New York, 1999), 290.

14. Ibid., 4; *Historical Statistics of the United States* (Washington, D.C., 1975), 243; *Statistical Abstract of the United States, 2001*, 426–427; *State Rankings, 2000*, 101; Melosi, *The Sanitary City*, 397; Molly Murray, "Rubbish: Del. Sign of Plenty," *News Journal* (Jan. 28, 1995).

15. Molly Murray, "The Problem Recycles: Landfills Teem with Trash," *News Journal* (April 18, 1999); *Environmental Problems in Delaware* (Dover, DE, 1970), 8–9, 3; Melosi, *The Sanitary City*, 422.

16. Boyer, *Governing Delaware*, 160; Celia Cohen, *Only in Delaware* (Newark, DE., 2002), 354; Molly Murray, "Hazardous Waste: Unknown Dangers by the Truckload," *News Journal* (April 17, 1994); John Bryson, interview (Oct. 27, 2004); Murray, "Del. Gives Go Ahead to Clean Dump," *News Journal* (April 26, 1991); Molly Murray, "Cleanup at Tyler Site Sufficient, EPA Says," *News Journal* (Feb. 6, 1996).

17. Mike Apgar, interview (Oct. 27, 2004); Mike Apgar, "We Can't Afford to Let This Happen Again," *Delaware Conservationist* (Summer, 1975), 19–21; Jeff Montgomery, "Tainting Prompts Public Well Tests," *News Journal* (Dec. 6, 2000); Molly Murray, "Army Creek: From Hazard to Haven" *News Journal* (July 16, 1994). For additional specific information on the environmental crisis at Army Creek see "Artesian Water Company v. the Government of New Castle County, Court of Chancery of Delaware, New Castle County, August 4, 1983."

18. Apgar, interview; Apgar, "We Can't Afford to Let This Happen Again," 19–21; Molly Murray, "Army Creek: From Hazard to Haven," *News Journal* (July 16, 1994).

19. Apgar, interview; Molly Murray, "Army Creek: From Hazard to Haven," *News Journal* (July 16, 1994).

20. Molly Murray, "Army Creek: From Hazard to Haven" *News Journal* (July 16, 1994); Jeff Montgomery, "Tainting Prompts Public Well Tests," *News Journal* (Dec. 6, 2000); n.a., "Delaware's Clean Water," *Delaware Conservationist* (Spring, 1985), 22.

21. *Historical Statistics of the United States: Colonial Times to 1970*, Part 1 (Washington, D.C., 1975), 460; Morgan, ed., State Rankings, 2001, 3.

22. Dot Abbott, "No-Till, Its Catching On," *Delaware Conservationist* vol. XXVII, no. 2 (1984), 8.

23. William F. Ritter, interview (Oct. 15, 2004); William F. Ritter, "Environmental Research of William F. Ritter," (Sept. 13, 2004), typed paper submitted to author.

24. McNeill, *Something New Under the Sun*, 27, 137; William F. Ritter," Water Quality in the Inland Bays: Agriculture—Friend or Foe," *Delaware Conservationist*, vol. XXVI, no. 2 (1983), 24; Molly Murray and Jeff Montgomery, "Poultry Industry Endangers Health of State Waterways," *News Journal* (July 28,

1998); Kevin C. Donnelly, interview (Sept. 7, 2005); J. Thomas Sims, "Agricultural Phosphorus and Water Quality: Environmental Challenges in the U.S. Atlantic Coastal Plain," Proceedings of the 2nd International Conference on Diffuse Pollution from Agriculture (Edinburgh, Scotland, 1997). A summary of William F. Ritter's research can be found in an unpublished paper, William F. Ritter et al., "Twenty Years of Water Quality Research in Sussex County, Delaware." Paper is in the hands of William F. Ritter, College of Agriculture, University of Delaware, Newark, DE. For more insight on the dangers of phosphorus by J. Thomas Sims see, for example, J. Thomas Sims, R. R. Simard, and B. C. Joern, "Phosphorous Loss in Agricultural Drainage: Historical Perspective and Current Research," *Journal of Environmental Quality* (1998),vol. XXVII, 277–293.

25. Molly Murray and Jeff Montgomery, "Poultry Industry Endangers Health of State Waterways," *News Journal* (July 28, 1998).

26. Jeff Montgomery and Molly Murray, "Waste Battle Squeezes Farmers," *News Journal* (July 28, 1998); Molly Murray, "Study: High Nitrate Levels Tainting Bays," *News Journal* (Aug. 6, 1992).

27. Rome, *The Bulldozer in the Countryside*, 23, 35, 256–257, 3; Boyer, *Governing Delaware*, 143. The assumption by the writer is that attitudes supporting the building of suburbs that were popular across the nation were also popular in Delaware.

28. Munroe, *History of Delaware*, 171; "Utopia Found in America," Robin K. Bodo, Del. Division of Historical & Cultural Affairs, (provided by Harry Themal); Robin Brown, "As They Like It," *News Journal* (May 17, 2004).

29. Munroe, *History of Delaware*, 243–244, 247; David Ames, "A Question of Balance," *Delaware Conservationist*, vol. XXXIII, no. 1 (1990), 32; Percentages of non-Wilmington residents of New Castle County are based on statistics in Munroe, *History of Delaware*, 269 (Appendix E).

30. Rome, *The Bulldozer in the Countryside*, 201.

31. Steinberg, *Down to Earth,* 222–22; Frank J. Scott's quote was taken from Steinberg.

32. Rome, *The Bulldozer in the Countryside,* 3, 87, 89, 103; Molly Murray, "Del. Unveils Septic System Aid Program," *News Journal* (June 18, 1991).

33. Williams, *The First State: An Illustrated History of Delaware* second ed., (Sun Valley CA, 1999), 166; *www. census gov/population factfinder.census.gov*.

34. John Schneider, "Small Bays—National Notability," *Delaware Conservationist*, vol XXXII, no. 2 (1989), 36; Jonathan Sharp, interview, (Nov. 23, 2005); Kathleen Jamison, "State of the State's Environment," *Delaware Conservationist*, vol. XXXIII, no. 1 (1990), 17.

35. David Small, "Loving Our Bays to Death," *Delaware Conservationist*, vol. XXXII, no. 2, (1989), 34; Molly Murray, "Sussex Natives Toe the Mark on Clean Water," *News Journal* (Oct. 31, 1990); Murray, "Eel Grass Fails to Survive," *News Journal*, (July 29, 1991).

36. Molly Murray, "Sewer Plant Discharge over Limit," *News Journal* (June 19, 1990); Murray, "Del. Finds Voluntary Phosphate Cutback Problematic," *News Journal* (Aug. 3, 1990); Murray, "Center Hopes Compromise Can Revive Bays," *News Journal* (Feb. 6, 1996).

37. Bob Stenger, interview (Sept. 9, 2004); Small, "Loving our Bays to Death," 34; Molly Murray, "Sewer Plant Discharge over Limit," *News Journal* (June 19, 1990).

38. Stenger, interview; Molly Murray, "Study: High Nitrate Levels Tainting Bays," *News Journal*, (August 6, 1992).

39. Donna S. Sharp, "A Bond with the Bay," *Delaware Conservationist*, vol. XXXII, no. 2 (1989), 22; Kathleen Jamison, "New Hope for Troubled Waters," *Delaware Conservationist*, vol. XXXII, no. 2 (1989), 10.

40. Molly Murray, "Stretches of Del. River 'Safe' for Swimming," *News Journal* (July 21, 1991); Murray, "Something Still Amiss in Del. Estuary,"

News Journal (Jan. 31, 1993); "Dirty Flows the Delaware," *Newsweek* (Nov. 2, 1970), 73; John Bryson, interview (Oct. 27, 2004); Tracy Bryant, "Delaware Bay: Healthy or Ailing," *Delaware Conservationist*, vol. XXVII, no. 4 (1985–86), 11.

41. Kathleen Jamison, "The Death and Life of Red Clay Creek," *Delaware Conservationist*, vol. XXX, no. 2 (Summer, 1988), 5-7; Manly P. Northam, *History: The National Vulcanized Fibre Company And The Vulcanized Fibre Industry* (Wilmington, DE, 1942), 1-2; "Revival of a Pretty Creek in Delaware," *Delaware Conservationist* (Summer, 1972), 20.

42. Jamison, "The Death and Life of Red Clay Creek," 4, 6, 7; Joan W. Priest, *Educating today, Preserving for tommorrow: A 40-Year History of the Delaware Nature Society* (Hockessin, DE, 2005), 40; Lorraine M. Fleming, *Delaware's Outstanding Natural Areas and Their Preservation* (Hockessin, DE, 1978), 80.

43. Robert Henry and Jeff Benoit, "Sea Level Rise: A Growing Threat to Delaware Beaches," *Delaware Conservationist*, vol. XXX, no. 1 (Spring, 1987), 4–5. Nineteenth century newspaper editor and historian Francis Vincent, *A History of the State of Delaware*, 59, described the process of gradual movement inland of the sea in coastal Delaware as simply a matter of water replacing sand, sand replacing marsh, and marsh replacing fast land.

44. Kelvin Ramsey and Marijke J. Reilly, The Hurricane of October 21–24, 1878 (Newark, DE, 2002), 21–38; Williams, *Delmarva's Chicken Industry*, 91–92; Williams, *The First State: An Illustrated History of Delaware*, 123; *Delaware's Storm of the Century* (published by DNREC and Sea Grant Program, University of Delaware), 1–4; Wendy Carey and Robert Dalrymple, "Northeasters," *Coastal Currents* (pamphlet published by University of Delaware Sea Grant), 1–2.

45. *Delaware's Storm of the Century*, 1–4; Carey and Dalrymple, "Northeasters," 1–2; Williams, *The First State: An Illustrated History of Delaware*, 123; Robert D. Henry, "So You Want to Build a cottage at the Beach!" *Delaware Conservationist* (Winter, 1974), 4–6.

46. Andrew D. Zimmerman, Frank R. Selby and William Cohen, *Life in a Refinery Town: The History and Impact of the Getty Oil Company in Delaware City, Delaware* (Newark, DE., 1983), 3, 95–96; Jeff Montgomery, "Over the years, plant gave a lot—and polluted a lot," *News Journal* (March 22, 2001); Carol E. Hoffecker, *Democracy in Delaware: The Story of the First State's General Assembly* (Wilminton, DE, 2004), 223–265.

47. Zimmerman et al., *Life in a Refinery Town*, 133, 135, 142, 165, 169; John Nickle, interview (Aug. 31, 2005); Boyer, *Governing Delaware*, 163; Molly Murray, "Texaco ordered to pay dumping fines," *News Journal* (Aug. 17, 1993); Murray, "Violations at Star," *News Journal* (May 8, 1994); Murray, "Polluters Fine Tops $2 Million," *News Journal* (Sept. 8, 1994); Murray, "Star Enterprise Odors Protested," *News Journal* (April, 25, 1996); Murray, "Del. Engineer Aims to Stifle Star Pollution," *News Journal* (Aug. 14, 1996).

48. Zimmerman et al., *Life in a Refinery Town*, 124–127; Boyer, *Governing Delaware*, 163.

49. Zimmerman et al., *Life in a Refinery Town*, 120–121; Boyer, *Governing Delaware*, 175–176; Phil Milford, "Delaware Ranked 'Least Healthy State in U.S.'" *News Journal* (Oct. 26, 1989); Jane Harriman, "Del. Rated at Bottom," *News Journal* (Jan. 20, 1993); Molly Murray, "Five Year Toxic Pollution Study Proposed," *News Journal* (June 12, 1996); Murray, "A Disheartening Search for Answers," *News Journal* (Aug. 14, 1995).

50. Williams, *Slavery and Freedom in Delaware*, 35–36.

51. Wilkinson, *E.I. du Pont, Botanist*, 66, 80, 90, 96, 119, 131; Norman B. Wilkinson, "E.I. du Pont: Botanist," *Delaware Today* (Oct.–Nov.), 1965), 18; Hoffecker, ed., "The Diaries of Edmund Canby...," *Del. Hist.*, vol. XVI, no.3 (Spring-Summer, 1975), 232. For the important role played by the du Ponts in the Delaware Nature Society's work see Joan W. Priest, *Educating today, Preserving for tomorrow: A 40-Year History of the Delaware Nature Society* (Hockessin, DE., 2005), 17, 23, 24, 25, 31, 39, 43, 70, 73, 75, 76.

52. Priscilla Thompson, "Creation of the Wilmington Park System Before 1896," *Del. Hist.*, vol. XVIII, no. 2 (Fall-Winter, 1978), 75–81; *Parks and Playgrounds of Wilmington* (Wilmington, DE, 1929), 9–29.

53. Thompson, "Creation of the Wilmington Park System Before 1896," 75–81; *Parks and Playgrounds in Wilmington*, 9.

54. *Journal of the Senate of the State of Delaware* (Dover, DE, 1909) 42, 45–47; Peter Huber, *Hard Green, Saving the Environment from the Environmentalists: A Conservative Manifesto* (New York, 1999), xi.

55. Zimmerman et al., *Life in a Refinery Town*, 76–77; *State of Delaware Intrastate Water Resources Survey* (Wilmington, DE, 1959), 20–9; *Environmental Problems in Delaware* (Dover, DE, 1970), 1–3; Michael Riska, "The State of Delaware's Plants and Animals," *Outdoor Delaware* (Spring–Summer, 2001), 5.

56. Ackerman, *Notes From The Shore*, 124; Murray, "Sewage Woes Could Cost Rehoboth Dearly," *News Journal* (Aug. 30, 1990).

57. Jacob Kreshtool, interview (Aug. 21, 2005); Molly Murray, "Clark Faces Hostile Environment," *News Journal* (June 30, 1991); Murray, "Crusader Coughs up a Bit of Doggerel," *News Journal* (Nov. 7, 1991).

58. Christopher A. G. Talou, "Front Lines," *Outdoor Delaware* (Winter, 1996), 2; Molly Murray, "Violations at Star," *News Journal* (May 8, 1994); Murray, "In Praise of 'Miss Horseshoe Crab,'" *News Journal* (Aug. 28, 1996); Russell W. Peterson, *Rebel with a Conscience* (Newark, DE, 1999), 184.

59. Norman G. Wilder, "Remembering Ted Harvey," *Delaware Conservationist*, (Fall, 1985) no. III, 27.

60. Boyer, *Governing Delaware*, 293 n. 58; Steven H. Amick, interview (July 21, 2006).

61. McNeill, *An Environmental History of the Twentieth-Century World*, 20; Dana Phillips, *The Truth of Ecology: Nature, Culture and Literature in America* (New York, 2003), 42; John D. Gates, "Is Ecology Just a Science?" *Morning News* (Jan. 22, 1971); Protecting Delaware's Natural Heritage (1999), 6; C. Jon Roush, "The Disintegrating Web: The Causes and Consequences of Extinction," *Delaware Conservationist*, vol. XXXIII, no. 1 (1990), 5.

62. Phillips, *The Truth of Ecology*, 44, 60, 70–72; Huber, *Hard Green*, 79, 161.

63. Larry Curtis, written response to criticisms of ecology, in my custody (no date).

64. Magoc, *So Glorious a Landscape*, 254; Roush, "The Disintegrating Web," 8; Melosi, The Sanitary City, 291; Til Purnell, interview, (April 15, 2003); Peterson, *Rebel with a Conscience*, 124; John D. Gates, "Is Ecology Just a Science," *Morning News* (Jan. 22, 1971). Perhaps as good a summary of the ecological perspective that appeared so routinely in late-twentieth-century publications dealing with Delaware's natural environment, although printed in 2002, is William A. McAvoy, "Disappearing Diversity: the loss of Native Flora as a reflection on Delaware's Environmental Health," *Outdoor Delaware*, (Spring–Summer, 2002), vol. XI, no. 1, 13–17; and *Delaware's Environmental Legacy*, 158, which was published in 1988.

CHAPTER EIGHT:
The Government Steps In, 1971–2000

1. Boyer, *Governing Delaware*, 32–33, 47, 61, 69; Paul Dolan and James R. Soles, *Government of Delaware* (Newark, DE, 1976), 23.

2. *Delaware's Environmental Legacy*, 136; n.a., *Protecting Delaware's Natural Heritage*, 38; Hoffecker, *Democracy in Delaware*, 198–202.

3. Hoffecker, *Democracy in Delaware*, 195–196; Celia Cohen, *Only in Delaware* (Newark, DE, 2002), 190.

4. Norman G. Wilder, "The Delaware Board of Game and Fish Commissioners; Its Objectives, History and Problems," (unpublished, 1968), 2–7.

5. *State of Delaware Intrastate Water Sources Sur-*

vey (Wilmington, DE, 1959), 18–11, 18–12, 18–14; "An Award and a Title" (Editorial), *Morning News* (Feb. 22, 1957); Rod Harmic, interview.

6. McNeill, *Something New Under the Sun*, 127; *Forty-Sixth Annual Report: Chief Engineer to the Water Department to the Board of Water Commissioners*, Fiscal 1914–1915 (Wilmington, DE, 1915), 20.
7. *Sewage Treatment Program* (Wilmington, DE, 1955), unpaginated.

8. Boyer, *Governing Delaware*, 144–145; *Protecting Delaware's Natural Heritage*, 39.

9. Ned Cooch, interview (Nov. 8, 2004); Peterson, *Rebel with a Conscience*, 125–126.

10. Roger Martin, ed., *Elbert N. Carvel* (Wilmington, DE, 1997), 29–30, 32; Peterson, *Rebel with a conscience*, 140; John A. Bivens, Jr., *First Annual Report: Delaware State Planning Office* (Dover, DE, 1962), 11; Rudolph F. Jass, *Annual Report, 1967: Delaware State Planning Office* (Dover, DE, 1967), 7.

11. Ned Davis, *Charles L. Terry: Governor of Delaware, 1965–69* (Wilmington, DE, 2000), 35; N.C. Vasuki, interview (Oct. 5, 2004); John A. Bivens, Jr., *Annual Report, Fiscal Year 1966: Delaware State Planning Office* (Dover, DE, 1967),11.

12. Lloyd Alexander, interview (July 18, 2005); "Wildlife Man of the Year" (Editorial), *Evening Journal* (Feb. 20, 1957); John Bryson, interview (Oct. 27, 2004); N.C. Vasuki, interview; June MacArtor, interview (July 29, 2005).

13. Gordon, "The American Environment," 31–32; Dawn Williams, "Delaware's Coastal Zone Act: A Story of Leadership and Community Action," (M.A. thesis, Duke University, Durham, NC, 1990), 7–10; *Morning News* (April 23, 1970).

14. Peterson, *Rebel with a Conscience* (Newark, DE, 1999), 31–46, 225, 313; Russell Peterson, interview (June 10, 2005).

15. Martin Ridge, ed., *Frederick Jackson Turner: Wisconsin's Historian of the Frontier* (Madison, WI, 1993), 11; Peterson, interview.

16. Peterson, *Rebel with a Conscience*, 123–125; Peterson, interview; Christopher L. Perry, ed., *Russell W. Peterson, Governor of Delaware, 1969–73* (Wilmington, DE, 1999), 57. For a look at voting in previous gubernatorial races see Cohen, *Only in Delaware*, 462, 464, 466, 468, 470, 472.

17. Sally Lindsay, "Showdown on Delaware Bay," *Saturday Review* (March 18, 1972), 36–37; Peterson, *Rebel with a Conscience*, 84; Cohen, *Only in Delaware*, 190, 165.

18. Peterson, interview; Peterson, *Rebel with a Conscience*, 118–119, 122; Hoffecker, *Democracy in Delaware*, 221; Cohen, *Only in Delaware*,168–169, 190.

19. Peterson, interview; Peterson, *Rebel with a Conscience*, 119–122; Cohen, *Only in Delaware*, 190.

20. Perry, ed., Russell W. Peterson, 62, 65; John C. Bryson, *Annual Report, 1975: Department of Natural Resources and Environmental Control* (Dover, DE, 1975), intro.; David Small, interview, (May 2, 2005); Peterson, *Rebel with a Conscience*, 225–226; Peterson, interview; Lynn Williams, interview (Nov. 8, 2004).

21. Lindsay, "Showdown on Delaware Bay," 34; Peterson, *Rebel with a Conscience*, 126–127; Perry, ed, *Russell Peterson*, 149; Cohen, *Only in Delaware*, 187–188.

22. Peterson, *Rebel with a Conscience*, 127–128.

23. Ibid., 130, 140; Perry, ed., *Russell W. Peterson*, 142; "Text of Governor Peterson's Address," *Morning News* (Jan. 14, 1971); "Future of the State" (Editorial), *Delaware State News*, (Jan. 14, 1971).

24. Perry, ed., *Russell W. Peterson*, 142; Peterson, *Rebel with a Conscience*, 130–139; Carol E. Hoffecker, *Delaware: A Bicentennial History* (New York, 1977) 207–208; Cohen, *Only in Delaware*, 188; David Swayze, interview (May 2, 2005); Lindsay, "Showdown on Delaware Bay," 36.

25. Peterson, *Rebel with a Conscience*, 139–142, 146, 149; Perry, ed., *Russell W. Peterson*, 151–152.

26. Dover Bureau, "Pollution Chief under Fire,"

Morning News (June 16, 1971); Lindsay, "Showdown on Delaware Bay," 36; Sussex Bureau, "Peterson Reaffirms Anti-industry Stand," *Morning News* (June 2, 1971); Peterson, *Rebel with a Conscience*, 131.

27. Peterson, *Rebel with a Conscience*, 142, 144–145; *Laws of the State of Delaware*, vol. 58, part 1, 490–499; Perry, ed., *Russell W. Peterson*,154–155; Hoffecker, *Democracy in Delaware*, 225.

28. John Schmaedeke, "House Passes Peterson Coast Bill," *Morning News* (June 22, 1971); Peterson, *Rebel with a Conscience*, 147–148.

29. Walt Rykiel, "Senate O K's Coast Zone Bill," *Morning News* (June 23, 1971); Peterson, *Rebel with a Conscience*, 148.

30. Cohen, *Only in Delaware*, 188–189; John Schmaedeke, "Peterson discloses financial crisis," *Morning News* (June 29, 1971); Peterson, *Rebel with a Conscience*, 177–202; Peterson, interview; Russell W. Peterson, "20 Years With the Coastal Zone Act," *Outdoor Delaware* (Winter, 1992), 19; Editorial, "Delaware Shows the Way," *New York Times* (July 5, 1971,) 18; Hoffecker, *Democracy in Delaware*, 223.

31. Michael Parkowski, interview (June 10, 2005); David Keifer, interview (June 19, 2005); Peterson, *Rebel with a Conscience*,177–178; Bill Wood, interview (July 20, 2005).

32. Peterson, *Rebel with a Conscience*, 230.

33. Daniel J. Fiorino, *Making Environmental Policy* (Berkeley, CA, 1995), 7, 24, 29, 34, 37.

34. Ibid., 128–129; Bjorn Lomborg, *The Skeptical Environmentalist: Measuring the Real State of the World* (Cambridge, U.K., 2001), 216, 222, 226, 229.

35. Swayze, interview; Molly Murray, "Tilt to Polluters Enrages Environmentalists," *News Journal* (Feb. 28, 1995); June McArtor, interview (July 29, 2005).

36. Martin, A *History of Delaware Through Its Governors*, 541–556; Hoffecker, *Democracy in Delaware*, 229, 232, 235–236; Peterson, *Rebel with a Conscience*, 182; Bryson, interview; Vasuki, interview.

37. Protecting Delaware's Natural Heritage, 58; *Laws of the State of Delaware*, vol. LIX, part I, 693; Austin P. Olney, *Annual Report, 1977, Delaware Department of Natural Resources and Environmental Control* (Dover, DE, 1977), unpaginated; "DRP Chronology," DSWA issued (Dover, DE, n.d.), 1; Bryson, interview; Vasuki, interview.

38. John C. Bryson, *Annual Report, 1976: Delaware Department of Natural Resources and Environmental Control* (Dover, DE, 1976), 7.

39. Swayze, interview; Joe Smyth, "Editorial," *Delaware State News* (April 27, 1977).

40. Peterson, *Rebel with a Conscience*, 187–188; Perry, ed. Russell W. Peterson, 160; Bryson, interview; Williams, *The First State: An Illustrated History of Delaware* (1999 edition), 161–162; *Delaware's Environmental Legacy*, 137, 142; Keifer, interview.

41. Til Purnell, interview; Molly Murray, "Del. Chamber of Commerce Names Toby Clark 'Environmentalist of the Year,'" *News Journal* (June 15, 1991).

42. *Protecting Delaware's Natural Heritage*, 85–88; Molly Murray, "Del's Farmland Preservation Efforts Draw Praise," *News Journal* (Jan. 13, 1999).

43. Michael de Courcy Hinds, "Delaware Tries to Erase Image of Its Poor Health," *New York Times* (Jan. 21, 1990); "Cancer, Dying in Delaware," *News Journal* (April 28, 2005); Cris Barrish, "3 Governors Task Forces Evaluated the Problem," *News Journal* (April 28, 2005); Boyer, *Governing Delaware*, 177–180; Ralph Moyed, "John Q. Public Takes the Rap in Cancer Report," *News Journal* (Aug. 16, 1995); Molly Murray, "A Disheartening Search for Answers," *News Journal* (Aug. 14, 1995).

44. Lomborg, *The Skeptical Environmentalist*, 229, 329, 330; Fiorino, *Making Environmental Policy*, 108, 128–129; Huber, *Hard Green*, xvii, 28, 141; Jeff Montgomery, "Delaware's Air a Toxic Mix of Chemicals, State Study Says," *News Journal* (Aug.

16, 2005); Peterson, *Rebel with a Conscience*, 244.

45. Harry F. Themal, "Growth is Vital but Must Be in Proper Balance," *News Journal* (June 7, 1999); Molly Murray, "Delay Urged for Coastal Zone Picks," *News Journal* (June 22, 1993); Editorial, *News Journal* (Oct. 19, 2000); Molly Murray, "Carper Plan may Change Coastal Regulations," *News Journal* (June 26, 1993); Molly Murray, "Environmentalists Ask Tulou to Abstain from Oil Ruling," *News Journal* (June 30, 1994); Molly Murray, "State Lagging in Enforcing Permits," *News Journal* (Sept. 15, 1996).

46. Jeff Montgomery, "Poultry May No Longer Rule the Roost," *News Journal* (Dec. 25, 1998); Molly Murray and Jeff Montgomery, "Poultry Industry Endangers Health of State Waterways," *News Journal* (July 28, 1998); "Title 3, Agriculture, Part II, Regulatory Provisions, Chapt.22, Nutrient Management," *http://www.delcode.state.de.us/title3/c022/sc01/index.htm.*

47. Molly Murray, "Beach Tests: A Summer of Confusion," *News Journal* (Aug. 20, 1989).

48. Peterson, *Rebel with a Conscience*, 196–198; Molly Murray, "93 Coastal Zone Rules Thrown Out," *News Journal* (May 20, 1994); Priest, *Educating today, Preserving for tomorrow: A 40-Year History of the Delaware Nature Society* , 55; *http://www.dnrec.state.de.us/development/CZA/CZbckgrd.htm*; Editorial, *News Journal* (Oct. 19, 2000).

49. John Burris, interview (July 18, 2005); Fiorino, *Making Environmental Policy*, 18.

50. *Delaware Annual Air Quality Report, 2000* (Dover, DE, 2001), 1, 8.

51. Ibid., 9, 10, 15, 19, 20, 23, 47, 28; John Burris, interview; Molly Murray, "A check on the Air," *News Journal* (Nov. 30, 1995); n.a., *Delaware Department of Natural Resources and Environmental Control: Program Self Assessment, 1995* (Dover, DE), 8–10; n.a., *Delaware's Environmental Legacy*, 40; Ray Malenfant, interview (Sept. 29, 2005); Jamison, *State of the State's Environment*, 19–21; Jeff Montgomery, "State Is Pollution Leader," *News Journal* (June 5, 2002); Montgomery, "Del. Toxic Emissions Lower in '02," *News Journal* (June 7, 2004); Jeff Montgomery, interview (Sept. 29, 2005).

52. *Protecting Delaware's Natural Heritage*, 81.

53. Priest, *Educating today, Preserving for tomorrow: A 40-Year History of the Delaware Nature Society*, 1–18; Lynn Williams, interview (Nov. 5, 2004); Gary Soulsman, "The Quiet Conservationist," *News Journal* (April 21, 2000); Patricia Rivera, "Conservationism Comes Naturally to Greenville Woman," *News Journal* (May 19, 2005); Fleming, *Delaware's Outstanding Natural Areas and Their Preservation*, 6.

54. Priest, *Educating today, Preserving for tomorrow: A 40-Year History of the Delaware Nature Society* , 18–21; Fleming, *Delaware's Outstanding Natural Areas And Their Preservation*, 7–9; *Protecting Delaware's Natural Heritage*, 83–84; Lorraine M. Fleming, interview, (Sept. 28, 2005); Ron Vickers, interview (Oct. 3, 2005).

55. Jamison, "State of The State's Environment," 28.

56. *Protecting Delaware's Natural Heritage*, 16; *Historical Statistics of the United States: Colonial Times to 1970*, Part I, 460; Kathleen O'Leary Morgan and Scott Morgan, *State Rankings, 2001* (Lawrence, KS, 2001), 3.

57. Molly Murray, "The Problem Recycles: Landfills Team with Trash," *News Journal* (April 18, 1999).

58. Ibid.

59. Ibid.; Pat Canzano, interview, (June 29, 2006).

60. Molly Murray, "The Problem Recycles," *News Journal*, (April 18, 1999); Canzano, interview; Dick Fleming, interview, (June 12, 2006).

61. Molly Murray, "Former Governor Urges Down to Earth Activism," *News Journal* (Jan. 11, 1990).

62. Molly Murray, "The Problem Recycles...," *News Journal* (April 18, 1999).

63. Jeff Montgomery, "Sprawl Threat is Called a

Crisis," *News Journal* (Oct. 17, 2000); Stacey Tiedge, "Resisting the Call of Sprawl," *News Journal* (Oct. 2, 2000).

64. Molly Murray, "Wetlands Plan draws Opposition," *News Journal* (Nov. 6, 1990); Murray, "Farmers Fight State Control of Wetlands," *News Journal* (April 16, 1992).

65. Murray, "Sussex Gets Earful on Land Use Plan," *News Journal* (March 18, 1997).

66. Pennsylvania Railroad, *Farming Possibilities of the Delaware, Maryland, Virginia Peninsula*, n.p., 1911, 47; Bryson, interview; Ed Kee, *Saving Our Harvest: The Story of the Mid-Atlantic Region's Canning and Freezing Industry* (Baltimore, MD, 2006), 401–402; Peterson, *Rebel with a Conscience*, 249.

67. Kevin C. Connelly, interview (Sept. 7, 2005); The Delaware Estuary Program, *The Delaware Estuary: Discover its Secrets— A Management Plan for the Delaware Estuary,* n.p., 1996, 64; Bryson, Annual Report, 1976: Delaware Department of Natural Resources and Environmental Control, 7.

68. *The Delaware Estuary*, 12; DNREC Self Assessment, 1995, 30; Jeff Montgomery, "Fish Kills Highlight Bay's Woes," *News Journal* (July 24, 2000); Molly Murray, "State Lagging in Enforcement Permits," *News Journal* (Sept. 15, 1996); Protecting Delaware's Natural Heritage, 17. The information about the nature of materials thrown into drainage ditches was gained by examining a number of drainage ditches, and by conversations with farmers and hunters.

69. DNREC Self Assessment, 1995, 23, 36; Molly Murray, "State Lagging," *News Journal* (Sept. 15, 1996); Molly Murray, "Waste discharge record improves," *News Journal* (Oct. 4, 1999).

70. Protecting Delaware's Natural Heritage, 15.

71. Boyer, *Governing Delaware*, 167; Protecting Delaware's Natural Heritage, 12–13, 15; Bill McAvoy, interview, May 15, 2006; McAvoy, summary sheet, May 15, 2006.

72. Protecting Delaware's Natural Heritage, 6, 8, 12, 54, 110–11, 114; *The Delaware Estuary*, 51; Robert Meadow, "Mystery of the Monster Phrag," *Outdoor Delaware* (winter, 2005), 19–21.

73. Ryan Cormier, "Elusive Cougar Reported near Bear Shopping Center," *News Journal* (Feb. 24, 2001); Molly Murray, "Dam beavers, Critics Say," *News Journal* (Aug. 8, 1991); Murray, "Sussex Landowners Get Another Shot at Deer," *News Journal* (Dec. 5, 1996); Murray, "Deer Damage to Crops Becomes Problem as Populations Increase," *News Journal* (Feb. 18, 1996).

74. James B. Trefethen, "The Return of the White-tailed Deer," *American Heritage*, vol XXI, no. 2 (Feb., 1970), 100–103.

75. Ibid., 103; Charles P. West, interview (Nov. 9, 2005); H. Lloyd Alexander, interview (July 18, 2005).

76. Bard S. McGinnes, *The White-tail Deer in Delaware* (Dover, 1950), 1, 7; "Delaware Deer Harvest," provided by Ken Reynolds of Delaware's Division of Fish and Wildlife; Molly Murray, "A Booming Season Likely for Deer hunters," *News Journal* (Sept. 9, 1992); Murray, "Sussex Landowners Get Another Shot at Deer," *News Journal* (Dec. 12, 1996); *Natural Resources and Environmental Control, 2000* (Dover, DE, 2000), unpaginated; James P. Sterba, "How Deer are Redesigning Our Forests," *National Wildlife*, (Oct.–Nov., 2005), 16.

77. Kenneth Reynolds, "Return of the Native," *Delaware Conservationist*, vol. XXXI, no. 3 (1988), 4–7; Molly Murray, "Turkeys Flock to Refinery," *News Journal* (Feb. 20, 1997). Generalization of gradual but slow growth in the wild turkey population is based on annual "Delaware Turkey Harvest," provided by Ken Reynolds of Delaware's Division of Fish and Wildlife.

78. Molly Murray, "Del. Bay Named a Reserve for Hungry, Migrating Birds," *News Journal* (April 27, 1990); Murray, "New Law Protects Horseshoe Crabs," *News Journal* (May 17, 1991); Murray, "Horseshoe Crabs Rapid decline," *News Journal* (Nov. 23, 1995); "Delaware's 2004 ASMFC Horseshoe Crab Compliance Report," Attachment D, Figure D-1, Figure D-2.

79. "Murray, "Cracking Crabs Mystery," *News Journal* (Oct. 18, 1999); Murray, "Nature's Highway," *News Journal* (May 7, 2000); Michael Riska, "The State of Delaware's Plants and Animals," *Outdoor Delaware* (Spring–Summer, 2001), 6; Murray, "New Law Protects Horseshoe Crabs," *News Journal* (May 17, 1991); Murray, "2 Year Ban on Horseshoe Crabs Proposed," *News Journal* (Nov. 2, 2005); Murray, "Crab-harvest Caps Stir Hot Debate," *News Journal* (Feb. 9, 2000).

80. Hess, *Birds of Delaware*, 12; Riska, "The State of Delaware's plants and animals," 7; "Table 4, Delaware aerial waterfowl count—December 17, 1974," and "Table 109, Delaware aerial waterfowl count—January 12, 2000," both provided by Ken Reynolds; Murray, "Canada Geese Numbers Plummet," *News Journal* (Nov. 1, 1995); Murray, "All of a Sudden, Geese everywhere," *News Journal* (Oct. 17, 1996); Murray, "Canada Geese Wearing out their Welcome," *News Journal* (Oct. 10, 1998).

81. Hess, *Birds of Delaware*, 23, 100; Murray, "Cleaner Delaware: Proof is in the Fishing," *News Journal* (Feb. 12, 1989); Murray, "Delaware is Battling the Birds," *News Journal* (Feb. 19, 1997).

82. H. Lloyd Alexander, "Endangered Species Update," *Delaware Conservationist* (Fall, 1979), 6; Molly Murray, "Bald Eagles Stay on State's Endangered List," *News Journal* (July 19, 1999); Murray, "Ospreys under Pressure," *News Journal* (May 10, 2004); Murray, "Ospreys Flourish at Wildlife Refuge," *News Journal* (July 3, 1999).

83. Murray, "Egrets and Herons are Declining," *News Journal* (July 29, 1996).

84. *Protecting Delaware's Natural Heritage*, 71–72, 14, 12.

85. Tom Horton, *An Island Out of Time: A Memoir of Smith Island in the Chesapeake* (New York, 1999) 70; Linda Nash, "The Changing Experience of Nature: Historical Encounters with a Northwest River," *Journal of American History*, vol. LXXXVI, no. 4, 1600.

86. Nash, "The Changing Experience of Nature," 1600; Mark Twain, *Life on the Mississippi* (New York, 1996), 119–120.

87. McNeill, *Something New Under the Sun*, 359.

Index

Page numbers in italics refer to figures, tables, and their captions.